Prairie Wetland Ecology

The Contribution of the
Marsh Ecology Research Program

edited by:
Henry R. Murkin
Arnold G. van der Valk
and
William R. Clark

Iowa State University Press / Am

Henry R. Murkin, PhD, is senior research scientist for the Institute for Wetland and Waterfowl Research (IWWR) of Ducks Unlimited Canada. Dr. Murkin's personal research interests are in wetland ecology and management.

Arnold G. van der Valk, PhD, is director, Iowa Lakeside Laboratory and professor in the Department of Botany, Iowa State University, Ames. Dr. van der Valk's research interests are in wetland restoration and landscape ecology.

William R. Clark, PhD, is professor in the Department of Animal Ecology and chair of the Ecology and Evolutionary Biology Program, Iowa State University, Ames. Dr. Clark's research interests are modeling animal populations in relation to landscape and ecosystem changes.

Iowa State University Press
2121 South State Avenue, Ames, Iowa 50014

Orders: 1-800-862-6657
Office: 1-515-292-0140
Fax: 1-515-292-3348
Web site: www.isupress.edu

Authorization to photocopy items for internal or personal use, or the internal or personal use of specific clients, is granted by Iowa State University Press, provided that the base fee of $.10 per copy is paid directly to the Copyright Clearance Center, 222 Rosewood Drive, Danvers, MA 01923. For those organizations that have been granted a photocopy license by CCC, a separate system of payments has been arranged. The fee code for users of the Transactional Reporting Service is 0-8138-2752-3/2000 $.10.

♾ Printed on acid-free paper in the United States of America

First edition, 2000

Library of Congress Cataloging-in-Publication Data

Prairie wetland ecology : the contribution of the Marsh Ecology Research Program / edited by Henry R. Murkin, Arnold G. van der Valk, and William R. Clark.—1st ed.
      p. cm.
   Includes bibliographical references and index.
   ISBN 0-8138-2752-3
     1. Wetland ecology. 2. Prairies. I. Murkin, Henry R. II. Valk, Arnold van der. III. Clark, William R. (William Richard). IV. Marsh Ecology Research Program.

QH541.5.M3 P73 2000
577.68—dc21
00-027666

The last digit is the print number: 9 8 7 6 5 4 3 2 1

To the memory of Peter D. Curry, conservationist
and true friend through good times and bad

Contents

Part III: Prairie Wetland Ecology

Part IV: Summary

Appendices:

Contributors

The numbers in parentheses after each name represent chapter numbers.

Bruce D.J. Batt (1,2)
c/o Ducks Unlimited, Inc.
One Waterfowl Way
Memphis, Tennessee 38120

Patrick J. Caldwell (1,10)
Institute for Wetland and Waterfowl Research
c/o Ducks Unlimited Canada
P.O. Box 1160
Stonewall, Manitoba
Canada R0C 2Z0

William R. Clark (11,12,13)
Department of Animal Ecology
Iowa State University
Ames, Iowa 50011

L. Gordon Goldsborough (8,12)
Department of Botany and University Field Station (Delta Marsh)
University of Manitoba
Winnipeg, Manitoba
Canada R3T 2N2

Sharon E. Gurney (8)
Water Quality Management Manitoba Environment, Suite 160
123 Main Street
Winnipeg, Manitoba
Canada R3C 1A5

John A. Kadlec (1,3,4,5,6)
College of Natural Resources
Utah State University
Logan, Utah 84322-5210

Henry R. Murkin (1,3,4,5,6,9,10,12,13)
Institute for Wetland and Waterfowl Research
c/o Ducks Unlimited Canada
P.O. Box 1160
Stonewall, Manitoba
Canada R0C 2Z0

Gordon G.C. Robinson (8)
Department of Botany
University of Manitoba
Winnipeg, Manitoba
Canada R3T 2N2

Lisette C.M. Ross (9)
Institute for Wetland and Waterfowl Research
c/o Ducks Unlimited Canada
P.O. Box 1160
Stonewall, Manitoba
Canada R0C 2Z0

Arnold G. van der Valk (3,4,5,6,7,12,13)
Department of Botany
353 Bessey Hall
Iowa State University
Ames, Iowa 50011-1020

Dale A. Wrubleski (12)
Institute for Wetland and Waterfowl Research
c/o Ducks Unlimited Canada
P.O. Box 1160
Stonewall, Manitoba
Canada R0C 2Z0

Preface

Our early understanding of prairie waterfowl and wetland ecology developed from research and conservation work inspired by the drought of the 1930s. Most waterfowl and wetland studies over the next 40 years were descriptive, but they highlighted the importance of conserving northern prairie wetlands as habitat for the wide array of wildlife species occupying these dynamic systems. It became evident, however, that purely descriptive studies were inadequate to advance our understanding of wildlife habitat requirements and the issues related to the general ecology and management of prairie wetlands. Weller (1978:280) stated:

> What is most needed to advance marsh management theory and practice are experimental data gathered concurrently by a team of specialists in marsh plants, limnology, hydrology, invertebrates and vertebrates. If such a group also had an isolated, artificial, multi-unit marsh complex designed exclusively for experimentation on the ecology of marshes and wildlife, the potential for understanding marsh systems and devising proper marsh management procedures for marsh wildlife would be unsurpassed.

In response to Weller's proposal, Bruce Batt, then Scientific Director of the Delta Waterfowl and Wetlands Research Station, and Patrick Caldwell of Ducks Unlimited Canada began discussions regarding a joint research program to address basic hypotheses on factors controlling the productivity of prairie wetlands. Their original idea of a "weekend project" was quickly discarded as the realities of the time, effort, and resources required by a long-term multidisciplinary experimental study came into focus. Nevertheless, they pressed on, and due to their efforts and the support of their respective boards, Ducks Unlimited Canada and the Delta Waterfowl and Wetlands Research Station embarked on the Marsh Ecology Research Program (MERP) in 1979.

Earlier work by Weller and his colleagues (Weller and Spatcher 1965; Weller and Fredrickson 1974) and the subsequent contribution by van der Valk and Davis (1978) documented the changes in prairie wetlands during the cyclical variations in annual precipitation (wet–dry cycles). These efforts highlighted the many uncertainties associated with the ecology of these dynamic systems. Early in MERP, a scientific team from a variety of disciplines was assembled to assess the priority research needs for these systems and to design a long-term, experimental study on the effects of water-level manipulations (wet–dry cycle) on the ecology of northern prairie wetlands. The resulting experimental design involved subjecting a series of 10 experimental cells to a simulated wet–dry cycle. Documenting changes in the primary producers, water chemistry, invertebrates, muskrat populations, and avian use as well as the nutrient budgets of the experimental cells were the major objectives of MERP. Ten years of fieldwork resulted in a broad array of new and diverse information on the ecology of prairie wetlands.

The objective of this book is to summarize the contributions of MERP to the current understanding of prairie wetland ecology. Part I describes the overall MERP study design and provides background information on the Delta Marsh where the study was located. A primary research objective of MERP was to assess the effects of water-level changes on the nutrient budgets during the wet–dry cycle in prairie wetlands. Part II summarizes the nutrient budgets for the various phases of the MERP study. In Part III, the advances through MERP are incorporated into reviews on individual components of the ecosystem during the wet–dry cycle. Finally, Part IV discusses the management implications of the MERP results and provides a general summary of the advances in our understanding of prairie wetland ecology provided by MERP. Three appendices also are included to provide a summary of the MERP field and laboratory sampling methods, additional nutrient budget data, and a list of MERP publications.

LITERATURE CITED

van der Valk, A.G., and C.B. Davis. 1978. Role of seed banks in the vegetation dynamics of prairie glacial marshes. Ecology 59:322-335.

Weller, M.W. 1978. Management of freshwater marshes for wildlife. In *Freshwater Wetlands: Ecological Processes and Management Potential*. (Eds.) R.E. Good, D.F. Whigham, and R.L. Simpson, pp.267-284. New York: Academic Press.

Weller, M.W., and L.H. Fredrickson. 1974. Avian ecology of a managed glacial marsh. Living Bird 12:269-291.

Weller, M.W., and C.E. Spatcher. 1965. Role of habitat in the distribution and abundance of marsh birds. Department of Zoology and Entomology Special Report #43. Agricultural and Home Economics Experiment Station, Iowa State University, Ames, Iowa.

Acknowledgments

A program of the size and scope of MERP required a broad base of financial support. The following individuals, companies, and foundations provided major financial contributions: Amoco Canada Petroleum, James F. Bell Foundation, Richard A.N. Bonnycastle, Peter D. Curry, Devonian Group of Charitable Foundations, Delta Waterfowl Foundation (formerly the North American Wildlife Foundation), Donner Foundation, Ducks Unlimited Canada, Ducks Unlimited Foundation, Ducks Unlimited Incorporated, William P. Elliott, Edson I. Gaylord, Richard Ivey Foundation, Laidlaw Foundation, John M.S. Lecky, Molson Foundation, Murphy Foundation, McLean Foundation, Senator Norman Patterson Foundation, Richardson Century Fund, and R. Howard Webster. The support of the boards of both Ducks Unlimited Canada and the Delta Waterfowl and Wetlands Research Station was instrumental in getting MERP underway and the fieldwork completed. Ducks Unlimited Canada provided support for the data analyses and preparation of this book.

Many individuals contributed to the success of MERP during the initial planning, fieldwork, data analyses, and preparation of this book. The following individuals served various terms on the MERP Scientific Team: Bruce D.J. Batt (Ducks Unlimited, Inc.), Patrick J. Caldwell (Ducks Unlimited Canada), William R. Clark (Iowa State University), Craig B. Davis (Iowa State University), John A. Kadlec (Utah State University), Richard M. Kaminski (Mississippi State University), Henry R. Murkin (Ducks Unlimited Canada), Jeffrey W. Nelson (Ducks Unlimited, Inc.), Gordon G.C. Robinson (University of Manitoba), and Arnold van der Valk (Iowa State University). Terry Neraasen (Ducks Unlimited Canada) and Dan McLaughlin (Concordia University) played important roles in the initial planning phase. David Cox of the Statistical Laboratory at Iowa State University was consulted extensively on statistical aspects of the experimental design. During the 10 years of fieldwork, more than 300 field assistants participated in the collection and processing of samples and data. Jonathan Scarth, Penny Alisauskas, Dale Wrubleski, Karen Tome, Sharon Gurney, Elaine Murkin, and Lisette Ross supervised various aspects of the fieldwork. Michael Stainton and Jack Boughen supervised and conducted the water-sample analyses. Arnold van der Valk and William Crumpton were responsible for the bulk of the plant tissue analyses. Craig Davis and William Crumpton made many important contributions to the plant-sampling design and analyses.

Ducks Unlimited Canada engineering staff designed and supervised construction of the experimental cell complex, which operated over a decade without a serious flaw ever being detected. The maintenance staff at the Delta station, most notably Russell Ward, Alf Lavallee, and Kevin Ward, provided a wide range of services to the project and field staff over the 10 years of fieldwork. Support of Peter Ward and the Bell family is gratefully acknowledged. Special thanks as well to Laura Gretsinger, Brenda Hales, Sandy Kennedy, and Pat Hope for the many ways they contributed to the needs of the field assistants and other MERP staff season after season.

The editors of this book would like to extend their appreciation to Stuart Morrison and Don Young of Ducks Unlimited Canada and Matthew Connolly of Ducks Unlimited, Inc. for their support throughout MERP and especially during the preparation of this summary book. We also would like to thank Bruce Batt and Pat Caldwell for their leadership through all the peaks and valleys associated with a long-term field study. In addition, I would like to thank Brian Gray and Michael Anderson for providing the time and support needed for "getting the book done." Arnold van der Valk and William Clark thank their respective department heads for providing time during the writing and editing process. Lisette Ross provided invaluable assistance during all aspects of the MERP study and especially during the preparation of this book (it wouldn't have happened without her).

The following individuals reviewed various chapters and sections of this book: Michael G. Anderson, Todd W. Arnold, Bruce D.J. Batt, Suzanne E. Bayley, Patrick J. Caldwell, William C. Crumpton, Kjell Danell, Craig B. Davis, James J. Dinsmore, Erik K. Fritzell, Susan M. Galatowitsch, Brian T. Gray, Sharon E. Gurney, Mark A. Hanson, Mickey E. Heitmeyer, Robert H. Kadlec, Richard M. Kaminski, James W. LaBaugh, Joseph S. Larson, Thomas R. McCabe, Robert K. Neely, Christopher Neill, Jeffery W. Nelson, Richard D. Robarts, Gordon G.C. Robinson, Frank C. Rohwer, David M. Rosenberg, Lisette C.M. Ross, Jennifer M. Shay, Eugene F. Stoermer, Douglas A. Wilcox, Thomas C. Winter, and Joy B. Zedler. Their input was an important contribution to the final product.

The editors and all members of the MERP Scientific Team would like to thank the hundreds of field assistants (MERPies) who spent countless hours collecting and processing samples and data under all possible conditions—they are truly the unsung heroes of MERP.

Finally, thanks to Tamara and Jonathan, my inspiration.

Henry R. Murkin

Part I

Introduction

1

Introduction to the Marsh Ecology Research Program

Henry R. Murkin
Bruce D.J. Batt
Patrick J. Caldwell
John A. Kadlec
Arnold G. van der Valk

ABSTRACT

The Marsh Ecology Research Program (MERP) was a long-term interdisciplinary study on the ecology of prairie wetlands. A scientific team from a variety of disciplines (hydrology, plant ecology, invertebrate ecology, vertebrate ecology, nutrient dynamics, marsh management) was assembled to design and oversee a long-term experiment on the effects of water-level manipulations on northern prairie wetlands. Ten years of fieldwork (1980–1989), combining a routine long-term monitoring program and a series of short-term studies, generated a wealth of new and diverse information on the ecology and function of prairie wetlands. The baseline years provided background information on the study area (Pederson 1981; Pederson and van der Valk 1984; Murkin et al. 1989; Wrubleski and Rosenberg 1990) and the opportunity to refine field techniques for the long-term monitoring program (Murkin et al. 1983; Wrubleski and Rosenberg 1984). The deep-flooding phase provided information on the effects of prolonged above-normal flooding on all components of the wetland system, including hydrology (Kadlec 1989), water chemistry (Kadlec 1986), vegetation (McKee et al. 1989; van der Valk 1994), algae (Hosseini and van der Valk 1989a, b), and invertebrates (Murkin and Kadlec 1986; Murkin et al. 1991) among others. The drawdown phase generated new information on prairie wetland response to drying and provided the opportunity to test various theories on drawdown response (Galinato and van der Valk 1986; Welling et al. 1988a; Kadlec 1989; van der Valk et al. 1989). The system response to the reflooding phase provided new insights on water depths and the wet–dry cycle (Gurney and Robinson 1988; Squires and van der Valk 1992; Kadlec 1993; Clark 1994). These are but a few of the MERP contributions to our understanding of prairie wetland ecology. The complete list of MERP publications is presented in Appendix 3. The objective of this book is to summarize the contributions of MERP to the current understanding of prairie wetland ecology and to provide the background for future work in these important ecosystems.

INTRODUCTION

Good et al. (1978) and Greeson et al. (1978) were among the first to summarize the state-of-our-understanding in freshwater wetland ecology. They emphasized the need for expanded interdisciplinary research in freshwater wetlands. Ten years later, van der Valk (1989), in *Northern Prairie Wetlands,* focused specifically on prairie wetlands and the influence of water-level changes on the dynamics and productivity of these systems. Primary production, secondary production, vegetation composition and distribution, animal use, and nutrient budgets are all affected as water levels rise and fall with variation in precipitation. Ranging from tiny potholes to huge lacustrine complexes, prairie wetlands are among the world's most productive ecosystems (Murkin 1989). Their importance as wildlife production areas has drawn considerable attention to their conservation and management (Weller 1978, 1981). This attention has increased in recent years with the need for information on the overall functions and values of wetlands within the prairie landscape (Murkin 1998). Understanding wetland values is critical to the development of comprehensive wetland conservation programs. In addition, the ongoing destruction of wetlands through drainage, urban sprawl, and other human-induced changes has added a great deal of urgency to the need for a better understanding of the ecology and management of all remaining wetland areas.

Recognizing the need for an improved understanding of prairie wetland ecology, the Marsh Ecology Research Program (MERP) was initiated in 1979 as a joint project of Ducks Unlimited Canada and the Delta Waterfowl and Wetlands Research Station (Murkin et al. 1984). The MERP was designed as a long-term replicated experiment addressing the impacts of various water levels on the main components of the prairie wetland ecosystem. The overall objective of MERP was to document the changes that occur within a prairie wetland during a simulated wet–dry cycle (van der Valk and Davis 1978). More specifically, MERP was designed to determine the storage (pools) and movement (fluxes) of nitrogen (N), phosphorus (P), and carbon (C) during the various stages of the cycle. Specific objectives for each of the major ecosystem components were as follows:

1. Hydrology—to estimate the terms of the water budget during all stages of the wet–dry cycle.

2. Water chemistry–to estimate the concentrations of N, P, C, and major ions in both surface and substrate pore-water pools and to monitor changes in these pools during the wet–dry cycle.

3. Macrophytes—to estimate annual above- and belowground macrophyte productivity and standing crops and subsequent changes related to the wet–dry cycle. Storage and movement of N, P, and C in macrophytes during the wet–dry cycle also were determined.

4. Macrophyte litter—to estimate the annual production of above- and belowground macrophyte litter and the loss and uptake of N, P, and C to and from the litter compartment.

5. Algae—to estimate algal standing crops and productivity (phytoplankton, epiphyton, epipelon, and metaphyton) during the various stages of the wet–dry cycle.

6. Invertebrates—to estimate standing crops of invertebrates (nekton, benthos, and emerging insects) during the various phases of the wet–dry cycle.

7. Vertebrates—to monitor use by waterfowl, coots, blackbirds, and muskrats in response to habitat changes during the wet–dry cycle.

The MERP experimental design and field-sampling protocols were developed to meet these objectives. Nutrient data were used to construct nutrient budgets for the experimental cells (see Part II chapters). Several additional objectives beyond those related specifically to nutrient storage and movement also were addressed. Examples are rate of recovery of plant communities following drawdown (Welling et al. 1988a), habitat use by various vertebrate groups during specific stages of the wet–dry cycle (Clark and Kroeker 1993), and the effect of the wet–dry cycle on individual invertebrate species (Ross and Murkin 1993). Many of these analyses are presented throughout this book and in detail in Part III (also see Appendix 3 for a complete listing of MERP publications to date).

STUDY DESIGN AND METHODS

Study Area

The MERP study area was located on the Delta Marsh in south central Manitoba, Canada (50°11′N, 98°19′W) (Figure 1.1). The Delta Marsh is a large lacustrine marsh (≈50,000 ha) along the southern shore of Lake Manitoba (see Chapter 2 for a detailed description

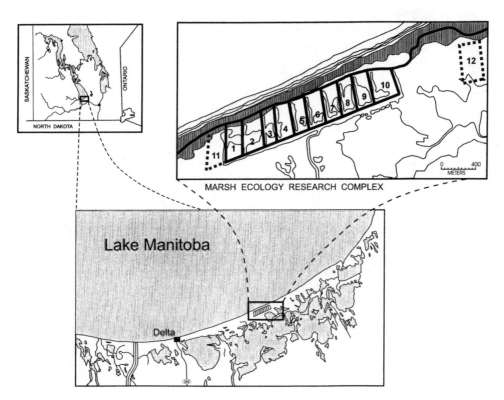

Figure 1.1   Location of the Marsh Ecology Research Program study area on the Delta Marsh, Manitoba.

of the physical and biological characteristics of the Delta Marsh). An important feature of the marsh is that its water level has been stabilized for many years (due to installation of a water-control structure on Lake Manitoba in 1961) resulting in lake marsh conditions described by van der Valk and Davis (1978) over much of its area.

The experimental complex consisted of 10 contiguous wetland cells (approximately 5 ha each) (cells 1 to 10 in Figure 1.1) created by building a series of dikes along the northern edge of the main marsh (Figure 1.2). Disturbance due to dike construction was confined to the dikes and a 10-m buffer strip on each side of the dike. No sampling was done in these buffer strips at any time during the study. Each diked cell was equipped with a stop-log control structure and electric pump with automatic controls to adjust and maintain (±2 cm) water levels. Besides the diked experimental cells, two undiked areas of similar size within the Delta Marsh (cells 11 and 12) were designated as reference cells (i.e., reference areas to the unmanipulated Delta Marsh) (Figure 1.1). The vegetation within the diked cells prior to the experimental treatments was comparable to the reference cells and main Delta Marsh (Pederson 1981).

Experimental Design

The general MERP objective was to monitor the response of various aspects of the wetland ecosystem to a series of water-level manipulations intended to simulate the wet–dry cycle of prairie wetlands. The schedule for water-level treatments and the number of replicates for each treatment are shown in Tables 1.1 and 1.2. The study was basically divided into four phases. The first two phases (baseline and deep flooding) were designed to provide background information on the experimental cells and to bring all cells to a similar condition (lake marsh stage) before the actual wet–dry cycle was applied during the last

Figure 1.2   Aerial photograph of the Marsh Ecology Research Program experimental cells, 1980.

two phases (drawdown and reflooding). The four phases of the experimental design were as follows:

Baseline year. During the first year of the study (1980), the control structures were left open and water levels within the cells were allowed to track the levels of the Delta Marsh. This method provided a year of baseline data before the actual water-level manipulations began. Two of the cells (3 and 7) were kept at baseline levels for a second year (1981) to allow for comparison of different durations of drawdown later in the study (see Table 1.2).

Deep flooding. Following the baseline year(s), all diked cells were flooded to ≈1 m above normal for a period of 2 years (Table 1.1). These levels were designed to flood the existing cattail (*Typha* spp.) stands to a depth of 1 m and were similar to actual natural high-water levels that occurred in the Delta Marsh in the 1950s (see Chapter 2). This deep flooding was intended to kill the existing emergent vegetation and set all the cells to the lake marsh stage described by van der Valk and Davis (1978). In part, deep flooding was also an attempt to reduce the variability in vegetation among the experimental cells prior to drawdown.

Drawdown. After the second year of above-normal flooding, the cells were drawn down (completely dewatered) by adjusting the water levels to 50 cm below the long-term level of the Delta Marsh (Table 1.1). Most cells were exposed to a 2-year drawdown. Two cells

Table 1.1. Schedule of water levels for MERP experimental cells, Delta Marsh

Year	Water levels
1980	10 cells at normal levels of Delta Marsh (247.50 m AMSL)
1981	8 cells flooded ≈1 m above normal (248.41 m AMSL)
	2 cells (3 and 7) at normal levels of Delta Marsh (247.50 m AMSL)
1982	10 cells flooded ≈1 m above normal (248.41 m AMSL)
1983	8 cells drawn down (247.00 m AMSL)
	2 cells (3 and 7) flooded ≈1 m above normal (248.41 m AMSL)
1984	10 cells drawn down (247.00 m AMSL)
1985–1989	4 cells (3,4,7, and 8) flooded at normal level (247.50 m AMSL)
	3 cells (1,5, and 9) flooded at medium level (247.80 m AMSL)
	3 cells (2,6, and 10) flooded at high level (248.10 m AMSL)

Table 1.2. Experimental cells within each of the water levels following drawdown (1985–89)

	Water level after drawdown (AMSL)		
Drawdown duration (years)	Shallow (247.50 m)	Medium (247.80 m)	High (248.10 m)
1	3,7	—	—
2	4,8	1,5,9	2,6,10

(3 and 7) were drawn down for 1 year to allow comparisons between drawdowns of 1- and 2-year durations. The drawdowns allowed vegetation to reestablish on the marsh substrates (Welling et al. 1988a, b).

Reflooding. It has been assumed that the rate at which prairie wetlands proceed through the various stages of the wet–dry cycle is determined by water depth, with deeper wetlands moving to the lake marsh stage more quickly than shallowly flooded wetlands (van der Valk and Davis 1978). Following drawdown, the cells were reflooded to three different levels (Table 1.2). The normal level was the long-term mean level (247.50 m above mean sea level [AMSL]) of the Delta Marsh and the medium (247.80 m) and high (248.10 m) levels were 30 and 60 cm above the long-term mean, respectively.

To ensure replicated treatments (on a hydrologic basis) during the reflooding experiments in 1985–1989, water-level treatments were assigned to the experimental cells as follows: the two low-water treatments with a 1-year drawdown (Table 1.2) were randomly assigned to the experimental complex, and the remaining treatments were then assigned to the other cells to ensure that each cell within a water-level treatment had a similar hydrologic setting (i.e., both low cells with 1-year drawdowns were flanked by cells with low and high treatments, both low 2-year drawdown cells by low and medium treatments, all medium treatment cells by high- and low-treatment cells, and the high-treatment cells by medium and low cells). This replicated hydrology is discussed in more detail in Kadlec (1983, 1993). The treatment assignments for the 1985–1989 period (Table 1.2) determined the water-level assignments during the early phases of the MERP experiments from 1981 to 1984 (Table 1.1).

Long-Term Monitoring Program

A standardized sampling program was developed to monitor the major components of the ecosystem (e.g., vegetation, invertebrates, algae) within the experimental wetlands. The strategy was to collect a comparable data set from year to year in each of the experimental cells for each of the major components being monitored. A brief summary of the long-term monitoring program is provided in Appendix 1. More detail can be obtained from Murkin and Murkin (1989) and the individual publications listed in Appendix 3.

Short-Term Studies

Besides the long-term monitoring program, short-term studies were an integral component of the overall MERP program (see Appendix 3 for a complete list of the MERP short-term graduate-student studies). Some of these studies provided background information needed to develop the long-term monitoring program (Pederson 1981; Wrubleski 1984). Some were integrated into the long-term monitoring program (Murkin 1983; Galinato 1985; Hosseini 1986; Welling 1987; Merendino 1989). Several aspects of the long-term monitoring program following drawdown began as short-term studies (Kroeker 1988; Squires 1991; Wrubleski 1991). Many of the short-term studies, however, were initiated because of questions raised by the long-term monitoring program. This approach was particularly true for studies of invertebrates (Nelson 1982; Neckles 1984; Bicknese 1987; Campeau 1990) and emergent macrophytes (Neill 1988, 1992).

THE WET–DRY CYCLE IN PRAIRIE WETLANDS

The dynamic nature of prairie wetlands is the result of the fluctuating water regime caused by the extreme variability of the prairie climate (Kantrud et al. 1989). Annual variations in spring runoff, summer precipitation, and evapotranspiration result in cyclical fluctuations in water levels within prairie basins. For example, the severe drought of the 1930s in western Canada and the Great Plains of the United States was followed by a period of record-high precipitation in the 1950s. The 1960s were initially dry but returned to wet conditions just prior to the 1970s. Conditions varied during the 1970s; however, a prolonged drought continued for much of the 1980s. Abundant precipitation returned in the early 1990s and water levels increased once again across most of the prairie landscape. These changes in water levels from dry to flooded conditions are often referred to as the wet–dry cycle (van der Valk and Davis 1978). These cycles and their associated hydrologic regimes (i.e., the depth and duration of flooding) are suggested to be the dominant factors controlling productivity of prairie wetlands. Any study examining the structure and function of these dynamic systems must therefore address the wet–dry cycle and associated hydrological processes (Kantrud et al. 1989).

van der Valk and Davis (1978) identified four distinct stages of the wet–dry cycle in what now is considered the model for prairie wetland ecosystem dynamics (Figure 1.3). During drought conditions, prairie wetlands undergo complete or partial drawdowns (dewatering) that expose all or part of the wetland bottom. During the dry marsh stage, seeds present in the seed bank germinate on the exposed substrate (van der Valk and Davis 1976). The resulting plant community is a mixture of emergent and annual species. Animals that rely on standing water (e.g., waterfowl and muskrats) must move elsewhere or develop mechanisms to withstand the dry conditions (e.g., some invertebrate groups) (Murkin 1989). Nutrient pools and fluxes (movements among pools) associated with water are limited to the remaining substrate pore-water component. With the lack of standing water, the hydrology of individual ponds and their interaction with the regional hydrology also are severely altered (Winter 1989).

When the drought ends, the prairie basins become flooded once again. Of the vegetation that developed on the exposed substrates during the dry period, the annual plants are eliminated by the flooding, whereas the emergent species expand through vegetative growth (van der Valk and Davis 1978). Submersed species germinate and become an important component of the plant community. This phase of the wet–dry cycle immediately following reflooding is called the regenerating stage (Figure 1.3). During this stage the wetland is recolonized by animals requiring flooded conditions. The expanding emergent vegetation serves as habitat to a variety of vertebrate and invertebrate species (Weller and Fredrickson 1974; Wrubleski 1991; Murkin et al. 1997). Invertebrate species that were dormant during the dry period reappear within the wetland (Neckles et al. 1990). The return of water also affects nutrient-cycling processes as pools of dissolved and suspend-

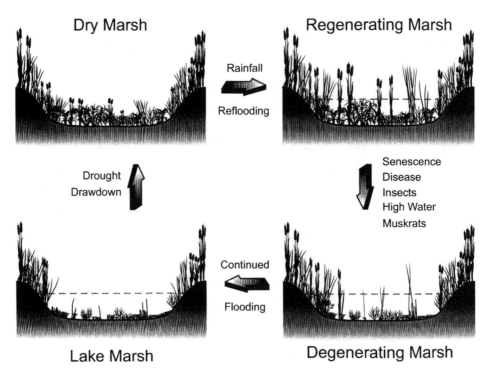

Figure 1.3   Schematic diagram of the wet–dry cycle in prairie wetlands (adapted from van der Valk and Davis [1978]).

ed nutrients within the surface water play important roles in overall wetland productivity and nutrient budgets. The return of water to the basin also has important implications for hydrologic processes within the wetland and surrounding landscape (LaBaugh et al. 1998).

With continued flooding, the emergent vegetation eventually begins to die back as a result of herbivory (primarily by muskrats), the direct effects of prolonged inundation on individual plant species, and disease. The depth of flooding influences the rate at which the vegetation is lost from the basin with deeper water normally resulting in faster die-off (van der Valk and Davis 1978). During this period the wetland enters the degenerating stage (Figure 1.3). Reduction in the standing crop of emergent vegetation also reduces the number and species of animals that depend on it for food and cover (Weller and Fredrickson 1974). The macrophytes also represent major pools of nutrients in the system; therefore, the loss of vegetation has significant implications for nutrient budgets. The changes in litter input also affect species that are adapted to feeding and living on the detrital resources provided by emergent vegetation (Murkin 1989).

Once the emergent vegetation is eliminated from the central portion of the basin and restricted to a narrow band around the fringe of the basin, the wetland enters the lake marsh stage (Figure 1.3). Emergent species cannot reestablish within the basin because their seeds cannot germinate under flooded conditions (van der Valk and Davis 1978). The lack of emergent vegetation and reduced litter production affects species that are dependent on living plants and associated detrital resources. The lack of vegetative cover within the basin affects species that use emergent vegetation as habitat (Weller and Fredrickson 1974). It also exposes the water surface to the wind and results in increased turbidity in the pond and thereby a general reduction in the growth and survival of submersed vegetation and algae (Nelson and Kadlec 1984). The wetland typically remains in the lake marsh stage until a drought or artificial drawdown occurs and begins a new cycle.

Under natural conditions, the various stages of the wet–dry cycle gradually change from one to the other and it is often difficult to determine when one stage ends and another begins. Changes in climatic conditions during the cycle may actually cause reversal (sudden dry period) or acceleration (prolonged rainy period) of the change from one stage to another.

Because of the long-term nature of the changes in climatic conditions, there have been few detailed studies of the wet–dry cycle on the prairies. Weller and Spatcher (1965) and Weller and Fredrickson (1974) followed the changes in bird communities during a wet–dry cycle in Iowa wetlands. They noted the loss of emergent vegetation with continued flooding and the impact of this loss on avian habitat use. These studies were followed by van der Valk and Davis (1976, 1978) who documented the vegetation response within the same Iowa wetlands during dry and flooded conditions and the role of seed banks in the vegetation dynamics of these systems. In Canada, Walker (1959, 1965) monitored the vegetation changes associated with fluctuating water levels in the Delta Marsh. High water resulted in loss of vegetation followed by recruitment from the seed bank with falling water levels. Millar (1973) documented the long-term changes in vegetation in a series of shallow marshes in Saskatchewan under alternating wet–dry conditions. Smith (1971) and Stout (1971) monitored waterfowl use of wetlands in Alberta and Saskatchewan during a complete wet–dry cycle; however, little information on vegetation or other habitat variables was presented.

These previous studies have provided background information on the wet–dry cycle and the ecology of prairie wetlands. Unfortunately, most have dealt primarily with a single aspect of the wetland ecosystem (e.g., vegetation [van der Valk and Davis 1978] and birds [Weller and Fredrickson 1974]) with little, if any, attempt to relate to other components of the system. Information on the interrelationships among nutrient budgets, macrophytes, algae, invertebrates, and the many other components of prairie wetlands is required for a more complete understanding of the structure and function of prairie wetland systems.

As Weller (1978) noted, the most important advances in wetland ecology theory will be made by a multidisciplinary scientific team working concurrently on a long-term study of wetland ecosystem dynamics. He also noted that the ideal study area would be "an isolated, multi-unit wetland complex designed exclusively for the experimentation on the ecology of marshes." The MERP is a direct response to Weller's challenge.

ACKNOWLEDGMENTS

Lisette Ross and Dale Wrubleski prepared the figures for this chapter. Craig Davis, Joseph Larson, Jeffrey Nelson, and Gordon Robinson provided comments on earlier drafts of the manuscript. This is Paper No. 94 of the Marsh Ecology Research Program, a joint project of Ducks Unlimited Canada and the Delta Waterfowl and Wetlands Research Station.

LITERATURE CITED

Bicknese, N.A. 1987. Factors influencing the decomposition of fallen macrophyte litter at the Delta Marsh, Manitoba, Canada. M.S. thesis, Iowa State University, Ames, Iowa.

Campeau, S. 1990. The relative importance of algae and detritus to freshwater wetland food chains. M.S. thesis, McGill University, Montreal, Quebec.

Clark, W.R. 1994. Habitat selection by muskrats in experimental marshes undergoing succession. Canadian Journal of Zoology 72:675-680.

Clark, W.R., and D.W. Kroeker. 1993. Population dynamics of muskrats in experimental marshes at Delta, Manitoba. Canadian Journal of Zoology 71:1620-1628.

Galinato, M.I. 1985. Seed germination studies of dominant mudflat-annual and emergent species of the Delta Marsh. M.S. thesis, Iowa State University, Ames, Iowa.

Galinato, M.I., and A.G. van der Valk. 1986. Seed germination traits of annuals and emergents during drawdown in the Delta Marsh, Manitoba, Canada. Aquatic Botany 26:89-102.

Good, R.E., D.F. Whigham, and R.L. Simpson (Eds.). 1978. *Freshwater Wetlands: Ecological Processes and Management Potential.* New York: Academic Press.

Greeson, P.E., J.R. Clark, and J.E. Clark (Eds.). 1978. *Wetland Functions and Values: The State of Our Understanding.* American Water Resources Association Technical Publication TPS79-2.

Gurney, S.E., and G.G.C. Robinson. 1988. VII. Small water bodies and wetlands. The influence of water level manipulation on metaphyton production in a temperate freshwater marsh. Verh. Internationale Verein Limnologie 23:1032-1040.

Hosseini, S.M. 1986. The effects of water level fluctuations on algal communities of freshwater marshes. Ph.D. dissertation, Iowa State University, Ames, Iowa.

Hosseini, S.M., and A.G. van der Valk. 1989a. Primary productivity and biomass of periphyton and phytoplankton in flooded freshwater marshes. In *Freshwater Wetlands and Wildlife.* (Eds.) R.R. Sharitz and J.W. Gibbons, pp.303-315. USDOE Symposium Series No. 61, Oak Ridge: USDOE Office of Scientific and Technical Information.

Hosseini, S.M., and A.G. van der Valk. 1989b. The impact of above-normal flooding on metaphyton in a freshwater marsh. In *Freshwater Wetlands and Wildlife.* (Eds.) R.R. Sharitz and J.W. Gibbons, pp.317-324. USDOE Symposium Series No. 61, Oak Ridge: USDOE Office of Scientific and Technical Information.

Kadlec, J.A. 1983. Water budgets for small diked marshes. Water Resources Bulletin 19:223-229.

Kadlec, J.A. 1986. Effects of flooding on dissolved and suspended nutrients in small diked marshes. Canadian Journal of Fisheries and Aquatic Sciences 43:1999-2008.

Kadlec, J.A. 1989. Effects of deep flooding and drawdown on freshwater marsh sediments. In *Freshwater Wetlands and Wildlife.* (Eds.) R.R. Sharitz and J.W. Gibbons, pp.127-143. USDOE Symposium Series No. 61. Oak Ridge: USDOE Office of Scientific and Technical Information.

Kadlec, J.A. 1993. Effect of depth of flooding on summer water budgets for small diked marshes. Wetlands 13:1-9.

Kantrud, H.A., J.B. Millar, and A.G. van der Valk. 1989. Vegetation of wetlands of the prairie pothole region. In *Northern Prairie Wetlands.* (Ed.) A.G. van der Valk, pp.132-187. Ames: Iowa State University Press.

Kroeker, D.W. 1988. Population dynamics of muskrats in managed marshes at Delta, Manitoba. M.S. thesis, Iowa State University, Ames, Iowa.

LaBaugh, J.W., T.C. Winter, and D.O. Rosenberry. 1998. Hydrologic functions of prairie wetlands. Great Plains Research 8:3-15.

McKee, K.L., I.A. Mendelssohn, and D.M. Burdick. 1989. Effect of long-term flooding on root metabolic response in five freshwater marsh plant species. Canadian Journal of Botany 67:3446-3452.

Merendino, M.T. 1989. The response of vegetation to seasonality of drawdown and reflood depth. M.S. thesis, Texas Tech University, Lubbock, Texas.

Millar, J.B. 1973. Vegetation changes in shallow marsh wetlands under improving moisture regime. Canadian Journal of Botany 51:1443-1457.

Murkin, E.J., and H.R. Murkin. (Eds.). 1989. Marsh Ecology Research Program long-term monitoring procedures manual. Delta Waterfowl and Wetlands Research Technical Bulletin 2, Portage la Prairie, Manitoba.

Murkin, H.R. 1983. Responses by aquatic macroinvertebrates to prolonged flooding of marsh habitat. Ph.D. dissertation, Utah State University, Logan, Utah.

Murkin, H.R. 1989. The basis for food chains in prairie wetlands. In *Northern Prairie Wetlands.* (Ed.) A.G. van der Valk, pp.316-338. Ames: Iowa State University Press

Murkin, H.R. 1998. Freshwater functions and values of prairie wetlands. Great Plains Research 8:3-15.

Murkin, H.R., and J.A. Kadlec. 1986. Responses by benthic macroinvertebrates to prolonged flooding of marsh habitat. Canadian Journal of Zoology 64:65-72.

Murkin, H.R., P.A. Abbott, and J.A. Kadlec. 1983. A comparison of activity traps and sweep nets for sampling nektonic invertebrates in wetlands. Freshwater Invertebrate Biology 2:99-106.

Murkin, H.R., B.D.J. Batt, P.J. Caldwell, C.B. Davis, J.A. Kadlec, and A.G. van der Valk. 1984. Perspectives on the Delta Waterfowl Research Station—Ducks Unlimited Canada Marsh Ecology Research Program. Transactions of the North American Wildlife and Natural Resources Conference 49:253-261.

Murkin, H.R., J.A. Kadlec, and E.J. Murkin. 1991. Effects of prolonged flooding on nektonic invertebrates in small diked marshes. Canadian Journal of Fisheries and Aquatic Sciences 48:2355-2364.

Murkin, H.R., E.J. Murkin, and J.P. Ball. 1997. Avian habitat selection and prairie wetland dynamics: a ten-year experiment. Ecological Applications 7:1144-1159.

Murkin, H.R., A.G. van der Valk, and C.B. Davis. 1989. Decomposition of four dominant macrophytes in the Delta Marsh, Manitoba. Wildlife Society Bulletin 17:215-221.

Neckles, H.A. 1984. Plant and macroinvertebrate responses to water regime in a whitetop marsh. M.S. thesis, University of Minnesota, Minneapolis, Minnesota.

Neckles, H.A., H.R. Murkin, and J.A. Cooper. 1990. Influences of seasonal flooding on macroinvertebrate abundance in wetland habitats. Freshwater Biology 23:311-322.

Neill, C. 1988. Control of primary productivity of emergent macrophytes in prairie marshes: effects of water depth and nutrient availability. M.S. thesis, University of Massachusetts, Amherst, Massachusetts.

Neill, C. 1992. Relationships between emergent plant production and seasonal flooding in prairie whitetop (*Scholochloa festucacea*) marshes. Ph.D. dissertation, University of Massachusetts, Amherst, Massachusetts.

Nelson, J.W. 1982. Effects of varying detrital nutrient concentrations on macroinvertebrate abundance and biomass. M.S. thesis, Utah State University, Logan, Utah.

Nelson, J.W., and J.A. Kadlec. 1984. A conceptual approach to relating habitat structure and macroinvertebrate production in freshwater wetlands. Transactions of the North American Wildlife and Natural Resources Conference 49:262-270.

Pederson, R.L. 1981. Abundance, distribution, and diversity of buried seed populations in the Delta Marsh, Manitoba, Canada. Ph.D. dissertation, Iowa State University, Ames, Iowa.

Pederson, R.L., and A.G. van der Valk. 1984. Vegetation change and seed banks in marshes: eco-logical and management implications. Transactions of the North American Wildlife and Natural Resources Conference 49:271-280.

Ross, L.C.M., and H.R. Murkin. 1993. The effect of above-normal flooding of a northern prairie marsh on *Agraylea multipunctata* Curtis (Trichoptera, Hydroptilidae). Journal of Freshwater Ecology 8:27-35.

Smith, A. G. 1971. Ecological factors affecting waterfowl production in the Alberta parklands. U.S. Bureau of Sport Fisheries and Wildlife Resource Publication 98.

Squires, L. 1991. Water depth tolerances of emergent species in the Delta Marsh, Manitoba. M.S. thesis, Iowa State University, Ames, Iowa.

Squires, L., and A.G. van der Valk. 1992. Water depth tolerances of the dominant emergent macro-phytes of the Delta Marsh, Manitoba. Canadian Journal of Botany 70:1860-1867.

Stout, J.H. 1971. Ecological factors affecting waterfowl production in the Saskatchewan parklands. U.S. Fish and Wildlife Service Resource Publication 99.

van der Valk, A.G. (Ed.). 1989. *Northern Prairie Wetlands*. Ames: Iowa State University Press.

van der Valk, A.G. 1994. Effects of prolonged flooding on the distribution and biomass of emergent species along a freshwater wetland coenocline. Vegetatio 110:185-196.

van der Valk, A.G., and C.B. Davis. 1976. The seed banks of prairie glacial marshes. Canadian Journal of Botany 54:1832-1838.

van der Valk, A.G., and C.B. Davis. 1978. The role of seed banks in the vegetation dynamics of prairie glacial marshes. Ecology 59:322-335.

van der Valk, A.G., C.H. Welling, and R.L. Pederson. 1989. Vegetation change in a freshwater wet-land: a test of a priori predictions. In *Freshwater Wetlands and Wildlife*. (Eds.) R.R. Sharitz and J.W. Gibbons, pp.207-217. USDOE Symposium Series No. 61, Oak Ridge: USDOE Office of Scientific and Technical Information.

Walker, J.M. 1959. Vegetation studies on the Delta Marsh, Delta, Manitoba. M.S. thesis, University of Manitoba, Winnipeg, Manitoba.

Walker, J.M. 1965. Vegetation changes with falling water levels in the Delta Marsh, Manitoba. Ph.D. dissertation, University of Manitoba, Winnipeg, Manitoba.

Weller, M.W. 1978. Management of freshwater marshes for wildlife. In *Freshwater Wetlands: Ecological Processes and Management Potential*. (Eds.) R.E. Good, D.F. Whigham, and R.L. Simpson, pp. 267-284. Academic Press, New York.

Weller, M.W. 1981. Freshwater Marshes: Ecology and Wildlife Management. University of Minnesota Press, Minneapolis, Minnesota.

Weller, M.W., and L.H. Fredrickson. 1974. Avian ecology of a managed glacial marsh. Living Bird 12:269-291.

Weller, M.W., and C.E. Spatcher. 1965. Role of habitat in the distribution and abundance of marsh birds. Department of Zoology and Entomology Special Report #43. Agricultural and Home Economics Experiment Station, Iowa State University, Ames, Iowa.

Welling, C.H. 1987. Reestablishment of perennial emergent macrophytes during a drawdown in a lacustrine marsh. M.S. thesis, Iowa State University, Ames, Iowa.

Welling, C.H., R.L. Pederson, and A.G. van der Valk. 1988a. Recruitment from the seed bank and the development of zonation of emergent vegetation during a drawdown in a prairie wetland. Journal of Ecology 76:483-496.

Welling, C.H., R.L. Pederson, and A.G. van der Valk. 1988b. Temporal patterns in recruitment from the seed bank during drawdowns in a prairie wetland. Journal of Applied Ecology 25:999-1007.

Winter, T.C. 1989. Hydrologic studies of wetlands in the northern prairie. In *Northern Prairie Wetlands*. (Ed.) A.G. van der Valk, pp.16-55. Ames: Iowa State University Press.

Wrubleski, D.A. 1984. Chironomid (Diptera: Chironomidae) species composition, emergence phenologies, and relative abundances in the Delta Marsh, Manitoba. M.S. thesis, University of Manitoba, Winnipeg, Manitoba.

Wrubleski, D.A. 1991. Chironomid recolonization of marsh drawdown surfaces following reflooding. Ph.D. dissertation, University of Alberta, Edmonton, Alberta.

Wrubleski, D.A., and D.M. Rosenberg. 1984. Overestimates of Chironomidae (Diptera) abundance from emergence traps with polystyrene floats. American Midland Naturalist 111:195-197.

Wrubleski, D.A., and D.M. Rosenberg. 1990. The Chironomidae (Diptera) of Bone Pile Pond, Delta Marsh, Manitoba, Canada. Wetlands 10:243-275.

2

The Delta Marsh

Bruce D.J. Batt

ABSTRACT

The Delta Marsh has existed for ≈2,300 to 3,000 years following the drying
of glacial Lake Agassiz, a period of isostatic uplift, and various periods during which the lake received water from the Assiniboine River alluvial fan. The
marsh remains similar in size and general configuration as it was 2500 to 3000
years B.P. Two scientific research institutions, the Delta Waterfowl and
Wetlands Research Station and the University of Manitoba Field Station, are
located within the marsh. Since the late 1930s these centers have studied
many aspects of the marsh's ecology, most notably the waterfowl, beach ridge
bird community, invertebrates, and vegetation.

The marsh has gone through a period of greatly fluctuating water levels in
the past, caused by natural cycles in precipitation, followed by an extended
period of artificially stabilized levels. Several studies have recorded resulting
changes in the marsh, especially the vegetative communities. Many anthropogenic initiatives have influenced the Delta Marsh, but the stabilization of
Lake Manitoba, to which the marsh is connected, has affected its fundamental nature. Since the regulation of lake levels in 1961, succession has taken
large parts of the Delta Marsh to the lake stage where it will probably remain
for the foreseeable future unless there is active intervention by wetland managers.

INTRODUCTION

The Delta Marsh is one of the most famous freshwater wetlands in North America. It is
located on the southern shore of Lake Manitoba (50°11′N, 98°19′W) ≈110 km northwest
of Winnipeg, Manitoba, Canada (see Figure 1.1 in Chapter 1). The village of Delta is situated near the center of its east-west extent. The first Europeans regularly transected the
marsh on one of the main routes used by trappers and traders from northwestern Canada
enroute to trading posts at the confluence of the Assiniboine and Red rivers where the city
of Winnipeg is located today. The marsh is perhaps most heralded for the seasonal spec-
tacles of waterfowl that gather there and that have been exploited by sport hunters since
the 19th century. In recognition of the Delta Marsh's historical prominence and its unique
international importance to North American continent's waterfowl populations, it has
been designated an Internationally Significant Wetland under the 1971 United Nations
Convention on Wetlands of International Importance (The Ramsar Convention).

From 1900 to the late 1920s, the village of Delta was served by a spur line of the Canadian Northern Railroad, which transported gypsum that had been shipped to Delta by barge across Lake Manitoba. The docking facilities and the rail line opened the areas associated with the marsh and the lake to other commercial and recreational opportunities that thrived for many years. Today, both Delta and the village of St. Ambroise at the eastern end of the marsh are served by paved trunk highways. A barrier ridge that separates Lake Manitoba from the marsh itself (Figure 2.1) provides both a beach and dry foundation for cottages, campgrounds, and hunting lodges. Hunting and fishing on the lake and marsh have been continuous since the late 1800s (Bell and Ward 1984). Trapping in the marsh once provided important seasonal income for local residents, but this activity is of minor importance today.

Other uses have been made of the marsh's natural resources. The main agricultural use is the harvest of hay, primarily of whitetop grass (*Scolochloa festucacea*) and common sedge (*Carex atherodes*), from ≈4,000 ha along the southern edge of the marsh. Minor other uses, such as the harvest of leopard frogs (*Rana pipiens*) for laboratory purposes and giant cane (*Phragmites australis*) for upholstery and roof thatching, have occurred from time to time.

The Delta Marsh is well-known to the scientific community because of the Delta Waterfowl and Wetlands Research Station and the University of Manitoba Field Station, which are located on the marsh edge (Figure 2.1). The Delta Waterfowl and Wetlands Research Station has operated continuously since 1938 and the University Station since 1966. Through their research and educational programs, these institutions have developed a record of scientific information that is unsurpassed at most other freshwater wetland research sites.

Many natural and artificial influences have molded the Delta Marsh. This chapter describes the physical setting and key long-term and recent changes in the marsh, current ecological characteristics, and the prospects for modern-day restoration and management programs.

PHYSICAL SETTING

The marsh is ≈15,000 ha and consists of a matrix of water and vegetation interspersed with large (>1,000 ha) open bays up to 3 m in depth and smaller (<1,000 ha) shallow bays that are usually <1 m in depth (Figure 2.2). It is physically separated from the lake by a forested barrier beach ridge. Smaller isolated openings or "sloughs" (1–5 ha, <0.5 m in depth) are found within otherwise extensive stands of vegetation. Natural channels and creeks exist and mark historical changes in drainage patterns between the lake and the marsh and between the marsh and the Assiniboine River, located ≈20 km to the south.

The soils are gleysols or regosols (Walker 1965) and exist dominantly as unconsolidated thick muck and peat deposits (Ehrlich et al. 1957; Walker 1959; Shay and Shay 1986) that have developed through the incomplete decomposition of organic material produced in the marsh. Kadlec (1989) provided detailed information on sediment physical charac-

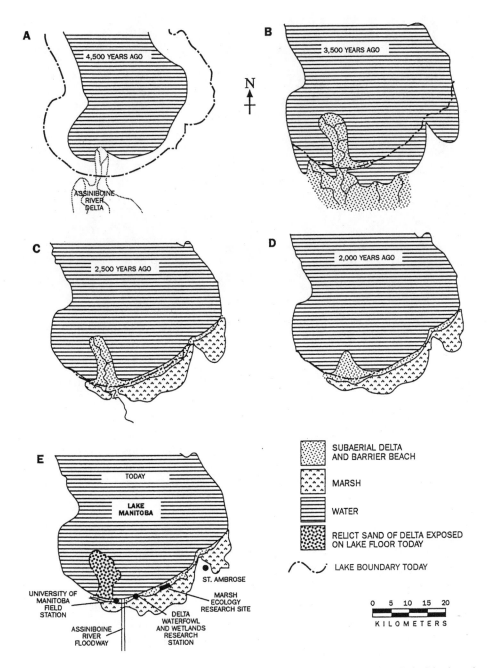

Figure 2.1   Approximate time and sequence of events leading to the formation of the Delta Marsh on the southern shore of Lake Manitoba. (A) The Assiniboine River began to flow into Lake Manitoba. (B) Deltaic deposits formed in the lake bottom. (C) Longshore drift shifted the deltaic deposits and built the barrier beach ridge. (D) About 2,000 years ago the marsh and lake stabilized at roughly the current configurations. (E) Today. After Teller and Last (1981).

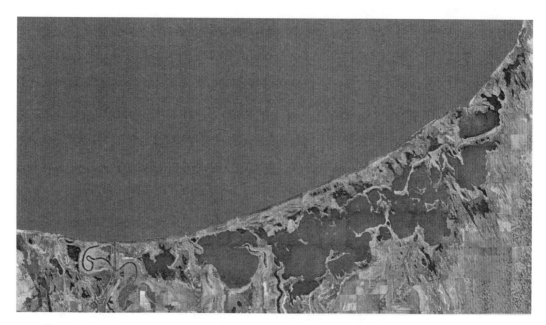

Figure 2.2   Aerial view of the Delta Marsh showing the complex of large windswept bays, smaller bays, sloughs, creeks, and channels that are interspersed with the uplands.

teristics and nutrient concentrations in the Delta Marsh. The waters are moderately brackish with conductivities averaging 2,600 mS/cm at 25°C (Anderson 1978). The pH varies from 8.2 to 9.0 and total alkalinity averages 337.6 mg/l as $CaCO_3$, largely bicarbonate ions (Anderson and Jones 1976). The cations are dominated by sodium and, in descending order, magnesium, calcium, and potassium. Bicarbonate and chloride are the predominant anions. More detailed information on the water chemistry of the marsh can be found in Kadlec (1986) and Goldsborough (1994).

GEOLOGICAL HISTORY

The Delta Marsh is the product of changes that have occurred since the last advance of glaciation, which was at its peak just prior to 20,000 B.P. (Pielou 1991). Glacial Lake Agassiz was formed ≈12,000 B.P. as the ice retreated northeastward. At its maximum extent, Lake Agassiz covered much of southern Manitoba, western Saskatchewan, North Dakota, and Minnesota. It variously advanced and retreated with the receding glacier, but drained eastward for the final time ≈9,500 B.P. (Teller and Thorliefson 1983). Lake Manitoba is a remnant basin of Lake Agassiz.

Teller and Last (1981) described a probable pattern of climactic variation and the relative degree of flooding of Lake Manitoba based on their analysis of lake sediment cores. For the Delta Marsh, the most important event occurred ≈4500 B.P. when the Assiniboine River began to discharge into the southern basin of Lake Manitoba (Figure 2.1), although Rannie et al. (1989) thought the first channels might have been cut ≈7000 years B.P. This condition persisted for between 2,500 and 2,000 years until the Assiniboine River quit flowing to Lake Manitoba (Rannie et al. 1989). Between 4500 and 3500 years B.P., Lake Manitoba rose and stabilized at about its present level. The only natural outflow from the lake is the Fairford River at the northern end of the lake.

The Assiniboine River carried a heavy sediment load that was deposited in a delta that extended several miles out into the southwestern corner of the lake. Prevailing northwest winds and water currents redistributed much of this coarse-grained sediment eastward and developed the present-day beach ridge that separates the Delta Marsh from Lake Manitoba (Figure 2.1b-d). This ridge was apparently completed in its present form ≈2,500 years B.P. (Sproule 1972) with only four natural channels for water flow between lake and marsh remaining today. The resultant sheltering of the Delta Marsh from the gross physical and hydrological features of Lake Manitoba enabled the development of the wetland system that still exists.

ECOLOGICAL CHARACTERISTICS

Climate

McGinn (1992) synthesized information from two climatological data collection sites on the Delta Marsh, one at the University of Manitoba Field Station (50°11′W, 98°23′W) and one at the Delta Waterfowl and Wetlands Research Station (50°11′N, 98°18′W). There also was a site at the Portage la Prairie airport, ≈30 km south of the marsh (49°54′W, 98°16′W). The University Field Station data were highly correlated with the other sites and were used for more detailed analyses over the 22–24-year period for which records were kept (1967–1991).

Mean annual temperature is 1.5 ± 0.9°C (±SE)with monthly means ranging from –19.8 ± 2.9°C in January to 19.1 ± 1.2°C in July. Precipitation occurs as both rain and snow. Over the period of record, rainfall has occurred in any month from March to December and averages 374.7 mm/year. Snowfall has occurred in any month from October through May and averages 135.0 mm of water equivalent per year. The mean annual precipitation is 498.6 ± 95.2 mm.

The proximity of Lake Manitoba modifies the climate of the Delta Marsh slightly over what is experienced at Portage la Prairie and other localities in the region. Temperatures during the spring, summer, and fall range from 0.4°C to 2.1°C cooler than Portage la Prairie but the most important effect of the lake is on the growing season (+3 days) and frost-free period (+6 days) compared with Portage la Prairie, which also is affected by the proximity of Lake Manitoba. Other climatological sites in the region, with no lake effect, have frost-free periods that are 10 to 30 days shorter (McGinn 1992). Precipitation on the Delta Marsh is not influenced by the lake although the prevailing northerly winds carry large volumes of snow from the lake to the wooded ridge and marsh where it settles, adding moisture along the northern edge of the marsh.

During the construction of the Marsh Ecology Research Program study site, two weather stations were established to gather temperature, precipitation, evaporation, and wind speed data for calculating water budgets (see Appendix 1) and for relating other eco-logical variables with weather. Data collection methods are described in Appendix 2 and in more detail in Murkin and Murkin (1989). A summary of these weather features is pre-sented in Table 2.1 and it is consistent with the information recorded by McGinn (1992). The more detailed data collected at these sites are used throughout this book wherever relationships with weather variables are examined.

Vegetation

The vegetation of the Delta Marsh has been described by Löve and Löve (1954) and Walker (1959, 1965) and major changes during the past 40 years by de Geus (1987). The dominant shoreline emergent species in the deep marsh zone (see Millar [1976] for marsh zone classification) are bulrush (*Scirpus lacustris*, made up of two subspecies: *S. lacustris* subsp. *validus* and *S. lacustris* subsp. *glaucus*) and cattail (*Typha* spp., made up of a com-plex of *T. latifolia*, *T. angustifolia*, and the hybrid *T. × glauca*) (Scoggan 1978–1979; Shay and Shay 1986; see Chapter 7) (Figure 2.3). Shorelines that are exposed to heavy wave action may be fringed with giant cane (*Phragmites australis*) but *Phragmites* is normally found at higher elevations than *Typha* and *Scirpus*. The shallow marsh zone is dominated by whitetop grass (*Scholochloa festucacea*) and, to a lesser extent, sedge species (*Carex* spp.). Further up the moisture gradient, a zone of *Phragmites* occurs that eventually gives way to a complex of other perennial and annual forbs. See Chapter 7 for a more detailed description of the vegetation of the Delta Marsh.

The exotic plant purple loosestrife (*Lythrum salicaria*) has become well established in the vicinity of the Delta Waterfowl and Wetlands Research Station and is present at a few other sites within the marsh (personal observation). The distribution of the plant appears to have been relatively stable during the last 25 years. However, there is probably a sig-nificant seed bank in marsh substrates that would respond if water levels occurred to

Table 2.1. Mean monthly weather data at the Marsh Ecology Research Program study site during the ice free period for the years 1980–1989

Month	Air temperature (°C)		Precipitation (mm)		Evaporation (mm)		Low wind (km/h)[a]		High wind (km/h)[b]	
	Mean	SE	Mean	SE	Mean	SE	Mean	SE	Mean	SE
May	12.5	1.9	44.4	20.4	167.4	39.8	6.2	1.5	17.7	0.9
June	16.9	2.0	80.2	37.7	185.1	22.7	4.8	1.5	17.5	1.0
July	19.8	0.6	77.4	38.4	190.4	13.1	3.6	1.4	14.9	1.1
August	18.5	1.8	88.1	55.4	163.3	29.1	3.8	1.2	15.8	0.5
September	12.1	0.9	60.8	14.7	113.1	33.0	4.6	1.7	16.0	1.3
October	5.6	1.0	41.5	23.8	71.6	31.2	3.9	2.0	17.9	1.3

[a]Measured at ground level.
[b]Includes data from 1986, 1987, 1988, and 1989 only, measured at 15 m above ground level.

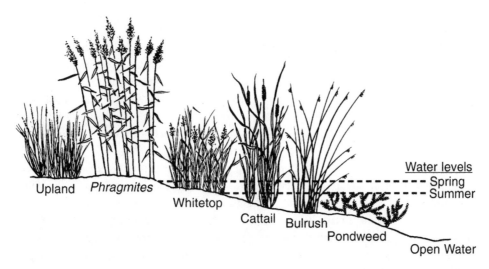

Figure 2.3 Typical shoreline vegetation of the Delta Marsh in sheltered areas. On windswept shorelines erosion and physical damage may eliminate vegetation between the open water and the *Phragmites*.

encourage germination (Thompson et al. 1987). This possibility must be incorporated into any future management plans for the Delta Marsh.

On the northern side of the marsh, the gradient continues to the wooded vegetative community of the barrier ridge. MacKenzie (1982) characterized the ridge between the village of Delta and the University of Manitoba Field Station. The dominant tree species were peach-leaved willow (*Salix amygdaloides*), Manitoba maple (*Acer negundo*), and green ash (*Fraxinus pennsylvanica* var. *subintegerrima*). Other trees included cottonwood (*Populus deltoides*), American elm (*Ulmus americana*), and hackberry (*Celtis occidentalis*). Shrubs included sandbar willow (*Salix interior*), red-berried elder (*Sambucus pubens*), red-osier dogwood (*Cornus stolonifera*), choke cherry (*Prunus virginiana*), and pin cherry (*Prunus pensylvanica*). The understory plants included common nettle (*Urtica dioica* var. *procera*), great burdock (*Arctium lappa*), wild cucumber (*Echinocystis lobata*), common hop (*Humulus lupulus*), wild morning-glory (*Convolvulus sepium*), and Virginia creeper (*Parthenocissus inserta*). To the south of the marsh, the vegetation grades through wet meadow to true prairie species (Löve and Löve 1954).

The submerged aquatic vascular plants of the eastern half of the marsh were surveyed by Anderson and Jones (1976). Eleven species were identified and covered ≈26.5% of the open-water area. Three species or species associations accounted for >60% of the submersed vegetation present: sago pondweed (*Potamogeton pectinatus*) (31%), sago pondweed and water milfoil (*Myriophyllum exalbescens*) (17%), and sheathed pondweed (*Potamogeton vaginatus*) (12%). Other species observed were Richardson's pondweed (*Potamogeton richardsonii*), bladderwort (*Utricularia vulgaris*), coontail (*Ceratophyllum demersum*), horned pondweed (*Zannichellia palustris*), and wigeon grass (*Ruppia occidentalis*).

Fauna

The vertebrate fauna of the Delta Marsh, in terms of abundance and diversity, is dominated by its avifauna. A large number of bird species use the marsh and associated woodlands for breeding and many species are transient during migratory movements. Historically, the most conspicuous of these birds have been waterfowl, the species and basic annual cycle patterns of which have been described by Hochbaum (1944, 1955) and Sowls (1955). The most common breeding species are mallard (*Anas platyrhynchos*), northern pintail (*Anas acuta*), gadwall (*Anas strepera*), blue-winged teal (*Anas discors*), American wigeon (*Anas americana*), northern shoveler (*Anas clypeata*), canvasback (*Aythya valisineria*), redhead (*Aythya americana*), and ruddy duck (*Oxyura jamaicensis*). Other common breeding waterbirds include eared grebe (*Podiceps nigricolis*), western grebe (*Aechmophorus occidentalis*), pied-billed grebe (*Podilymbus podiceps*), black-crowned night heron (*Nycticorax nycticorax*), American bittern (*Botaurus lentiginosis*), American coot (*Fulica americana*), black tern (*Chlidonias niger*), Forster's tern (*Sterna forsteri*), red-winged blackbirds (*Agelaius phoeniceus*), yellow-headed blackbirds (*Xanthocephalus xanthocephalus*), common grackle (*Quiscalus quiscula*) and sedge and marsh wrens (*Cistothorus platensis* and *Cistothorus palustris*, respectively). Harriers (*Circus cyaneus*) and short-eared owls (*Asio flammeus*) are common raptors.

The breeding bird community on the forested beach ridge is dominated by nine species (MacKenzie et al. 1982): mourning dove (*Zenaidura macroura*), eastern kingbird (*Tyrannus tyrannus*), western kingbird (*Tyrannus verticalis*), least flycatcher (*Empidonax minimus*), gray catbird (*Dumatella carolinensis*), American robin (*Turdus migratorius*), warbling vireo (*Vireo gilvus*), yellow warbler (*Dendroica petechia*), and northern oriole (*Icterus galbula*). From midsummer until mid-September, the skies above the marsh and the trees and power lines along the periphery are home to several millions of migrant tree swallows (*Tachycineta bicolor*) that apparently stage there to take advantage of emerging masses of insects, especially chironomids (Chironomidae) from both the lake and the marsh.

In terms of its potential impact on the prairie marsh ecosystem (Errington et al. 1963; Weller and Spatcher 1965), the muskrat (*Ondatra zibethicus*) is the most significant mammal in this system. Muskrat populations at Delta have fluctuated widely as a result of cyclic fluctuations in water levels. The ecology of this species and the mechanisms influencing its population fluctuations are discussed in Chapter 11. Other common mammals include white-tailed deer (*Odocoileus virginianus*), mink (*Mustela vison*), striped skunk (*Mephitis mephitis*), raccoon (*Procyon lotor*), boreal red-backed voles (*Clethrionomys gapperi*), short-tailed weasel (*Mustela erminea*), long-tailed weasel (*Mustela frenata*), woodchuck (*Marmota monax*), Richardson ground squirrel (*Spermophilus richardsonii*) and Franklin ground squirrel (*S. franklinii*). Several bat species, including silver-haired (*Lasionycteris noctivagans*) and hoary bats (*Lasiurus cinereus*), are found associated with the beach ridge (Barclay 1986).

Little research has been conducted on the fish community of the marsh. LaPointe (1986) examined fish movements between Lake Manitoba and the marsh through Cram Creek, just west of the University of Manitoba Field Station (Figure 2.1). Fifteen species were recorded moving to and from the marsh during the ice-free period between early May and the end of August. The most abundant species were the common white sucker (*Castostomus commersoni*), northern pike (*Esox lucius*), yellow perch (*Perca flavescens*), carp (*Cyprinus carpio*), and brown bullhead (*Ictalurus nebulosus*). Ten other species were caught and were representative of the fish community of Lake Manitoba. LaPointe (1986) was unable to conclude anything about the true relative abundance of the species observed because he did not sample when the creek was frozen and fish could move freely under the ice. He also was unable to sample during wind sets into or out of the marsh. In a second part of his study, LaPointe (1986) determined that fish predation reduced macroinvertebrate abundance and biomass and affected invertebrate community structure. Several temporal, structural, spatial, and taxonomic variables were shown to influence invertebrate response to predation.

The aquatic invertebrates of the Delta Marsh have been comprehensively studied (Murkin and Kadlec 1986a; Hann 1991; Murkin et al. 1991, 1992) especially with regard to interactions with waterfowl (Murkin and Kadlec 1986b; Peterson et al. 1989; Wrubleski 1989). Certain groups have received more detailed attention, especially the Chironomidae (Wrubleski 1984; Wrubleski and Roback 1987; Wrubleski and Rosenberg 1990) and the Crustacea (Smith 1968; Hann and Lonsberry 1991; Hann and Goldsborough 1997).

RECENT DEVELOPMENTS

The Delta Marsh is intimately linked to the fluctuations of water levels of Lake Manitoba. Several natural connections between the lake and the marsh (Figure 2.4) ensure that water levels in the marsh follow those of the lake. Anthropogenic interventions have been made to the flow of water between the lake and the marsh since the early 1900s (Technical Committee for the Development of the Delta Marsh 1969). Physical structures have been placed, at one time or another, on all the major connections between the two water bodies. Each was intended to stabilize marsh levels to achieve the waterfowl, furbearer, or fisheries management goals of the day. Nevertheless, the impact of most of these endeavors was relatively short term because erosion and disrepair effectively restored the connections between the lake and the marsh.

Managers also have removed structures on occasion to increase the free passage of fish, which seems logical from a fisheries management perspective. However, there is probably competition between fish and other vertebrates, such as waterfowl (Murkin and Batt 1987) and this management may reduce waterfowl use. Carp also are well established in Lake Manitoba and are extremely abundant in the lake's peripheral marshes. Carp have long been known to dramatically change the productivity of marshes (Tryon 1954; Robel 1961; King and Hunt 1967). Their omnivorous feeding habits reduce standing crops of plant and animal foods. They also reduce primary production as a result of the turbidity that they create through their feeding activities. Allowing the free passage of carp is contrary to most established marsh management objectives.

Another complicating factor relative to fish impacts was the completion of the

Figure 2.4   A wooded ridge separates Lake Manitoba from the Delta Marsh. Connections between the lake and the marsh occur at several locations and allow free passage of water, fish, and other biota between the two water bodies.

Assiniboine River Floodway in 1969 to divert floodwaters from the Assiniboine River north to Lake Manitoba across the Delta Marsh (Figure 2.1). This floodway effectively restored the connection between the river and the lake that had been terminated ≈2,300 years previously by natural forces. This connection has allowed the movement of fish between the previously distinct watersheds and new species are now found in Lake Manitoba and the marsh (Stewart et al. 1985). Although no measurements are available, the physical presence of the floodway undoubtedly impacted local surface and subsurface water movements as well.

Stabilization of Lake Manitoba water levels has had greater impact on the Delta Marsh than any other geologic, climatological, or anthropogenic intervention in the last several centuries because of the link between lake and marsh water levels. The first records of Lake Manitoba water levels were established in 1923. Until 1961, levels fluctuated within an upper and lower range of 2.1 m (Figure 2.5). This fluctuation caused some hardship for users of the lake edge during high-water years because pastures and hay meadows were made inaccessible, beach developments were threatened, and the village of Delta was flooded. During low-water years, the loss of free passage between the lake and the marshes around the lake was thought to reduce fish stocks because of reduced access to the peripheral marshes. In 1961, the Fairford River was dammed and a control structure was installed at its origin on Lake Manitoba to stabilize lake water levels. The river also was channelized to allow increased rates of flow when needed. These developments provided positive control of Lake Manitoba waters at a target level of 247.6 m above sea

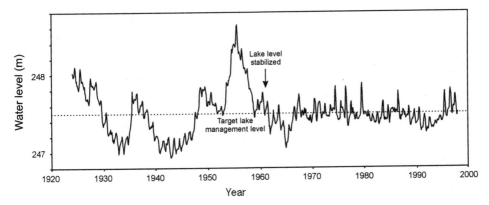

Figure 2.5   Water level data (m) for Lake Manitoba, as recorded at the Steeprock gauging station by the province of Manitoba.

level, operating within a range of ±0.3 m (Anonymous 1973; Crowe 1974). As a result both the lake and marsh have been effectively stabilized (Figure 2.5) within a very narrow range of variation.

Prior to the installation of the Fairford River Dam, from 1923 to 1961, annual mean levels ranged from a low of 247.06 m in 1943 to a high mean level of 248.42 m in 1955. During those 38 years there were two extraordinarily high periods in 1923 (248.07 m) and 1955, and two very low periods in 1933 (247.11 m) and 1943. Since 1961 there has been very little fluctuation in mean annual water levels. Thus, the natural hydrological fluctuations of water levels in the Delta Marsh, indeed in all marshes connected to Lake Manitoba, have been dramatically dampened. This dampening effect is true for both the long-term fluctuations discussed previously and the wind setups (seiches) that occur on the lakeshore in response to prevailing northerly winds (McGinn 1992). Wind setups are typically short in duration and only have localized influence on the marsh levels where openings between lake and marsh still exist.

Numerous studies have addressed the dynamic relationships of water regime and marsh vegetation in wetlands (Bourn and Cottam 1939; Kadlec 1962; Harris and Marshall 1963; Weller and Fredrickson 1974; Weller and Spatcher 1965; Meeks 1969; van der Valk and Davis 1976). The model that emerged from these and other studies was synthesized by Weller (1978) and van der Valk and Davis (1978) and is discussed more comprehensively in Chapter 1 and elsewhere in this book. The critical point relative to the Delta Marsh is that periodic flooding and drawdown are natural events that force a marsh to cycle through productive periods over time. The response by vegetation to wet and dry periods has long been exploited by marsh managers, and Weller (1978) concluded that "Stability seems deadly to a marsh system…". A refinement of the model of vegetative succession was developed by van der Valk (1981), but the conclusion regarding the effect of stability on marsh vegetation was identical. The long-term maintenance of wetlands that continuously cycle through wet and dry periods was demonstrated by McAndrews et al. (1967), who determined that a single prairie pothole wetland had been in existence for 10,000 years and countless wet–dry cycles.

The recent 30-year-plus period of essentially stable water levels on the Delta Marsh has caused many areas within the marsh to advance to, and remain at, the relatively unproductive lake marsh stage (van der Valk and Davis 1978). Predictably, it will remain in this stage unless climatic variation increases to such an extent that abundance or paucity of water overwhelms the engineering capability of the Fairford Dam. Today much of the marsh exists as a series of open-water bays surrounded by cattail-dominated (*T.* × *glauca*) shorelines and wet-meadow zones. These areas lack the habitat diversity that is the typical goal of marsh management (Weller 1978, 1981; Kaminski and Prince 1981; Murkin et al. 1982).

FUTURE PROSPECTS

de Geus (1987) quantified vegetative changes in the Delta Marsh from 1948 to 1982 and concluded that *Typha* (primarily × *glauca*) had become the dominant plant in the marsh, replacing *Phragmites*, and would probably continue to expand vegetatively if stable water levels persist. In the absence of active management or climatic change such as has not been experienced in the recent several centuries, the major extent of the Delta Marsh will continue to exist as a series of shallow, muck-bottomed, winter-kill lakes surrounded by cattail. This state will probably be followed by a gradual infilling of the marsh through the accumulation of organic sediments and the continued expansion of cattail. The dominant *T.* × *glauca* is vigorous and tolerant of deep water (Waters and Shay 1990), a factor that might contribute to infilling with accumulated organic material. This process will probably be accelerated in the western portion of the marsh by silts deposited by the overflow of the Assiniboine River Diversion directly into the marsh as has occurred on several occasions since it was constructed. Clearly, if wildlife management is a priority objective for the Delta Marsh system, the restoration of a managed wet–dry cycle is necessary. This restoration can only be established by one of two options: (1) the planned regulation of Lake Manitoba at higher or lower levels, or (2) a separation of the marsh from the lake and management of each system independently.

Careful regulation of the high- and low-water cycle on Lake Manitoba would be preferred because all the wetlands associated with the lake would benefit. However, due to investments in the agricultural, residential, and recreational infrastructure that now surround Lake Manitoba, it is unlikely that this option would be considered for political and economic reasons. Separation from the lake is a more limited and achievable approach. It would allow artificial management of water levels through wet and dry periods and would allow for the more surgical management of the fisheries and wildlife resources.

All the raw materials needed to return one of the world's most famous wetlands to a more naturally productive and dynamic ecosystem remain in place. Leadership and unprecedented mutual will and cooperation by local residents and public and private resource management agencies and organizations have been lacking. The results of the research reported in this book provide the basis for such management.

PRACTICAL BASIS FOR THE MARSH ECOLOGY RESEARCH PROGRAM

The Marsh Ecology Research Program (MERP) was conceived by staff of the Delta Waterfowl and Wetlands Research Station and Ducks Unlimited Canada, following discussions during the late 1970s. As discussed in Chapter 1, they observed that the technical basis of marsh management schemes for waterfowl had not advanced very far in recent decades because of the lack of reliable data from replicated, experimental studies. At the Delta Waterfowl and Wetlands Research Station, there was also concern over the possible changes that might be developing because of the constant water levels imposed on the Delta Marsh by regulation of Lake Manitoba. Ducks Unlimited also was aware that their inventory of developed wetlands in Canada was continuing to grow and they wanted to ensure that their management plans were based on the best information available. Both organizations recognized a need for increased numbers of experienced individuals who could become the future wetlands managers and scientists needed to develop and implement more sophisticated and broadly based management programs. Beyond these factors, waterfowl research through the Delta Waterfowl and Wetlands Research Station on breeding areas had focused primarily on the biology of the birds themselves. The station was interested in expanding their research activities to develop a better understanding of the key ecosystem, which provides many of the birds' requirements that influence recruitment and survival during the breeding period.

The technical basis for the MERP is described in Chapter 1. The site for the MERP experimental complex was chosen because of several features. First, the area had representative portions of shallow- and deep-marsh zones characteristic of the Delta Marsh and other northern prairie marshes. Second, the area is remote and completely within the private ownership of the Delta Waterfowl and Wetlands Research Station, thus ensuring access for researchers, while affording effective control of trespass and disturbance. Third, the site is served by an existing road system. This road system was crucial, given the large number of individuals working on the site each day; the large numbers and volumes of water, soil, vegetation, and animal samples to be handled; and the logistics of transporting and maintaining sampling devices. Finally, the existing road provided the northern boundary of the experimental cells, thereby reducing construction costs.

The water levels used in the experiments were chosen to reflect conditions that had occurred naturally during the previous 5 decades. Thus, the experimental site and design has provided information that is relevant to any future management efforts on the Delta Marsh, specifically, and wetland management in general.

ACKNOWLEDGMENTS

I thank Henry Murkin, Gordon Robinson, Bill Clark, Arnold van der Valk, and two anonymous reviewers for comments on earlier drafts of this paper. I hope they agree with my reconciliation of the suggestions they provided. Lisette Ross provided important assis-

tance with many logistic matters. Roger Pederson, Bill Tedford and Shane Gabor provided several items of reference materials. Ron Hempel and Dale Wrubleski provided assistance with the figures. I also thank Gordon Robinson for permission to cite information
from The University of Manitoba Field Station Annual Reports. Gordon Goldsborough
provided the figure showing Lake Manitoba water levels. This is Paper No. 95 of the
Marsh Ecology Research Program, a joint project of Ducks Unlimited Canada and the
Delta Waterfowl and Wetlands Research Station.

LITERATURE CITED

Anderson, M.G. 1978. Distribution and production of sago pondweed (*Potamogeton pectinatus* L.)
on a northern prairie marsh. Ecology 59:154-160.
Anderson, M.G., and R.E. Jones. 1976. Submerged aquatic vascular plants of east Delta Marsh.
Wildlife Report of Delta Marsh Management Study, Manitoba Department of Renewable
Resources and Transportation Services, Winnipeg, Manitoba.
Anonymous. 1973. Lake Manitoba regulation. Manitoba Water Commission Report. Manitoba
Water Commission, Winnipeg, Manitoba.
Barclay, R.M.R. 1986. Foraging strategies of silver-haired (*Lasionycteris noctivagans*) and hoary
(*Lasiurus cinereus*) bats. Myotis 23-24:161-166.
Bell, C.H., and P. Ward. 1984. Delta Waterfowl Research Station. In *Flyways: Pioneering Waterfowl
Management in North America*. (Eds.) A.S. Hawkins, R.C. Hanson, H.K. Nelson, and H.M.
Reeves, pp.321-328. United States Department of Interior, Fish and Wildlife Service.
Washington: U.S. Government Printing Office.
Bourn, W.S., and C. Cottam. 1939. The effect of lowering marsh levels on marsh wildlife.
Transactions of the North American Wildlife and Natural Resources Conference 4:343-
350.
Crowe, J. 1974. Lake Manitoba—the third great lake. Research Branch Information Series 5,
Manitoba Department of Mines, Resources and Environment Management, Winnipeg, Manitoba.
de Geus, P.M. 1987. Vegetation changes in the Delta Marsh, Manitoba between 1948–1980. M.S.
thesis, University of Manitoba, Winnipeg, Manitoba.
Ehrlich, W.A., E.A. Poyser, and L.E. Pratt. 1957. Report on the reconnaissance soil survey of
Carberry map sheet area. Soils Report 7, Manitoba Soils Survey, Winnipeg, Manitoba.
Errington, P.L., R. Siglin, and R. Clark. 1963. The decline of a muskrat population. Journal of
Wildlife Management 27:1-8.
Goldsborough, L.G. 1994. Weather and water quality data summary (1994), University Field Station
(Delta Marsh). In *University Field Station (Delta Marsh) Annual Report 29*. (Ed.) L.G.
Goldsborough, pp.11-19. Winnipeg: University of Manitoba.
Hann, B.J. 1991. Invertebrate grazer periphyton interactions in a eutrophic marsh pond. Freshwater
Biology 26:87-96.
Hann, B.J., and L.G. Goldsborough. 1997. Responses of a prairie wetland to press and pulse addi-
tions of inorganic nitrogen and phosphorus: invertebrate community structure and interactions.
Archiv für Hydrobiologie 140:169-194.
Hann, B.J., and B. Lonsberry. 1991. Influence of temperature on hatching of eggs of *Lepidurus
couesii* (Crustacea, Notostraca). Hydrobiologia 212:61-66.
Harris, S.W., and W.H. Marshall. 1963. Ecology of water-level manipulations on a northern marsh.
Ecology 44:331-343.

Hochbaum, H.A. 1944. *Canvasback on a Prairie Marsh.* Washington: American Wildlife Institute.

Hochbaum, H.A. 1955. *Travels and Traditions of Waterfowl.* Minneapolis: University of Minnesota Press.

Kadlec, J.A. 1962. Effects of a drawdown on a waterfowl impoundment. Ecology 43:267-281.

Kadlec, J.A. 1986. Effects of flooding on dissolved and suspended nutrients in small diked marshes. Canadian Journal of Fisheries and Aquatic Sciences 43:1999-2008.

Kadlec, J.A. 1989. Effects of deep flooding and drawdown on freshwater marsh sediments. In *Freshwater Wetlands and Wildlife.* (Eds.) R.R. Sharitz and J.W. Gibbons, pp.127-143. USDOE Symposium Series No. 61, Oak Ridge: USDOE Office of Scientific and Technical Information.

Kaminski, R.M., and H.H. Prince. 1981. Dabbling duck and aquatic macroinvertebrate responses to manipulated wetland habitat. Journal of Wildlife Management 44:1-15.

King, D.R., and G.S. Hunt. 1967. Effect of carp on vegetation in a Lake Erie marsh. Journal of Wildlife Management 31:181-188.

LaPointe, G.D. 1986. Fish movement and predation on macroinvertebrates in a lakeshore marsh. M.S. thesis, University of Minnesota, St Paul, Minnesota.

Löve, A., and D. Löve. 1954. Vegetation of a prairie marsh. Bulletin of the Torrey Botanical Club 81:16-34.

MacKenzie, D.I. 1982. The dune-ridge forest, Delta Marsh, Manitoba: overstory vegetation and soil patterns. Canadian Field-Naturalist 96:61-68.

MacKenzie, D.I., S.G. Sealy, and G.D. Sutherland. 1982. Nest-site characteristics of the avian community in the dune-ridge forest, Delta Marsh, Manitoba: a multivariate analysis. Canadian Journal of Zoology 60:2212-2223.

McAndrews, J.H., R.E. Stewart, Jr., and R.C. Bright. 1967. Paleoecology of a prairie pothole: a preliminary report. In *Midwestern Friends of the Pleistocene Guidebook.* (Eds.) L. Clayton and T.F. Freers, pp.101-113. North Dakota Geological Society Annual Field Conference 18, Miscellaneous Series 30.

McGinn, R.A. 1992. Climatology of the Delta Marsh. In *University Field Station (Delta Marsh) Annual Report 27.* (Ed.) L.G. Goldsborough, pp.65-77. Winnipeg: University of Manitoba.

Meeks, R.L. 1969. The effect of drawdown date on wetland plant succession. Journal of Wildlife Management 33:817-821.

Millar, J.B. 1976. Wetland classification in western Canada: a guide to marshes and shallow open water wetlands in the grasslands and parklands of the Prairie Provinces. Canadian Wildlife Service Report Series 37.

Moore, D.D. 1969. Delta Marsh management study report. Ducks Unlimited Canada Report, Winnipeg, Manitoba.

Murkin, E.J., and H.R. Murkin. 1989. Marsh Ecology Research Program long-term monitoring procedures manual. Delta Waterfowl and Wetland Research Station Technical Bulletin 2, Portage la Prairie, Manitoba.

Murkin, E.J., H.R. Murkin, and R.D. Titman. 1992. Nektonic invertebrate abundance and distribution at the emergent vegetation-open water interface in the Delta Marsh, Manitoba, Canada. Wetlands 12:45-52.

Murkin, H.R., and B.D.J. Batt. 1987. The interactions of vertebrates and invertebrates in peatlands and marshes. Memoirs of Entomological Society of Canada 140:15-30.

Murkin, H.R., and J.A. Kadlec. 1986a. Responses by benthic macroinvertebrates to prolonged flooding of marsh habitat. Canadian Journal of Zoology 64:65-72.

Murkin, H.R., and J.A. Kadlec. 1986b. Relationship between waterfowl and macroinvertebrate densities in a northern prairie marsh. Journal of Wildlife Management 50:212-217.

Murkin, H.R., R.M. Kaminski, and R.D. Titman. 1982. Responses of dabbling ducks and aquatic invertebrates to an experimentally manipulated cattail marsh. Canadian Journal of Zoology 60:2234-2242.

Murkin, H.R., J.A. Kadlec, and E.J. Murkin. 1991. Effects of prolonged flooding on nektonic invertebrates in small diked marshes. Canadian Journal of Fisheries and Aquatic Sciences 48:2355-2364.

Peterson, L.P., H.R. Murkin, and D.A. Wrubleski. 1989. Waterfowl predation on benthic macroinvertebrates during fall drawdown on a northern prairie marsh. In *Freshwater Wetlands and Wildlife*. (Eds.) R.R. Sharitz and J.W. Gibbons, pp.681-689. USDOE Symposium Series 61, Oak Ridge: USDOE Office of Scientific and Technical Information.

Pielou, E.C. 1991. *After The Ice Age: The Return of Life to Glaciated North America*. Chicago: University of Chicago Press.

Rannie, W.F., L.H. Thorliefson, and J.T. Teller. 1989. Holocene evolution of the Assiniboine River paleochannels and Portage la Prairie alluvial fan. Canadian Journal of Earth Science 26:1834-1841.

Robel, R. 1961. The effects of carp populations on the production of waterfowl food plants on a western waterfowl marsh. Transactions of the North American Wildlife and Natural Resources Conference 26:147-159.

Scoggan, H. J. 1978–1979. *The flora of Canada*. Parts 1–4. National Museum of Natural Science, Ottawa, Ontario.

Shay, J.M., and T.C. Shay. 1986. Prairie marshes in western Canada, with specific reference to the ecology of five emergent macrophytes. Canadian Journal of Botany 64:443-454.

Smith, T.G. 1968. Crustacea of the Delta Marsh region, Manitoba. Canadian Field Naturalist 82:120-139.

Sowls, L.K. 1955. *Prairie Ducks*. Washington: Wildlife Management Institute.

Sproule, T.A. 1972. A paleoecological investigation into the postglacial history of the Delta Marsh. M.S. thesis, University of Manitoba, Winnipeg.

Stewart , K.W., I.M. Suthers, and K. Leavesley. 1985. New fish distribution records in Manitoba and the role of a man-made interconnection between two drainages as an avenue of dispersal. Canadian Field Naturalist 99:317-326.

Technical Committee for the Development of the Delta Marsh. 1969. The Delta Marsh, its values, problems and potentialities. Manitoba Department of Mines and Natural Resources, Winnipeg.

Teller, J.T., and W.M. Last. 1981. Late Quarternary history of Lake Manitoba, Canada. Quarternary Research V 16:97-116.

Teller, J.T., and L.H. Thorliefson. 1983. The Lake Agassiz-Lake Superior connection. In *Glacial Lake Agassiz*. (Eds.) J.T. Teller and L. Clayton, pp.17-18. Geological Association of Canada, Special Paper 26.

Thompson, D.Q., R.L. Stuckey, and E.B. Thompson. 1987. Spread, impact, and control of purple loosestrife (*Lythrum salicaria*) in North America. U.S. Fish and Wildlife Service Report 2.

Tryon, C.A. 1954. The effect of carp exclosures on growth of submerged aquatic vegetation in Pymatuning Lake, Pennsylvania Journal of Wildlife Management 18:251-254.

van der Valk, A.G. 1981. Succession in wetlands: a Gleasonian approach. Ecology 62:688-696.

van der Valk, A.G., and C.B. Davis. 1976. The seed banks of prairie glacial marshes. Canadian Journal of Botany 54:1832-1838.

van der Valk, A.G., and C.B. Davis. 1978. The role of seed banks in the vegetation dynamics of prairie glacial marshes. Ecology 59:322-335.

Walker, J.M. 1959. Vegetation studies in the Delta Marsh, Delta, Manitoba. M.S. thesis, University of Manitoba, Winnipeg, Manitoba.

Walker, J.M. 1965. Vegetation changes with falling water levels in the Delta Marsh, Manitoba. Ph.D. dissertation, University of Manitoba, Winnipeg, Manitoba.

Waters, I., and J.M. Shay. 1990. A field study of the morphometric response of *Typha glauca* to a water depth gradient. Canadian Journal of Botany 68:2339-2343.

Weller, M.W. 1978. Management of freshwater marshes for wildlife. In *Freshwater Wetlands: Ecological Processes and Management Potential*. (Eds.) R.E. Good, D.F. Whigham, and R.L. Simpson, pp.267-284. New York: Academic Press.

Weller, M.W. 1981. *Freshwater Marshes, Ecology and Wildlife Management.* University of Minnesota Press, Minneapolis, Minnesota.

Weller, M.W., and L.H. Fredrickson. 1974. Avian ecology of a managed glacial marsh. Living Bird 12:269-291.

Weller, M.W., and C.S. Spatcher. 1965. Role of habitat in the distribution and abundance of marsh birds. Special Report Number 43, Iowa Agriculture and Home Economics Experiment Station, Ames, Iowa.

Wrubleski, D.A. 1984. Species composition, emergence phenologies, and relative abundances of Chironomidae (Diptera) from the Delta Marsh, Manitoba, Canada. M.S. thesis, University of Manitoba, Winnipeg, Manitoba.

Wrubleski, D.A. 1989. The effect of waterfowl feeding on a chironomid (Diptera: Chironomidae) community. In *Freshwater Wetlands and Wildlife*. (Eds.) R.R. Sharitz and J. W. Gibbons, pp.691-696. USDOE Symposium Series No. 61. Oak Ridge: USDOE Office of Scientific and Technical Information.

Wrubleski, D.A., and S.S. Roback. 1987. Two species of *Procladius* (Diptera: Chironomidae) from a northern prairie marsh: descriptions, phenologies and mating behaviour. Journal of the North American Benthological Society 6:198-212.

Wrubleski, D.A., and D.M. Rosenberg. 1990. The Chironomidae (Diptera) of Bone Pile Pond, Delta Marsh, Manitoba, Canada. Wetlands 10:243-275.

Part II

Nitrogen, Phosphorus, and Carbon Budgets During the MERP Experiments

3

The MERP Nutrient Budgets

John A. Kadlec
Arnold G. van der Valk
Henry R. Murkin

Abstract

Nitrogen (N), phosphorus (P), and carbon (C) nutrient budgets of the Marsh Ecology Research Program (MERP) cells were constructed using seasonal input–output budgets and estimates of intrasystem nutrient pools and fluxes. The three nutrient budgets (N, P, and C) have a nearly identical structure (Figure 3.1) and differ primarily in their pathway of macrophyte uptake of N and P (from the pore water) and of C (by photosynthesis). Nutrient pools or compartments (aboveground and belowground macrophytes, algae, invertebrates, suspended solids, dissolved inorganic and organic nutrients in surface water and pore water) were measured as were inputs and outputs of nutrients due to pumping, seepage, and precipitation. Intrasystem net nutrient fluxes were estimated from changes in the size of nutrient pools. The N fluxes to and from the atmosphere were assumed to be equal and are not included in the N model. Because there was more than one unknown flux in some of the mass balance equations, the direction and partitioning of some fluxes were constrained on the basis of previous studies of nutrient cycling in wetlands or by using best professional judgment. Three seasonal budgets were developed for each year: late spring–early summer (1 June-1 August); late summer–fall (1 August-1 October); and winter–early spring (1 October-1 June).

Introduction

Although changes in the vegetation and fauna of prairie wetlands during wet–dry cycles had been described by the late 1970s (Weller and Spatcher 1965; Weller and Fredrickson 1974; van der Valk and Davis 1978a), little was known about how water-level fluctuations during these cycles affected ecosystem processes such as primary production and nutrient cycling (Davis and van der Valk 1978a, b; van der Valk and Davis 1978b). Consequently, the major objective of the Marsh Ecology Research Program (MERP) was to assess the effects of water-level changes on the nitrogen (N), phosphorus (P), and carbon (C) budgets during a simulated wet–dry cycle. Known changes in the abundance of emergent vegetation during wet–dry cycles (Weller and Fredrickson 1974; van der Valk and Davis 1978a, b) and a preliminary understanding of the links between the vegetation nutrient pool and the other pools within wetlands (Neely and Baker 1989) made it possible to hypothesize about the changes in the sizes of nutrient pools and fluxes (nutrient movements among these pools) that would occur during the MERP study. The primary predictions related to changes in nutrient budgets during the various phases of the experiments within MERP were as follows:

1. The release of nutrients into the water column from plants killed by 2 years of deep flooding (see Table 1.1 in Chapter 1), especially during the first year of flooding.

2. The release of nutrients from litter and sediments during the 1- and 2-year drawdowns (see Table 1.1 in Chapter 1) because of accelerated decomposition of litter, especially in cells drawn down for 2 years.

3. A reduction in the size of the macrophyte nutrient pools in cells in the medium and high treatments (see Table 1.2 in Chapter 1) during the reflooding years (1985–1989), especially by the fifth year of flooding at these water levels (1989).

In this chapter, the mass balance approach used to develop the N, P, and C budgets for the MERP experimental cells is described. A brief overview of the structure of these budgets is presented first, followed by a description of how fluxes within these budgets were calculated. A schematic overview of these general budgets is presented in Figure 3.1. More detailed budgets used to calculate seasonal fluxes of N, P, and C to and from the macrophytes pools are shown in Figure 3.2. Acronyms used in these budgets are defined in Table 3.1, and the equations used to calculate the fluxes among pools are presented in

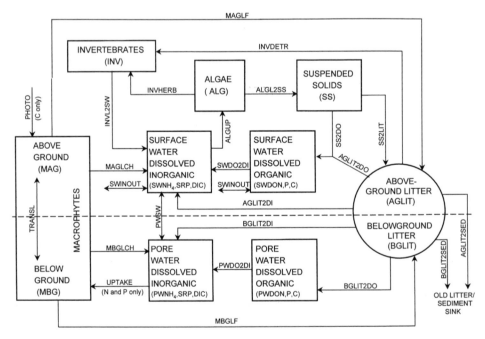

Figure 3.1   The MERP mass balance nutrient budgets. See Table 3.1 for definition of acronyms and Table 3.2 for equations used to calculate fluxes.

A. 1 June to 1 August. LATE SPRING/EARLY SUMMER

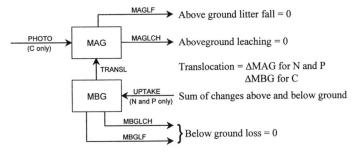

B. 1 August to 1 October. LATE SUMMER/FALL

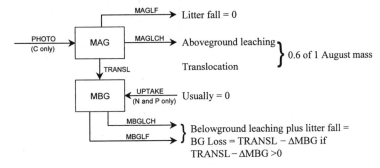

C. 1 October to 1 June. WINTER/EARLY SPRING

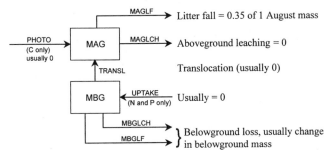

Figure 3.2   Submodels of seasonal fluxes to and from macrophyte compartments. See Table 3.1 for definition of acronyms.

Table 3.2. The results of these nutrient budgets during the various stages of the MERP study and comparisons of predicted (hypothesized) and actual changes in nutrient pools and fluxes and their ecological implications are examined in Chapters 4 and 5. Chapter 6 incorporates the MERP nutrient budget results into an overall discussion of nutrient budgets during the wet–dry cycle in prairie wetlands.

Table 3.1. Definitions of pool and flux acronyms used in the MERP nutrient budgets (Figures 3.1 and 3.2)

Pools or fluxes	Acronym[a]
Pools	
Macrophytes	
aboveground	MAG
belowground	MBG
Algae	ALG
Invertebrates	INV
Surface water dissolved organic nutrients	SWDON, SWDOP or SWDOC
Surface water dissolved inorganic nutrients	$SWNH_4$, SWSRP or SWDIC
Pore water dissolved organic nutrients	PWDON, PWDOP or PWDOC
Pore water dissolved inorganic nutrients	$PWNH_4$, PWSRP or PWDIC
Suspended solids	SS
Litter pool	
aboveground	AGLIT
belowground	BGLIT
Aboveground Fluxes	
Net carbon fixation (photosynthesis)	PHOTO
Macrophyte leaching	MAGLCH
Algal uptake	ALGUP
Algal losses to suspended solids	ALGL2SS
Invertebrates losses to surface water	INVL2SW
Invertebrate nutrient uptake from litter	INVDETR
Invertebrate nutrient uptake from algae	INVHERB
Litter leaching of inorganic nutrients to surface water	AGLIT2DI
Litter leaching of dissolved organic nutrients to surface water	AGLIT2DO
Suspended solids to dissolved organic nutrients	SS2DO
Conversion of dissolved organic to dissolved inorganic nutrients	SWDO2DI
Hydrologic inputs–outputs to surface water	SWINOUT
Above- and belowground fluxes	
Translocation	TRANSL
Macrophyte litter-fall	MAGLF
Loss of aboveground particulate litter to old litter/sediments	AGLIT2SED
Sedimentation of suspended particles to litter pool	SS2LIT
Pore water–surface water exchange	SWPW
Belowground fluxes	
Belowground macrophyte uptake	UPTAKE
Belowground leaching	MBGLCH
Belowground litter-fall	MBGLF
Litter leaching of inorganic sediments	BGLIT2DI
Litter leaching of organic nutrients	BGLIT2DO
Mineralization of dissolved organic nutrients	PWDO2DI
Particulate litter inputs to old litter/sediments	BGLIT2SED

[a]Net change in the mass of a nutrient pool during a season is indicated by a Δ before an acronym in the models.

Table 3.2. Equations and constraints for calculating net nitrogen, phosphorus, and carbon fluxes from changes in nutrient pools in above- and belowground macrophytes.

Flux	Season	Constraints	Equation
Macrophytes			
UPTAKE (N or P only)	Early summer	IF $(\Delta MAGx + \Delta MBGx) > 0$	UPTAKE $= \Delta MAGx + \Delta MBGx$, otherwise 0
	Late summer	IF $(\Delta MAGx + \Delta MBGx - MAGLCH) > 0$	UPTAKE $= \Delta MAGx + \Delta MBGx - MAGLCH$, otherwise 0
	Winter	IF $\Delta MBGx > 0$	UPTAKE $= \Delta MBGx$, otherwise 0
PHOTO (C only)	Early summer		PHOTO $= \Delta MAG + \Delta MBG$
	Late summer	IF $(TRANSL - \Delta MBG) < 0$	PHOTO $= \Delta MBGx - TRANSL$, otherwise 0
	Winter		PHOTO $= 0$
TRANSL (N or P)	Early summer	MAGLCH $= 0$	TRANSL $= \Delta MAGx$
	Late summer	MAGLCH ≥ 0	TRANSL $= \Delta MAGx - MAGLCH$
	Winter		TRANSL $= 0$
TRANSL (C only)	Early summer		TRANSL $= \Delta MBG$
	Late summer	IF $(1 - MAG \text{ leach rate}^a) \cdot \Delta MAG < \Delta MBG$	TRANSL $= \Delta MBG$, otherwise $(MAG \text{ leach rate} - 1) \cdot \Delta MAG$
	Winter	IF $\Delta MBG > 0$	TRANSL $= \Delta MBG$, otherwise 0
MAGLCH	Late summer		MAGLCH $= -(MAG \text{ leach rate}^a) \cdot \Delta MAG$
MBGLCH			MBGLCH $= (MBG \text{ leach rate}^b) \cdot MBGLOSS^c$
MAGLF	Winter		MAGLF $= -\Delta MAGx$
MBGLF			MBGLF $= (1 - MBG \text{ leach rate}) \cdot MBGLOSS$
Algae and invertebrates			
ALGUP		IF $(\Delta ALGx + INVHERB) > 0$	ALGUP $= \Delta ALGx + INVHERB$, otherwise 0
ALGL2SS		IF $(\Delta ALGx + INVHERB) < 0$	ALGL2SS $= -(\Delta ALGx + INVHERB)$, otherwise 0
INVL2SW		IF $\Delta INVx < 0$	INVL2SW $= -\Delta INVx$, otherwise 0
INVDETR		IF $\Delta INVx > 0$	INVDETR $= 0.5^d \cdot (\Delta INVx)$, otherwise 0
INVHERB		IF $\Delta INVx > 0$	INVHERB $= 0.5^d \cdot (\Delta INVx)$, otherwise 0
Dissolved nutrients and suspended solids			
SWINOUT			SWINOUT $=$ Pumping inputs/outputs + precipitation inputs + seepage inputs/outputs
SWDO2DI		IF $\Delta SWDOx < 0$	SWDO2DI $= -\Delta SWDOx$, otherwise 0
PWDO2DI		IF $\Delta PWDOx < 0$	PWDO2DI $= -(\Delta PWDOx)$, otherwise 0
PWSW		IF $LIT2SW^e < 0$	PWSW $= -LIT2SW$, otherwise 0
SS2LIT			SS2LIT $= ALGL2SS - \Delta SSx$

Continued

Table 3.2.   (*Continued*)

Flux	Season	Constraints	Equation
Litter/sediment pool			
AGLIT2DO/SS2DO		IF ΔSWDOx > 0	AGLIT2DO/SS2DO = ΔSWDOx, otherwise 0
AGLIT2DI		IF LIT2SW[e] > 0	AGLIT2DI = LIT2SW, otherwise 0
BGLIT2DO		IF ΔPWDOx > 0	BGLIT2DO = ΔPWDOx, otherwise 0
BGLIT2DI			BGLIT2DI = ΔPWxxx – MBGLCH + UPTAKE (N&P only) + SWPW – PWDO2DI
AGLIT2SED			AGLIT2SED = AGDECOMPx[f] – AGLIT2DO – AGLIT2DI – INVDETR
BGLIT2SED			BGLIT2SED = BGDECOMPx[g] – BGLIT2DO

Note: See Table 3.1 definitions of pool and flux acronyms. Separate models were developed for N, P, and C. An x represents N, P, or C and xxx represents NH_4, SRP, or DIC.

[a]Fraction of aboveground nutrient mass lost via leaching.

[b]Fraction of belowground nutrient mass lost via leaching.

[c]MBGLOSS: Early summer   MBGLOSS = 0
                      Late summer   If (TRANSL + ΔMBG) < 0, then MBGLOSS = TRANSL + ΔMBG, otherwise 0
                      Winter   If ΔMGB < 0, then MBGLOSS = –ΔMBG, otherwise 0

[d]Fraction of the invertebrate diet derived from litter and algae.

[e]LIT2SW = ΔSWxxx – SWINOUT – INVL2SW – SWDO2DI + ALGUP – MAGLCH.

[f]AGDECOMPx = AGk • AGLITx, AGk being the aboveground litter decomposition constant and AGLITx being the mass of a nutrient in the aboveground litter. AGLIT2DI is constrained to be + and less than (AGDECOMPx – AGLIT2DO – AGLIT2SED). AGk is 0.08 during a season.

[g]BGDECOMPx = BGk • BGLITx where BGk is the belowground decomposition constant and BGLITx is the mass of nutrient in the belowground litter. BGk is 0.08 per season.

MASS BALANCE AND NUTRIENT BUDGETS

The quantitative description of the inputs, outputs, and internal cycling of materials in an ecosystem is called an ecosystem mass balance (Whigham and Bayley 1979; Nixon and Lee 1986). If the material being measured is N, P, or C, then the mass balance is termed a nutrient budget (Mitsch and Gosselink 1993). Mass balances provide a useful framework for organizing information about nutrient inputs and outputs and changes in nutrient pools and fluxes among these pools within an ecosystem. The advantages of using mass balance techniques include the following: (1) the method is holistic and readily integrates information over large areas (the MERP experimental cells averaged 5 ha) and extended periods of time (seasons and years as in the MERP study); (2) the calculations are conceptually simple (see Table 3.2); (3) the method is based on a fundamental precept of science (i.e., conservation of mass); and (4) for large ecosystem studies such as MERP, the data required for the calculations can be obtained from routine monitoring programs (Brezonik

et al. 1987; Evans et al. 1997; see Appendix 1). The MERP experimental cells presented
a unique opportunity to apply the mass balance approach to examine the effects of changing water levels on the nutrients pools and fluxes in a prairie wetland system.

Nutrients inputs into wetlands are primarily through precipitation and surface and
groundwater inflows (Kadlec 1983; Neely and Baker 1989; Winter 1989). Primary outputs
are through surface and groundwater outflows. Long-term loss of nutrients to the sediments also is considered an output and, in some cases, exports to the atmosphere may be
important (e.g., loss of N through denitrification) (Neely and Baker 1989). Intrasystem
cycling is the flux or exchange of nutrients among pools within the wetland (Mitsch and
Gosselink 1993). Pools are the amount of N, P, or C contained in any one compartment of
the ecosystem (e.g., aboveground macrophyte biomass) at a point in time. Fluxes or movement of nutrients among these pools are inferred from changes in pool sizes between time
periods. Fluxes are the result of seasonal or short-term physical, chemical, and biological
processes, most of which are well-known (e.g., uptake of nutrients by plants, translocation
of nutrients in emergent vegetation, and litter decomposition) (Howard-Williams 1985;
Bowden 1987).

MERP Nutrient Budgets

The MERP mass balance approach and nutrient budgets were based primarily on estimates of inputs and outputs to the experimental cells and changes in nutrient pool sizes
(i.e., vascular plants or macrophytes, algae, invertebrates, surface water, pore water) within the cells during the growing season. Appendix 1 and Murkin and Murkin (1989)
describe the sampling protocols used for all aspects of the MERP study. Input–output
estimates are available from the water budgets of the cells combined with the nutrient concentrations of the water entering or leaving the systems (Kadlec 1983, 1986a).

The general approach used to calculate the MERP nutrient budgets was similar to those
used in other wetlands by Mitsch and Reeder (1991), Grigal (1991), and Bowden et al.
(1991), except for the level of detail of some processes, fluxes, or pools. In common with
Mitsch and Reeder (1991) and Bowden et al. (1991), a detailed hydrologic budget (Kadlec
1983, 1986a, 1993) was used to determine waterborne inputs and outputs. In Bowden et
al. (1991), Grigal (1991), and the MERP, measurements of macrophyte peak standing crop
and nutrient concentrations were used to estimate the mass of nutrients within the live
macrophyte pool, whereas Mitsch and Reeder (1991) used a detailed primary production
submodel. The MERP fluxes were determined by changes in pool sizes between sampling
periods, whereas Bowden et al. (1991) used field and laboratory studies of processes in
the N cycle to estimate fluxes.

Because the MERP sample unit was an experimental cell, pools and fluxes were determined for an entire cell. Besides the 10 diked cells in the MERP experimental complex,
there were two undiked reference cells nearby in the Delta Marsh of approximately the
same size and shape (see Figure 1.1 in Chapter 1). With the exception of the way in which
inputs and outputs of dissolved inorganic and organic nutrients in surface and groundwater were calculated (see Inputs-Outputs section to follow), the budget methodology for
diked and undiked cells was identical. Field data were interpolated as required to provide

seasonal nutrient pool estimates (kg/ha) on 1 June, 1 August, and 1 October each year. Thus, for each experimental cell, nutrient fluxes were estimated for three periods each year: late spring–early summer (1 June-1 August); late summer–fall (1 August-1 October); and winter–early spring (1 October-1 June). The three periods chosen were very distinct, especially with respect to macrophyte growth and production. Late spring–early summer was the period of maximum macrophyte growth. Late summer–fall was the period of maximum standing crops of macrophytes followed by senescence as fall proceeded. Winter–early spring was the period of minimal macrophyte production, representing the dormant winter period and the early spring period when the marsh begins to thaw but before much macrophyte growth occurs. These periods are also distinct with respect to algal and invertebrate production (Murkin and Kadlec 1986; Murkin et al. 1991; Robinson et al. 1997a, b).

Overall the N, P, and C budget models are very similar (Figure 3.1), except that C inputs to macrophytes was due to photosynthesis by aboveground leaves and stems, whereas N and P input to macrophytes was due to uptake by belowground tissues from pore water. Because adequate data for C were available only for 1985–1989, C budgets are available only for the reflooding years. The N exchanges with the atmosphere were not measured regularly in the MERP cells and are not included in the N model. Some preliminary sampling in the MERP cells indicated that N fixation and denitrification rates were very low and comparable in magnitude (J.C. Cornwell, University of Maryland Center for Environmental and Estuarine Studies, unpublished data). Thus, it was assumed for these budget determinations that they cancelled each other out. Neely and Baker (1989) presented a review of N fixation and denitrification in prairie wetlands. They concluded that available information is variable and there is a need for more reliable estimates of the magnitude of these processes in prairie wetland systems. Nevertheless, the lack of direct measurements of N fixation and denitrification was a shortcoming of the MERP study.

Inputs–Outputs

Detailed analyses of the hydrology of each of the experimental cells (Kadlec 1983, 1993) and monthly analyses of nutrient concentrations in surface and pore water allowed calculation of inputs and outputs via pumping and seepage and input via precipitation (Kadlec 1986a,b). The major inputs of water into the cells were precipitation, pumping, and seepage, whereas the major outputs were evapotranspiration, pumping, and seepage (see Appendix 1 for a detailed description of the methods used to measure hydrologic inputs and outputs). Changes in storage were measured using a combination of staff gauges, continuous-water-level recorders, and contour maps (10-cm interval) of each cell. Both pumping (pumps were metered) and precipitation (there were two weather stations with standard rain gauges at the site) were measured directly. Although a variety of methods was used to calculate evapotranspiration, adjusted class A evaporation pan data from the study area were used for the evapotranspiration estimates for water-budget determinations. Seepage was determined by difference in the water budgets and verified by measurements of water-level changes during periods of no evapotranspiration (see Chapter 12) and seepage well readings (Kadlec 1983).

For the overall budgets, the monthly inputs and outputs were summed over each season. Samples of pump water and precipitation (from samples collected at the weather stations) were collected for determination of nutrient concentrations (see Appendix 1 for the detailed procedures of water sample collection). For water seeping into or out of the cells, surface-water nutrient concentrations were used in the input–output calculations. The water chemistry of these marshes, as indicated by N and P concentrations and budgets, strongly suggested that seepage was primarily through dikes and the north beach ridge separating the experimental cells from Lake Manitoba (Kadlec 1986a, b; 1993; unpublished data). Inward seepage was significant only in the drawdown years when water levels in the diked cells were lower than in the surrounding marsh (Kadlec 1993).

The water budgets for the undiked reference marshes (cells 11 and 12) were much less precise than for the diked experimental cells. For these marshes, differences between precipitation and evaporation, with an adjustment for water-level change, were assumed to be met by surface flow from the surrounding marsh. Computationally, this amount was equivalent to seepage in the diked cells. For purposes of nutrient budgets, the surface flow was considered to contain surface-water nutrient concentrations.

Intrasystem Cycling

Pools

Nutrient pool sizes within the MERP experimental cells were calculated by multiplying nutrient concentrations in the material in a pool by the total mass of material in that pool. The MERP budgets have nine pools (boxes and circles on Figure 3.1). In the eight cases represented by boxes, the pools were measured directly, whereas the litter/sediment pool (circle indicating the current year's litter production plus old litter/sediment sink, which accounted for the remaining litter from previous growing seasons) is a residual or net accumulation or storage pool that was not measured.

Macrophytes

Aboveground macrophyte standing crops were sampled once a year near peak biomass in late July-early August (see Appendix 1). These data were used to calculate the 1 August estimates of the mass of nutrients in the macrophyte aboveground pool. The 1 October aboveground-pool estimates were set at 40% of August 1 estimates based on reported differences between standing-live and standing-dead nutrient concentrations (Nelson et al. 1990; van der Valk et al. 1991). With the onset of winter, the majority of the remaining nutrients was transferred to the litter sink via winter litter fall (Davis and van der Valk 1978a,b). Spring nutrient mass in the aboveground macrophyte pool due to survival of overwintering buds and shoots was assumed to be only 5% of the previous summer peak (Bernard and Gorham 1978).

Belowground macrophyte standing crop was sampled twice a year; once in spring (late May-early June) near seasonal minimum biomass when belowground reserves were reduced by early spring growth, and again in late fall (late September-early October) near presumed maximum biomass following nutrient and carbohydrate translocation to roots and rhizomes in late summer and fall (Linde et al. 1976; van der Valk and Davis 1978b; Neely and Baker 1989). The 1 August belowground biomass in the MERP experimental cells was assumed to be the mean of the June and September-October estimates (Schierup 1978).

Algae and Invertebrates

The pools of N, P, and C in algal biomass had four components: phytoplankton, periphyton, metaphyton, and epipelon (see Chapter 9 and Appendix 1). The MERP algal monitoring varied somewhat over the course of the study. During the deep-flooding years, the best estimate of phytoplankton N, P, and C was the "suspended" component of the surface-water samples. The ratio of P:N:C suggested that the particulate material suspended in the water column was primarily of algal origin (Kadlec, unpublished data). Some data on pre-drawdown periphyton, phytoplankton, and metaphyton were available from Hosseini and van der Valk (1989a,b), and detailed post-drawdown monitoring of algal productivity and biomass was carried out by Gurney and Robinson (1989) and Robinson et al. (1997a, b). Metaphyton (mats of filamentous algae) was routinely harvested and analyzed for N, P, and C as part of the macrophyte sampling and analyses (see Appendix 1). Interpolation from data collected at the closest time were used to estimate N, P, and C algal pools on 1 June, 1 August, and 1 October of each year.

   Invertebrate standing crops were measured on a variety of schedules based on the community in question (benthic, epiphytic, and nektonic) (see Appendix 1). Invertebrate sampling was stratified by vegetation type and biomass estimates were extrapolated to the entire cell by using the known area of each vegetation type in each cell. The biomass estimates of the sampling periods nearest to 1 June, 1 August, and 1 October were used to develop a combined pool of the invertebrate communities within the experimental cells. Tissue samples of the dominant taxa sampled were analyzed for N, P, and C concentrations to allow conversion of biomass to nutrient mass for each time period.

Dissolved Nutrients and Suspended Solids

Surface-water and pore-water volume determinations combined with water chemistry analyses provided data on the four dissolved nutrient pools (see Figure 3.1) as well as the suspended-solids pool within the surface water (see Appendix 1; Kadlec 1986a,b). Surface-water samples were taken by submersing a collection bottle in mid-water column, whereas the pore-water samples were pumped from pore-water wells installed in the experimental cells (see Appendix 1 for a detailed description of sampling methods). Pore-water pools were estimated for a sediment layer 30 cm in thickness based on volume, sediment bulk density (weight of solids per unit volume), and pore-water nutrient concentrations.

Litter/sediment pool

In the MERP nutrient budgets, the litter/sediment pool consisted of the litter from the current growing season (circle in Figure 3.1) plus the accumulated litter in the sediments from previous growing seasons (old litter/sediment sink in Figure 3.1). Only fluxes into and out of the litter/sediment pool were considered. No estimates of the total size of this compartment were included in the budgets. The litter and sediment pool accumulated inputs of macrophyte and algal litter and acted as a source of dissolved N or P for surface and pore water. The behavior of this pool is based on field studies of the dynamics of the litter pool in prairie wetlands (Davis and van der Valk 1978a,b; Murkin et al. 1989; Neckles and Neil 1994) and in the MERP cells (van der Valk et al. 1991; Wrubleski et al. 1997a, b). A substantial body of evidence (Kadlec 1979; Barko et al. 1991; Bowden et al. 1991) suggests that the combined litter and sediment pool in wetlands is relatively stable and contains a very large proportion (probably >90%) of the total N and P in most wetland ecosystems. This proportion may not be accurate for newly created or restored wetlands during the first few years of development; however, we assumed this was the case for established wetlands such as the Delta Marsh. Its mass therefore would be essentially unaffected by seasonal and other short-term changes in inputs and outputs. Because the litter/sediment pool was determined by difference from mass balance calculations, it is also affected by measurement errors, so changes in this pool also incorporate total system error in flux and pool estimates.

Fluxes

Fluxes were not measured directly but were estimated from changes in pool sizes by using mass balance calculations. In Figure 3.1, fluxes are shown as arrows. The various fluxes and associated acronyms that were used in the budgets are defined in Table 3.1 and the equations used to calculate them in Table 3.2. The direction of the arrows in Figure 3.1 indicates the direction of the expected movement of a nutrient. For some fluxes, movement may be in either direction depending on time of year, the water levels, or both. Known seasonal patterns and other constraints on fluxes were used in the budgets to establish the direction of fluxes (Table 3.2; Figure 3.2). In calculating the fluxes, non-negativity constraints were imposed where negative fluxes were conceptually impossible (Table 3.2); however, negative values rarely occurred in these situations, thereby supporting the various assumptions in the budgets.

   In the budget determinations, because many fluxes are interrelated and the results of one flux estimate was needed to calculate another, fluxes were calculated in the following order:

1. Fluxes to and from aboveground and belowground macrophytes.
2. Invertebrate–algal fluxes (invertebrate consumption of algae, invertebrate consumption of litter, death of invertebrates, and algal uptake).
3. Fluxes in surface and pore water (release of dissolved organic nutrients from aboveground or belowground litter and its subsequent mineralization).
4. Fluxes associated with suspended particles in surface water.
5. Fluxes of dissolved organic and inorganic material from the litter/sediment pool to surface water.
6. Surface and pore water dissolved inorganic mass balances (all the inputs and outputs associated with changes in ammonia (NH_4), soluble reactive phosphorus (SRP), or dissolved inorganic carbon (DIC) pools).

7. Fluxes from the various inputs and outputs from the litter/sediment pool were accumulated to provide an estimate of net flux to this pool.

As mentioned, many assumptions were made based on our current understanding of nutrient pools and fluxes in prairie wetlands. Sensitivity analyses of these assumptions and restrictions given in Table 3.2 were conducted by manipulating the appropriate variables over their theoretically possible ranges in the mass balance equations. These analyses indicated that the assumptions and constraints had little affect on the overall budget outputs. The nutrient budgets in the MERP cells were controlled largely by macrophyte uptake and releases and changes in surface water caused by pumping water into and out of the cells. All other compartments and fluxes were minor in comparison so assumptions regarding changes in these components had little effect on the overall budgets.

Macrophytes

In the MERP nutrient budgets, a relatively small number of processes account for the observed changes in aboveground and belowground nutrient pools associated with living macrophytes (Figures 3.1 and 3.2). These include fluxes resulting in net increases in mass (root uptake of N and P from the sediments or C from air or water during photosynthesis), fluxes between aboveground and belowground pools (translocation), and fluxes resulting in losses from the macrophyte pools (leaching and litter fall). Macrophyte herbivory is usually considered to be minimal in prairie marshes (Murkin 1989) although muskrats may have some impact at high population levels (Fritzell 1989; see Chapter 11). Macrophyte consumption by muskrats was estimated to be between 1 and 11% of the annual macrophyte production in the MERP experimental cells (see Chapter 11). In the MERP nutrient-budget determinations, muskrat herbivory is not included as a separate flux from litter fall. In summary, inputs to the macrophyte pools were partitioned among root uptake (belowground pool), photosynthesis (aboveground pool), and translocation. Losses from the individual pools were partitioned among translocation, leaching, and litter fall. The importance of each of the factors varies with season (Figure 3.2).

The most straightforward fluxes to estimate were root uptake, photosynthesis, and translocation. These fluxes were estimated directly from changes in the size of the standing crops of nutrients in the above- and belowground macrophyte pools during the appropriate season (Table 3.2). Root uptake of N and P was calculated from the changes in aboveground and belowground masses. During the period of rapid growth in late spring–early summer, any increase in aboveground mass over belowground losses would be considered uptake. Later in the growing season (late summer–fall), a correction for aboveground leaching losses was included in these calculations. In winter–early spring when there is little or no aboveground activity, any increase in belowground mass would be attributed to uptake for N and P. There would be no or minimal photosynthetic uptake of C during winter–early spring.

Translocation should be primarily upward from belowground tissues to aboveground tissues during the late spring–early summer; downward during late summer–fall; and zero during winter–early spring (Figure 3.2). Most of the N and P in new-growth macrophytes in the spring is from belowground reserves (Jervis 1969; Linde et al. 1976; Schierup 1978); therefore, the bulk of aboveground macrophyte N and P mass at this time comes from translocation from belowground tissues (Table 3.2; Figure 3.2). In late summer–early fall, the pattern for N and P reverses and translocation will be downward. Therefore, the loss of N and P mass in the standing macrophytes during this period, corrected for leaching losses, is considered as translocation (litter fall during this period is considered to be near zero, see below) (Table 3.2). No translocation of N and P takes place in the winter. There are some differences in the translocation patterns for C compared with N and P (Table 3.2). Because C-fixation by photosynthesis occurs above ground and there is little, if any, C uptake by roots (Hwang and Morris 1992), most translocation of C is expected to be downward. Any positive change in C in the belowground pool was considered the result of translocation. In some situations in late summer–fall, when the loss of aboveground C exceeds the increase in belowground mass, the loss of belowground mass corrected for leaching losses was considered as translocation (once again it is assumed that litter fall is minimal during this period).

Leaching losses from aboveground and belowground macrophyte pools were calculated from changes in the size of the pools (corrected for translocation in the case of belowground leaching) during the period in question. Aboveground leaching was assumed to occur primarily after aboveground tissues senesced in the late summer–early fall (Figure 3.2) and was calculated as a portion of the loss of aboveground mass during that period (Table 3.2). The portions of aboveground mass lost to leaching for the macrophyte species occurring in the MERP experimental cells were determined by litter-bag studies conducted in the MERP experimental cells and in the Delta Marsh (Murkin et al. 1989; Nelson et al. 1990; van der Valk et al. 1991; Wrubleski et al. 1997a,b). Belowground leaching losses were a portion of the change in total belowground mass after correction for translocation (see footnote to Table 3.2). Because belowground tissues are in contact with pore water, immediate leaching losses can be significant. Benner et al. (1991) showed a 20% loss of biomass in dead belowground *Spartina alterniflora* tissues in the first month, which they attributed primarily to leaching. The MERP belowground leaching rates also averaged 20% within 2 days of placing root and rhizome samples below ground (Wrubleski et al. 1997b). Thus, the MERP budgets used a proportion of 0.20 leaching losses from belowground litter.

Although there is litter-fall throughout the year, the bulk occurs due to fragmentation during winter storms (Davis and van der Valk 1978a,b). Therefore, the loss of aboveground macrophyte nutrient mass to fallen litter (Figure 3.2) was assumed to occur only in winter–early spring. It was equal to 35% of the previous midsummer (1 August) peak because of the assumptions used to calculate 1 October (40% of the aboveground standing crop on 1 August) and 1 June (5% of aboveground standing crop the previous year) pools of N, P, and C in aboveground biomass. Thus, nutrient fluxes attributed to winter–early spring litter-fall merely reflect assumptions made about changes in macrophyte aboveground nutrient pools and are not based on direct measurements of litter-fall. Nevertheless, the magnitude of this flux as estimated is consistent with those observed in field studies of the transfers of nutrients from macrophytes to litter (Nelson et al. 1990;

van der Valk et al. 1991). Losses from the belowground macrophyte pools, after correction for translocation, were due to a combination of belowground leaching and particulate loss (Figure 3.2). The allocation of belowground macrophyte losses to litter-fall was simply the proportion of the losses (0.80) not assigned to leaching.

Algae and Invertebrates

Changes in the pools of N, P, and C in algae and invertebrates were assumed to be closely linked (Figure 3.1; Table 3.2). In both compartments, there is undoubtedly much more rapid turnover than is reflected in the seasonal time scale used in these models (Murkin 1989; Robinson et al. 1997a, b). Thus, the MERP budgets undoubtedly underestimate the fluxes associated with the algae and invertebrates pools. Increases in pool sizes of algal N, P, and C (Figure 3.1) were assumed to be due to uptake from the surface-water dissolved-inorganic nutrient pool and losses allocated first to invertebrate consumption and then to death (loss to the surface-water suspended-solids pool) (Table 3.2).

Increases in the pools of N, P, and C in invertebrates, which included both benthic and planktonic invertebrates, were attributed to consumption of litter, algae, or both (Figure 3.1). There is some uncertainty regarding the degree to which invertebrates ultimately depend on litter versus algae as a nutrient source (Murkin 1989). In the MERP budgets, increases in invertebrate pools of N, P, or C were apportioned equally between consumption of litter (detritivory) and algae (herbivory) (Table 3.2). All losses from the invertebrate pool in the budgets were assumed to be due to mortality. It also was assumed that the rapid decomposition of invertebrates releases nutrients from the invertebrate pool directly to surface water.

Dissolved Nutrients and Suspended Solids

The surface-water hydrologic flux (SWINOUT in Figure 3.1) combined seepage, precipitation, and pumping inputs and outputs to the surface-water nutrient pools (Table 3.2). Part of the changes in pool sizes of surface-water nutrients was due to these water movements to and from the experimental cells. These movements were measured directly (see Appendix 1) and combined with concentration data to estimate hydrologic inputs and outputs to the surface-water nutrient pools. In addition, several other fluxes affected both the surface-water nutrient pools (organic and inorganic) and the suspended-solids nutrient pools. Inputs to the dissolved organic pool include dissolved organic material from the breakdown of suspended solids material and the decomposition of aboveground macrophyte litter. The MERP budget analyses were unable to distinguish between the input from suspended solids breakdown and macrophyte litter breakdown (Figure 3.1) so they are reported as a combined input (calculated from the increase [if any] in the surface-water dissolved-organic pool over the period in question) (Table 3.2). Losses from the surface-water dissolved-organic pool were due to mineralization of the dissolved-organic matter (calculated from the decrease [if any] in the surface-water dissolved-organic pool size over the period in question) (Table 3.2). Besides this mineralization input to the surface-water dissolved-inorganic pool, other inputs included macrophyte leaching, invertebrate losses, and pore water–surface water exchange (Table 3.2). Output from the dissolved-inorganic pool included uptake by algae (and other microbes).

Water movements were very minor in the sediments (Kadlec, unpublished data); therefore, no hydrologic flux was calculated for the pore-water nutrient pools (i.e., net changes in the dissolved pore-water pools were due primarily to microbial activity). Inputs to the pore-water organic pools included primarily the breakdown of belowground macrophyte litter (Figure 3.1). Losses from the pore-water dissolved-organic pool was assumed to be primarily mineralization of organic to inorganic material. Inputs to the pore-water inorganic pool beside mineralization of the organic pool included macrophyte leaching and breakdown of belowground macrophyte litter. Losses from the pore-water inorganic pool were due primarily to uptake by macrophytes.

Litter/Sediment Pool

Inputs to the litter/sediment pool included litter fall from both above- and belowground macrophytes. Macrophyte litter can be lost through decomposition (production of dissolved inorganic, dissolved organic, and particulate material) and invertebrate consumption. Furthermore, dissolved organic compounds can be decomposed to release a variety of dissolved inorganic compounds. In general, the literature for prairie marshes is inadequate to estimate the relative amounts of these fluxes of dissolved material from the litter/sediment pool. Losses to the surface- and pore-water pools were discussed previously and in the calculations shown in Table 3.2. To calculate particulate losses from the litter pool to the old litter/sediment sink (long-term loss of material from the system), the initial mass of nutrients in the aboveground and belowground litter pool from the current growing season was determined from the estimated rates of litter-fall and decomposition (see footnote to Table 3.2). Decomposition rates were derived from litter-bag studies in the MERP experimental cells (van der Valk et al. 1991). Decomposition was assumed not to take place in winter. For each time period, a total particulate flux to the old litter/sediments was calculated as the product of the overall loss rate and the estimated mass of the litter pool corrected for losses to the organic and inorganic surface-water pools and invertebrate consumption. Belowground litter-pool loss to the litter/sediment pool (Table 3.2) was calculated in the same manner as aboveground losses. Belowground decomposition rates were determined from buried litter-bag studies within the experimental cells (Wrubleski et al. 1997b).

ACKNOWLEDGMENTS

Lisette Ross developed the early drafts of the figures in this chapter. This is Paper No. 96 of the Marsh Ecology Research Program, a joint project of Ducks Unlimited Canada and the Delta Waterfowl and Wetlands Research Station.

LITERATURE CITED

Barko, J.W., D. Gunnison, and S.R. Carpenter. 1991. Sediment interactions with submersed macro-phyte growth and community dynamics. Aquatic Botany 41:41-65.

Benner, R., M.L. Fogel, and E.K. Sprague. 1991. Diagenesis of belowground biomass of *Spartina alterniflora* in salt-marsh sediments. Limnology and Oceanography 36:1358-1374.

Bernard, J. M., and E. Gorham. 1978. Life history aspects of primary production in sedge wetlands. In *Freshwater Wetlands: Ecological Processes and Management Potential*. (Eds.) R.E. Good, D.F. Whigham, and R.L. Simpson, pp.39-51. New York: Academic Press.

Bowden, W. B. 1987. The biogeochemistry of nitrogen in freshwater wetlands. Biogeochemistry 4:313-348.

Bowden, W.B., C.J. Vorosmarty, J.T. Morris, B.J. Peterson, J.E. Hobbie, P.A. Steudler, and B. Moore. 1991. Transport and processing of nitrogen in a tidal freshwater wetland. Water Resources Research 27:389-408.

Brezonik, P.L., L.A. Baker, and T.E. Perry. 1987. Mechanisms of alkalinity generation in acid-sensitive waters. In *Sources and Fates of Aquatic Pollutants*. (Eds.) R.A. Hites and S.J. Eisenreich, pp.229-260. Washington: American Chemical Society.

Davis, C.B., and A.G. van der Valk. 1978a. Decomposition of standing and fallen litter of *Typha glauca* and *Scirpus fluviatilis*. Canadian Journal of Botany 56:662-675.

Davis, C.B., and A.G. van der Valk. 1978b. Litter decomposition in prairie glacial marshes. In *Freshwater Wetlands: Ecological Processes and Management Potential*. (Eds.) R.E. Good, D.F. Whigham, and R.L. Simpson, pp.99-113. New York: Academic Press.

Evans, H.E., P.J. Dillon, and L.A. Molot. 1997. The use of mass balance investigations in the study of the biogeochemical cycle of sulfur. Hydrological Processes 11:765-782.

Fritzell, E.K. 1989. Mammals in prairie wetlands. In *Northern Prairie Wetlands*. (Ed.) A.G. van der Valk, pp.268-301. Ames: Iowa State University Press.

Grigal, D.F. 1991. Elemental dynamics in forested bogs in northern Minnesota. Canadian Journal of Botany 69:539-546.

Gurney, S.E., and G.G.C. Robinson. 1989. Algal primary production. In *Marsh Ecology Research Program: Long-term Monitoring Procedures Manual. Technical Bulletin 2*. (Eds.) E.J. Murkin and H.R. Murkin, pp.18-22. Portage la Prairie:Delta Waterfowl and Wetlands Research Station.

Hosseini, S.M., and A.G. van der Valk. 1989a. The impact of prolonged, above-normal flooding on metaphyton in a freshwater marsh. In *Freshwater Wetlands and Wildlife*. (Eds.) R.R. Sharitz and J.W. Gibbons, pp.317-324. USDOE Symposium Series 61. Oak Ridge: USDOE Office of Scientific and Technical Information.

Hosseini, S.M., and A.G. van der Valk. 1989b. Primary productivity and biomass of periphyton and phytoplankton in flooded freshwater marshes. In *Freshwater Wetlands and Wildlife*. (Eds.) R.R. Sharitz and J.W. Gibbons, pp.303-315. USDOE Symposium Series 61. Oak Ridge: USDOE Office of Scientific and Technical Information.

Howard-Williams, C. 1985. Cycling and retention of nitrogen and phosphorus in wetlands: a theoretical and applied perspective. Freshwater Biology 15:391-431.

Hwang, Y.H., and J.T. Morris. 1992. Fixation of inorganic carbon from different sources and its translocation in *Spartina alterniflora* Loisel. Aquatic Botany 43:137-147.

Jervis, R.A. 1969. Primary production in the freshwater marsh ecosystem of Troy Meadows, New Jersey. Bulletin of the Torrey Botanical Club 96:209-231.

Kadlec, J.A. 1979. Nitrogen and phosphorus dynamics in inland freshwater wetlands. In *Waterfowl and Wetlands: An Integrated Review*. (Ed.) T.A. Bookhout, pp.17-41. Madison: The Wildlife Society.

Kadlec, J.A. 1983. Water budgets for small diked marshes. Water Resources Bulletin 19:223-229.

Kadlec, J.A. 1986a. Input-output nutrient budgets for small diked marshes. Canadian Journal of Fisheries and Aquatic Science 43:2009-2016.

Kadlec, J.A. 1986b. Effects of flooding on dissolved and suspended nutrients in small diked marshes. Canadian Journal of Fisheries and Aquatic Science 43:1999-2008.

Kadlec, J.A. 1993. Effect of depth of flooding on summer water budgets for small diked marshes. Wetlands 13:1-9.

Linde, A.F., T. Janisch, and D. Smith. 1976. Cattail—the significance of its growth, phenology, and carbohydrate storage to its control and management. Wisconsin Department of Natural Resources Technical Bulletin 94. 27pp.

Mitsch, W.J., and J.G. Gosselink. 1993. *Wetlands, 2nd edition.* New York: Van Nostrand Reinhold.

Mitsch, W.J., and B.C. Reeder. 1991. Modeling nutrient retention of a freshwater coastal wetland - estimating the roles of primary productivity, sedimentation, resuspension and hydrology. Ecological Modeling 54:151-187.

Murkin, E.J., and H.R. Murkin (Eds.). 1989. Marsh ecology research program: long-term monitoring procedures manual. Technical Bulletin 2, Delta Waterfowl and Wetlands Research Station, Portage la Prairie, Manitoba.

Murkin, H.R. 1989. The basis for food chains in prairie wetlands. In *Northern Prairie Wetlands.* (Ed.) A.G. van der Valk, pp.316-338. Ames: Iowa State University Press.

Murkin, H.R., and J.A. Kadlec. 1986. Responses by benthic macroinvertebrates to prolonged flooding of marsh habitat. Canadian Journal of Zoology 64:65-72.

Murkin, H.R., J.A. Kadlec, and E.J. Murkin. 1991. Effects of prolonged flooding on nektonic invertebrates in small diked marshes. Canadian Journal of Fisheries and Aquatic Sciences 48:2355-2364.

Murkin, H.R., A.G. van der Valk, and C. B. Davis. 1989. Decomposition of four dominant macrophytes in the Delta Marsh, Manitoba. Wildlife Society Bulletin 17:215-221.

Neckles, H. A., and C. Neil. 1994. Hydrologic control of litter decomposition in seasonally flooded prairie marshes. Hydrobiologia 286:155-165.

Neely, R.K., and J.L. Baker. 1989. Nitrogen and phosphorus dynamics and the fate of agricultural runoff. In *Northern Prairie Wetlands.* (Ed.) A.G. van der Valk, pp.92-131. Ames: Iowa State University Press.

Nelson, J.W., J. A. Kadlec, and H. R. Murkin. 1990. Seasonal comparisons of weight loss for two types of *Typha glauca* Godr leaf litter. Aquatic Botany 37:299-314.

Nixon, S.W., and V. Lee. 1986. Wetlands and water quality. Wetlands Research Program Technical Report Y-86-2. U.S. Army Corps Engineers Waterway Experiment Station, Vicksburg, Mississippi.

Robinson, G.G.C., S.E. Gurney, and L.G. Goldsborough. 1997a. Response of benthic and planktonic algal biomass to experimental water-level manipulation in a prairie lakeshore wetland. Wetlands 17:167-181.

Robinson, G.G.C., S. E. Gurney, and L.G. Goldsborough. 1997b. The primary productivity of benthic and planktonic algae in a prairie wetland under controlled water-level regimes. Wetlands 17:182-194.

Schierup, H.-H. 1978. Biomass and primary production in a *Phragmites communis* Trin. swamp in North Jutland, Denmark. Verh. Internat. Verein. Limnol. 20:94-99.

van der Valk, A.G., and C.B. Davis. 1978a. The role of the seed banks in the vegetation dynamics of prairie glacial marshes. Ecology 59:322-335.

van der Valk, A.G., and C.B. Davis. 1978b. Primary production of prairie glacial marshes. In

Freshwater Wetlands: Ecological Processes and Management Potential. (Eds.) R.E. Good, D.F. Whigham, and R.L. Simpson, pp.21-37. New York: Academic Press.

van der Valk, A.G., J.M. Rhymer, and H.R. Murkin. 1991. Flooding and the decomposition of litter of 4 emergent plant species in a prairie wetland. Wetlands 11:1-16.

Weller, M.W., and L. H. Fredrickson. 1974. Avian ecology of a managed glacial marsh. Living Bird 12:269-291.

Weller, M.W., and C. S. Spatcher. 1965. Role of habitat in the distribution and abundance of marsh birds. Iowa Agriculture and Home Economics Experiment Station, Special Report 43, Ames, Iowa.

Whigham, D.F., and S.E. Bayley. 1979. Nutrient dynamics in freshwater wetlands. In *Wetland Functions and Values: The State of Our Understanding.* (Eds.) P.E. Greeson, J.R. Clark, and J.E. Clark, pp.468-478. American Water Resources Association Technical Publication TPS79-2.

Winter, T.C. 1989. Hydrologic studies of wetlands in the northern prairies. In *Northern Prairie Wetlands.* (Ed.) A.G. van der Valk, pp.16-54. Ames: Iowa State University Press.

Wrubleski, D.A., H.R. Murkin, A.G. van der Valk, and C.B. Davis. 1997a. Decomposition of litter of three mudflat annual species in a northern prairie marsh during drawdown. Plant Ecology 129:141-148.

Wrubleski, D.A., H.R. Murkin, A.G. van der Valk, and J.W. Nelson. 1997b. Decomposition of emergent roots and rhizomes in a northern prairie marsh. Aquatic Botany 58:121-134.

4

The Baseline and Deep-Flooding Years

John A. Kadlec
Henry R. Murkin
Arnold G. van der Valk

ABSTRACT

Pumping water into the cells did not increase concentrations of nutrients in surface water during the first year of flooding, but did increase the mass of nutrients in the surface-water pool because of the increased volume of water. The large decreases in mass of nitrogen (N) and phosphorus (P) in the aboveground macrophyte pool due to flooding was responsible for the major changes in intrasystem nutrient cycling between the baseline and deep-flooding years. During deep flooding, both N and P concentrations and mass in pore water increased. The belowground macrophyte N and P mass did not decrease as rapidly as the aboveground mass, and even during the second year of deep flooding they were still 71% and 60% of the baseline N and P pools, respectively. Deep flooding resulted in a progressive increase in mass of N and P in metaphyton and invertebrates, both reached their peaks in the second year of flooding. By the second year of deep flooding, the mass of N and P in all measured nonmacrophyte pools was slightly greater than that in the macrophyte pools. The mean total mass of N and P in all measured pools by the second year of flooding was only 10% and 15% lower for N and P, respectively, than in the baseline years. These losses are probably due to burial of litter in the old litter/sediment pool and to microbial processes, including denitrification. Macrophyte litter decomposition seems to be a major factor regulating the size of the pools and fluxes during the deep-flooding years. Draining the cells at the end of the deep flooding resulted in more N and P being lost than were initially pumped into the cells.

INTRODUCTION

The initial deep-flooding experiments of the Marsh Ecology Research Program (MERP) study from 1981 to 1983 (Figure 4.1, also see Table 1.1 in Chapter 1) were designed to set the experimental cells to the lake marsh stage of the prairie wetland wet–dry cycle (van der Valk and Davis 1978). Prolonged increases in water levels in prairie wetlands that result from consistent surplus of water supplied by precipitation can cause elimination of nearly all of their emergent macrophyte zones (Weller and Spatcher 1965; Weller and Fredrickson 1974; Kantrud et al. 1989). The large volumes of water entering prairie

Figure 4.1   View of a MERP experimental cells during the baseline year (1980), first year of deep flooding (1981), and the second year of deep flooding (1982).

marshes during periods of above-normal flooding, the concomitant increase in the size of the surface-water pool, and reduction in the size of the macrophyte pool should have major effects on the overall input–output nutrient budgets of these wetlands and on intrasystem storage pools and fluxes of nutrients among these pools. Prior to this phase of the MERP experiments, little was known about the magnitude and significance of changes in nutrient budgets and cycles during high-water periods or during development of the lake marsh stage (Neely and Baker 1989).

The closest parallel to the deep flooding of prairie marshes during high-water years is the flooding of new reservoirs (Cook and Powers 1958; Hecky et al. 1984; Kadlec 1986a). Both new waterfowl impoundments and larger reservoirs show their peak production, often reflected at all trophic levels, immediately following initial flooding (Kadlec 1962; Whitman 1974). This phenomenon is sufficiently predictable to have been dubbed "the reservoir effect". Hecky et al. (1984) attributed this effect primarily to the death and decay of newly inundated vegetation and associated litter. This submersed plant litter is thought to be the source of nutrients that support the observed high levels of aquatic primary production. Litter is also a direct source of detrital food resources and habitat for invertebrates. Thus, secondary production in newly flooded habitats is suggested to be high because both herbivores and detritivores benefit from increased primary production (mostly algal production) and the litter and associated nutrient resources made available when the existing plants inundated by the flooding are killed and subsequently decompose (Murkin 1989).

This initial pulse of increased production in prairie wetlands following flooding is short lived and with continued flooding, productivity begins to decline. These wetlands eventually enter the lake marsh stage, which is normally considered the period of lowest productivity during the wet–dry cycle of prairie marshes (van der Valk and Davis 1978). With reduced macrophyte production and eventual loss of nutrient release from the decomposition of existing plant litter, available resources for secondary consumers are reduced. Algal communities during the lake marsh stage are dominated by phytoplankton that do not reach the biomass levels of the other algal communities common to earlier stages of the wet–dry cycle of prairie wetlands (Goldsborough and Robinson 1996; Robinson et al. 1997a,b).

The flooding of the MERP cells to 1 m above normal for 2 years was predicted not only to have a major impact on the input–output nutrient budgets of the experimental cells (due to the large volume of water pumped into the cells) but also to have significant effects on intrasystem nutrient cycling (pools and internal fluxes). The specific predictions made about the effects of deep flooding on the nutrient budgets of the MERP experimental cells were as follows:

1. The living macrophyte pools would be reduced both above and below ground due to mortality caused by the high water levels.

2. The decomposition of the aboveground macrophyte litter would be complete after 2 years of flooding, thereby releasing the nutrients associated with the dead standing litter

to other pools and resulting in a substrate free of litter at the initiation of the drawdown experiments to follow.

3. The death and decomposition of emergent macrophytes would release substantial amounts of N and P into surface-water and pore-water nutrient pools.

4. The release of nutrients from the macrophyte pools was predicted to support increased algal and invertebrate production with their highest production during the first year of deep flooding.

5. Eventually, with the lower macrophyte production and decline in the size of the litter pools through decomposition, the overall productivity of the cells would be lower during the second year of deep flooding (lake marsh conditions).

To test these predictions and associated hypotheses, both the overall input–output nutrient budgets of the experimental cells (i.e., nutrient inputs and outputs into each cell due to pumping, precipitation, seepage, and evapotranspiration) and their intrasystem nutrient cycles (i.e., the mass of N and P in pools and fluxes among these pools) during the baseline years (1980—all cells, 1981—cells 3 and 7) and both years of deep flooding (1 m above normal) were estimated using the nutrient-budget framework and flux calculations outlined in Chapter 3. Pool sizes and fluxes among years and flooding treatments were compared using analysis of variance, and in those cases where the assumptions of parametric procedures were violated, Kruskal–Wallis nonparametric tests (Ott 1988) were used.

Losses of nutrients at the end of the deep-flooding years due to pumping of water out of the cells to initiate the drawdown stage of MERP also were examined. These losses are of special interest because the long-term productivity of managed marshes may be reduced if the mass of nutrients pumped out of them exceeds the amount initially pumped into them, which could happen if the concentration of nutrients in surface water is higher when water is being pumped out because of the decomposition of macrophytes killed by deep flooding and other fluxes to the surface-water nutrient pools.

HYDROLOGY

During the 2 years of deep flooding, water levels in the MERP experimental cells were increased to about 1 m over the substrate in the existing cattail stands, an elevation of 248.4 m above mean sea level (AMSL). Because cells 3 and 7 were to be drawn down for only 1 year later in the study (see Table 1.1 in Chapter 1), they were flooded 1 year later than the other experimental cells. Eight diked cells were flooded in 1981 and 1982, and cells 3 and 7 were flooded in 1982 and 1983. Although this lag complicated the hydrology of the cells by having adjacent cells at different water levels during 1 or both years (Table 4.1), it permitted separation of year effects from water-level effects. The two cells (3 and 7) whose deep flooding were delayed by 1 year had very different hydrologic budgets from the other eight cells (Table 4.1). They had more seepage into them in 1981 when they were at normal water levels and the adjacent cells were flooded, and more seepage out of them in 1983 when adjacent cells were drawn down. In spite of these differences in hydrologic budgets, the mean size of nutrient pools and nutrient fluxes in the deep-flood-

Table 4.1. Summary annual water budgets, MERP experimental cells, 1980–1983, by hydrologic category.

Hydrologic category	Flooded area – ha (SE)	Volume – 1,000 m³/experimental cell (SE)[a]			
		Pumped in[b]	Precipi- tation	Evapotrans- piration	Net seepage
NNN[c]	2.73	0	10.1	−14.0	+3.0
N = 10 (1980), 2[d] (1980–1983)	(0.20)	(0)	(0.6)	(1.0)	(0.5)
NHN	3.16	0	8.6	−11.7	−5.9
n = 2 (1981)	(0.72)	(0)	(2.0)	(2.6)	(0.1)
HHH	6.13	29.3	22.0	−28.5	−21.7
n = 2 (1980), 8 (1981)	(0.18)	(1.9)	(0.9)	(0.8)	(0.8)
NHH	6.14	51.1	25.1	−28.6	−39.6
n = 6 (1981), 2 (1982)	(0.20)	(3.6)	(1.5)	(0.9)	(1.2)
DHD	5.89	70.6	16.4	−26.0	−53.0
n = 2 (1983)	(0.42)	(1.5)	(1.2)	(1.8)	(0.01)

Note: Means and SE were calculated using the annual mean values for each budget component for each cell in each hydrologic category.

[a]The standard error represents the statistical summation of the data and does not reflect the error of measurement inherent in the methods used to obtain the data.

[b]Pumping required to maintain experimental water levels. Does not include initial pumping to experimental levels.

[c]Letters refer to water levels in cell and adjacent cells (i.e., NNN indicates cell at normal water levels (N) between two cells of normal water levels; NHH indicates a cell at high levels (H) between a cell at normal levels and another at high levels; D indicates drained marsh (1983 only).

[d]Reference cells (cells 11 and 12).

ed cells were generally very similar and no significant year effects were found (see Tables A2.1 and A2.2 in Appendix 2). Consequently, when considering intrasystem cycling (pools and fluxes), the experimental cells were lumped into baseline, 1-year- and 2-year-flooded treatments.

Compared with the mean volume of water present in the cells during the baseline years ($\approx$9,000 m³), a large volume of water was pumped into the cells ($\approx$50,500 m³) to raise their water level to the deep-flooding level (248.4 m). Detailed information about the amount of water pumped into individual cells to establish a water level of 248.4 m and other aspects of their hydrology can be found in Kadlec (1983). Out of the total volume of water pumped into the experimental cells (50,500 m³), $\approx$8,100 m³ seeped into the dikes and beach ridge (on the northern edge of the cells) while the cells were being pumped up.

During the summer of the deep-flooding years, water had to be pumped into the cells to maintain the experimental levels of 248.4 m to offset losses due to seepage and evapo-transpiration above inputs due to precipitation (Table 4.1). In general, annual mean losses of water due to evapotranspiration exceeded inputs due to precipitation by $\approx$25% (Table

4.1). The experimental cells were not all exactly the same size, nor did they have the same topography (see Appendix 1). Consequently, the flooded area varied among cells and so did monthly and annual differences in precipitation and evapotranspiration rates. Both precipitation inputs and evapotranspiration losses were calculated using the area in each cell that was covered with standing water, which during the deep-flooding years was the area of the entire cell. These inputs and losses account for the observed variation in volumes of water added by precipitation or lost by evapotranspiration in Table 4.1.

When the cells were deeply flooded, pumping for most of the summer was needed to offset losses due to seepage (Table 4.1). Seepage losses in a cell depended primarily on water levels in adjacent cells. When water levels in adjacent cells are also high (hydrologic class HHH in Table 4.1), then seepage during the summer was ≈21,700 m³. When a high cell had an adjacent cell with normal water levels (NHH), seepage increased to 39,600 m³. When a high cell was flanked by two drawdown cells (DHD), as occurred in 1983 when only two cells (3 and 7) were still flooded, seepage increased to 53,000 m³ (Table 4.1). Therefore, the two cells (3 and 7) that were not flooded deeply until 1982 had much more water pumped into them to maintain a water level of 248.4 m than the other eight cells during the second year of deep flooding. Seepage rates varied among cells not only because of water levels in adjacent cells but also because of local differences in dike-seepage rates (Kadlec 1983). Although some downward movement of water into the sediments and underlying groundwater was anticipated (Kadlec 1983), vertical flow was minimal (Kadlec 1986a) and, therefore, seepage losses were due to horizontal flow through the dikes.

When the cells were drained and pumped dry in the fall of 1982 or 1983, an average of 65,700 m³ of water was pumped out per cell. This amount included ≈15,400 m³ of water stored in dikes and surficial sediments as the water level was lowered to the drawdown level of 247.0 m (Kadlec, unpublished data). At this elevation, nearly all of the marsh bottom in each cell was above the water table (i.e., there was no standing water in the cells during this period).

N AND P BUDGETS

Inputs–Outputs

Because of differences in seepage into and out of the experimental cells (Table 4.1), the effect of pumping on the overall input–output nutrient budgets of the cells in both years of deep flooding was highly variable and depended on the cell, year, and the forms of N and P (Kadlec 1986b). Input–output nutrient budgets for ammonia (NH₄), total dissolved nitrogen (TDN), soluble reactive phosphorus (SRP), and total dissolved phosphorus (TDP) in deep-flooded cells are presented in Table 4.2. The major input of the NH₄ and SRP when the cells were deep flooded was in precipitation. Inputs of total dissolved N during the deep-flooding years came primarily from pumped inputs, whereas total dissolved P inputs were actually slightly higher in precipitation than in pumped inputs. The major losses of dissolved N and P were by seepage, with the highest losses occurring in 1983 when only cells 3 and 7 were flooded.

Table 4.2. Mean annual input–output nutrient budgets (kg/ha/yr) for the flooded MERP experimental cells.

Nutrient	Year	Treatment	Pumped	Precipi-tation	Seepage	ΔStorage
NH$_4$	1981	Flooded Year One	1.22	4.22	−1.02	3.10
	1982	Flooded Year Two	0.53	2.36	−0.87	1.02
	1982	Flooded Year One	0.50	2.53	−0.95	6.02
	1983	Flooded Year Two	0.39	2.21	−2.66	−7.55
Total dissolved N	1981	Flooded Year One	17.6	8.98	−15.6	15.5
	1982	Flooded Year Two	17.7	5.01	−16.3	0.19
	1982	Flooded Year One	15.1	5.34	−11.6	8.67
	1983	Flooded Year Two	11.1	5.28	−44.2	−39.0
Soluble reactive P	1981	Flooded Year One	0.51	3.68	−1.86	3.35
	1982	Flooded Year Two	0.47	0.54	−0.65	0.17
	1982	Flooded Year One	0.22	0.57	−0.36	1.36
	1983	Flooded Year Two	0.33	0.51	−2.03	−1.41
Total dissolved P	1981	Flooded Year One	0.56	0.70	−0.75	0.95
	1982	Flooded Year Two	0.89	0.61	−0.97	−0.05
	1982	Flooded Year One	0.51	0.68	−1.64	1.69
	1983	Flooded Year Two	0.68	0.78	−2.76	−2.27
Suspended N	1981	Flooded Year One	10.1	2.61	—[a]	−2.73
	1982	Flooded Year Two	4.37	1.06	—	1.57
	1982	Flooded Year One	3.77	1.03	—	−8.49
	1983	Flooded Year Two	0.82	2.23	—	0.35
Suspended P	1981	Flooded Year One	1.34	0.36	—	−0.54
	1982	Flooded Year Two	0.68	0.18	—	−1.48
	1982	Flooded Year One	0.75	0.15	—	0.27
	1983	Flooded Year Two	0.48	0.36	—	0.10

Note: The first year of flooding for eight cells was 1981 and for two other cells (3 and 7) was 1982.
[a] No losses possible through seepage.

The mean mass of total dissolved N (2.48 and 58.8 kg/ha of NH$_4$ and dissolved organic nitrogen (DON), respectively) and P (1.08 and 1.03 kg/ha of SRP and dissolved organic phosphorus (DOP), respectively) pumped into the cells when they were initially flooded and the mean amount drained out of the cells when they were drawn down indicates that experimental cells lost an average of 1.52, 15.2, 3.81, and 2.28 kg/ha of NH$_4$, DON, SRP and DOP, respectively, in excess of what was pumped into them when they were filled. These estimated net losses do not include additional nutrients lost due to pumping the cells from 247.5 to 247.0 m to initiate the drawdown phase of MERP at the end of the deep-flooding years. Thus, these nutrient losses are due solely to draining enough water out of the cells to return water levels to baseline conditions (247. 5 m). These losses represent ≈10% and 25% of the mean total mass of N and P, respectively, in all measured nutrient pools during the baseline years.

Intrasystem Cycling

Pools

Tables 4.3 and 4.4 summarize information on the mass of N and P in the major nutrient pools during the baseline and deep-flooding years. As mentioned, there were no significant effects of delaying the flooding of cells 3 and 7 by 1 year; therefore, these effects are not considered in subsequent discussions. Detailed year-by-year nutrient-pool size estimates for N and P during the baseline and deep-flooding years can be found in Tables A2.1 and A2.2 in Appendix 2.

Macrophytes
During the baseline year, the largest nutrient pools in the cells by far were the above- and belowground macrophyte pools, ≈158 kg/ha of N and 21 kg/ha of P (Tables 4.3 and 4.4).

Table 4.3.   Mean (SE) nitrogen (N) pool sizes (kg/ha) in the MERP experimental cells by flooding treatment, 1980–1983.

		Treatment	
Pool	Baseline	Flooded year one	Flooded year two
Macrophytes			
Aboveground	72.42a[a]	12.27b	17.39b
	(4.33)	(3.98)	(2.83)
Belowground	85.44a	78.02ab	60.30b
	(3.14)	(7.53)	(9.06)
Algae	0.85a	5.01b	15.34c
	(0.32)	(0.17)	(2.87)
Invertebrates	0.73a	0.99b	3.16c
	(0.08)	(0.05)	(0.17)
Surface water			
NH_4	0.84a	1.81a	2.36b
	(0.99)	(0.60)	(2.48)
Dissolved organic N	10.54a	26.55b	30.72b
	(5.49)	(6.34)	(3.43)
Suspended solids	4.02a	2.24a	2.21a
	(1.83)	(0.80)	(0.52)
Pore water			
NH_4	8.41a	14.50a	23.29b
	(6.42)	(3.79)	(3.50)
Dissolved organic N	20.39a	29.18b	30.93b
	(6.49)	(4.92)	(4.93)

Note: All values are means over the growing season, except aboveground macrophytes, which are mean peak standing crops ($n = 10$).
[a]Within a row, means with different letters are significantly different ($P < 0.05$).

Table 4.4.   Mean (SE) phosphorus (P) pool sizes (kg/ha) in the MERP
experimental cells by flooding treatment, 1980–1983.

		Treatment	
Pool	Baseline	Flooded year one	Flooded year two
Macrophytes			
Aboveground	10.28a[a]	1.66b	3.43b
	(0.99)	(0.53)	(0.53)
Belowground	10.94a	7.80b	6.52b
	(0.60)	(0.76)	(1.02)
Algae	0.12a	0.71b	2.16c
	(0.05)	(0.02)	(0.70)
Invertebrates	0.10a	0.10a	0.32b
	(0.03)	(0.01)	(0.02)
Surface water			
Soluble reactive P	0.62a	0.55a	1.45a
	(0.18)	(0.12)	(0.33)
Dissolved organic P	0.27a	0.76b	0.70b
	(0.04)	(0.04)	(0.03)
Suspended solids	0.86a	0.37a	0.38a
	(0.29)	(0.05)	(0.05)
Pore water			
Soluble reactive P	1.20a	3.18b	5.26b
	(0.14)	(0.22)	(0.56)
Dissolved organic P	0.43a	0.75ab	1.02b
	(0.04)	(0.17)	(0.21)

Note: All values are means over the growing season, except aboveground macrophytes,
which are mean peak standing crops ($n = 10$).
[a]Within a row means with different letters are significantly different ($P < 0.05$).

All the other pools collectively contained ≈30% of the N and <20% of the P in the macrophyte pools. Deep flooding greatly reduced aboveground macrophyte biomass (Figure
4.1) and associated N and P pools decreased by >80% during the first year of deep flooding (Table 4.3 and 4.4). Deep flooding had much less effect on belowground biomass in
1981 and 1982 (van der Valk 1994) and consequently on the mass of belowground macrophyte nutrient pools, which were reduced only 10% for N and ≈30% for P in the first year
of deep flooding. During the second year of deep flooding, even though mean aboveground macrophyte biomass was slightly lower, there was a small increase in the mass of
N and P in aboveground biomass. This increase was due primarily to an increase in submersed aquatic species that have a higher concentration of N and P than emergent species
(van der Valk 1994). Belowground biomass continued to decline during the second year
of flooding, although it remained the largest single nutrient pool for both N and P during
the deep-flooding period (as during the baseline years). Although the aboveground macrophyte pool had declined nearly 75% for N and 70% for P by the second year of flooding,

mean belowground N and P macrophyte pools had declined only 30% and 40%, respectively. Roots and rhizomes of some emergent species were able to survive the first 2 years of deep flooding by going dormant. However, most of the surviving belowground tissues died during the winter following the second year of deep flooding (van der Valk 1994). Consequently, if the experimental cells had been exposed to a third year of flooding, we hypothesize that there would have been much reduced pools of belowground macrophyte N and P.

Algae and Invertebrates

Standing crops of N and P in algae (primarily metaphyton) increased more than fivefold in the first year of deep flooding and increased again more than threefold in the second year of flooding (Tables 4.3 and 4.4). These changes reflect changes in metaphyton biomass reported by Hosseini and van der Valk (1989). As noted, the mean mass of N and P suspended solids after deep flooding actually decreased >65% below that found in the baseline year (Tables 4.3 and 4.4). Suspended solids pumped into the MERP cells were mostly planktonic algae (see Chapter 3). The decline in suspended solids (i.e., phytoplankton) during the flooding years may be due to competition for light and nutrients caused by floating mats of metaphyton, which covered large areas of the flooded cells.

Standing crops of N and P in macroinvertebrates (Tables 4.3 and 4.4) had the smallest measured pools in the cells. Their mean biomass during the first year of flooding was similar to that during the baseline years for both N and P, but increased about threefold during the second year of deep flooding because chironomid densities increased in areas where the emergent vegetation was eliminated. Detailed information on invertebrates during the deep-flooding years can be found in Murkin and Kadlec (1986) and Murkin et al. (1991). There is some potential for export of N and P from the cells via emerging insects (Vallentyne 1952); however, we assumed that because most emerging insects (primarily chironomids in the MERP cells) remained in the marsh to mate and lay eggs that the mass of N and P in adult insects leaving the cells was probably very small. Regardless, the effect of invertebrate nutrient pools and associated fluxes on the experimental-cell nutrient budgets during the baseline and deep-flooding years was minimal compared with the other nutrient pools in the cells.

Dissolved Nutrients and Suspended Solids

There was little change in the surface-water nutrient pools in the undiked reference cells over the period of study (Tables A2.1 and A2.2 in Appendix 2). Flooding resulted in a substantial increase in the mass of dissolved inorganic and organic N and P in the surface-water pools of the deep-flooded cells during the first year of flooding (Tables 4.3 and 4.4), primarily due to the increase in water volume. The only exception was the mass of inorganic P (SRP), which during the first year of flooding was similar to that during the baseline years. Concentrations of dissolved nutrients in surface water did not change significantly during the first year of flooding (Kadlec 1986a); therefore, these initial changes in mean mass of nutrients in surface water were primarily due to increases in water volume. During the second year of flooding, the mass of dissolved N and P in the surface water increased due to an increase in concentrations during the second year. The mass of N and P, however, in the suspended solids pool in the surface water dropped by nearly 50% for N and >50% for P during the first year of flooding and remained at about the same level during the second year of deep flooding (Tables 4.3 and 4.4). As mentioned, this outcome is probably due to the loss of phytoplankton due to competition for light and nutrients by the expanding metaphyton community during the deep-flooding years.

As with surface water, the nutrient pools in the pore water of the reference cells changed little over the period of study (Tables A2.1 and A2.2 in Appendix 2). The mass of dissolved N and P in pore water increased during the first year of flooding and again during the second year of flooding (Tables 4.3 and 4.4). The greater mass of N and P stored in pore water in flooded cells was the result of both an increase in area flooded (and hence volume of pore water) and an increase in concentration of dissolved species of N and P (Kadlec 1986a). The mass of dissolved organic N and P in pore water had a proportionally smaller increase (≈50%) in pool size after flooding than did the mass of dissolved organic nutrients in the surface-water pool (>100%). The inorganic pore-water pools of NH_4 and SRP, however, were much larger than those of the surface-water pool, both during the baseline and deep-flooding years (Tables 4.3 and 4.4). The pore-water dissolved-organic pools of N and P were comparable in size to the surface-water pools, especially during the deep-flooding years.

Litter Pool/Sediment Sink

As stated in Chapter 3, no estimates of total litter/sediment pool size were included in the budgets because of the accumulation of litter over many years. Only fluxes to and from the current year's litter pool and to the old litter/sediment sink were determined on an annual basis.

Fluxes

Macrophytes

Because of the large size of the macrophyte nutrient pools (Tables 4.3 and 4.4), N and P fluxes (Tables 4.5 and 4.6) in the experimental cells were driven primarily by the changes in these pools. Annual nutrient cycling in the cells during the baseline years, as indicated by the larger fluxes, was dominated by plant uptake, translocation, leaching, and above- and belowground litter-fall (Figure 4.1). During the first year of deep flooding, there was much reduced macrophyte uptake (−85% for N and −94% for P) of nutrients, whereas above- and belowground litter-fall was reduced by only 10% for N and 33% for P (Tables 4.5 and 4.6). There was little change in macrophyte fluxes during the second year of flooding; however, fluxes associated with aboveground litter-fall were reduced due to the reduced macrophyte pools sizes caused by the deep flooding.

Algae and Invertebrates

Algal uptake declined during the first year of flooding and then, due primarily to the development of extensive metaphyton mats, increased during the second year of flooding to levels much higher than the baseline years (Table 4.5). With the increased metaphyton stand-

Table 4.5. Mean (SE) annual (1 Oct.–30 Sept.) nitrogen (N) fluxes (kg/ha/yr) in the MERP experimental cells, 1980–1983.

Flux	Treatment		
	Baseline + reference[a]	Flooded year one[b]	Flooded year two[c]
Macrophytes			
Uptake	96.31a[d]	13.94b	15.96b
	(6.50)	(3.53)	(4.66)
Translocation	57.81a	5.48b	11.56b
	(4.16)	(2.74)	(1.84)
Aboveground leaching	26.42a	3.68b	5.22b
	(1.88)	(1.19)	(0.85)
Belowground leaching	5.36a	3.92a	4.84a
	(0.91)	(0.89)	(1.36)
Aboveground litter-fall	26.89a	21.77a	4.29b
	(2.32)	(1.26)	(1.39)
Belowground litter-fall	21.43a	15.68a	19.35a
	(3.62)	(3.56)	(5.44)
Algae and invertebrates			
Algal uptake	6.61a	1.94a	11.46b
	(3.28)	(0.97)	(3.94)
Algal litter-fall[e]	4.01a	18.08b	14.14b
	(1.29)	(2.98)	(3.16)
Invertebrate loss	0.44a	0.23b	0.23b
	(0.07)	(0.07)	(0.12)
Invertebrate uptake:	0.57a	1.83b	3.74c
litter and algae combined	(0.11)	(0.14)	(0.22)
Dissolved nutrients/suspended solids			
Mineralization			
Aboveground SWDO to SWDI	7.77a	13.81b	9.29a
	(0.75)	(3.23)	(1.70)
Belowground PWDO to PWDI	6.16a	7.85a	6.68a
	(2.21)	(1.79)	(1.71)
Pore–surface water exchange	−30.86a	−18.91b	−4.50c
	(3.26)	(2.82)	(4.04)
Litter/sediment pool			
Litter leaching			
Dissolved inorganic to SW	0a	1.08a	0.31a
	(0.63)	(0.82)	(0.31)
Dissolved organic to SW	2.08a	21.98b	2.70a
	(0.83)	(5.51)	(0.63)
Dissolved organic to PW	12.73a	12.95a	7.60a
	(2.52)	(3.20)	(1.49)
Dissolved inorganic to PW	55.09a	6.77b	7.30b
	(7.43)	(5.41)	(7.31)
Aboveground litter to sediments	34.48a	21.80b	2.70c
	(0.83)	(5.50)	(0.63)
Belowground litter to sediments	7.79a	11.11b	11.03b
	(1.23)	(1.44)	(1.92)

[a] Baseline, $n = 10$ (1980), $n = 2$ (1981); Reference, $n = 2$ (1980–1983).
[b] $n = 8$ (1981), $n = 2$ (1982).
[c] $n = 8$ (1982), $n = 2$ (1983).
[d] Within a row means with different letters are significantly different ($P < 0.05$).
[e] Includes suspended solids to litter.

Table 4.6. Mean (SE) annual (1 Oct.–30 Sept.) phosphorus (P) fluxes (kg/ha/yr) in the MERP experimental cells, 1980–1983.

	Treatment		
Flux	Baseline + reference[a]	Flooded year one[b]	Flooded year two[c]
Macrophytes			
Uptake	15.14a[d]	0.94b	4.29b
	(1.02)	(0.23)	(1.25)
Translocation	9.04a	0.77b	2.32b
	(0.65)	(0.38)	(0.37)
Aboveground leaching	4.11a	0.50b	1.03b
	(0.29)	(0.16)	(0.17)
Belowground leaching	1.43a	1.12a	0.65b
	(0.24)	(0.25)	(0.18)
Aboveground litter-fall	3.96a	2.76a	0.58b
	(0.34)	(0.16)	(0.19)
Belowground litter-fall	5.71a	4.46a	2.61a
	(0.96)	(0.46)	(0.53)
Algae and invertebrates			
Algal uptake	0.76a	0.24a	1.53a
	(0.38)	(0.12)	(0.53)
Algal litter-fall[e]	0.60a	1.92ab	2.14b
	(0.19)	(0.32)	(0.47)
Invertebrate loss	0.12a	0.02a	0.02a
	(0.02)	(0.01)	(0.01)
Invertebrate uptake: litter and	0.14a	0.18b	0.38c
algae combined	(0.02)	(0.01)	(0.02)
Dissolved nutrients and			
suspended solids			
Mineralization			
Aboveground SWDO to SWDI	0.53a	0.28b	0.52a
	(0.05)	(0.07)	(0.10)
Belowground PWDO to PWDI	0.29a	0.66a	1.03a
	(0.10)	(0.16)	(0.26)
Pore–surface water exchange	−4.61a	−0.38a	−0.75b
	(0.49)	(0.06)	(0.67)
Litter/and sediment pool			
Litter leaching			
Dissolved inorganic to SW	0.07a	0.55a	0.66a
	(0.02)	(0.41)	(0.66)
Dissolved organic to SW	0.18a	0.87b	0.09a
	(0.07)	(0.21)	(0.01)
Dissolved organic to PW	0.33a	0.76a	1.26a
	(0.07)	(0.19)	(0.25)
Dissolved inorganic to PW	8.89a	2.16b	3.92b
	(1.19)	(1.02)	(2.51)
Aboveground litter to sediments	4.39a	2.75b	2.38b
	(0.28)	(0.28)	(0.23)
Belowground litter to sediments	2.73a	1.33b	1.23b
	(0.21)	(0.14)	(0.17)

[a]Baseline, $n = 10$ (1980), $n = 2$ (1981); Reference, $n = 2$ (1980–1983).
[b]$n = 8$ (1981), $n = 2$ (1982).
[c]$n = 8$ (1982), $n = 2$ (1983).
[d]Within a row means with different letters are significantly different ($P < 0.05$).
[e]Includes suspended solids to litter.

ing crops, algal litter fall increased during the flooding years. Although the invertebrate fluxes were small compared with the other fluxes in the system, invertebrate uptake increased during both years of deep flooding (Tables 4.5 and 4.6).

Dissolved Nutrients and Suspended Solids

Although pore-water mineralization rates appeared to be little affected by the deep flooding, the surface-water mineralization of N increased during the first year of flooding before returning to baseline levels during the second year (Table 4.5). The P mineralization, in contrast, declined during the first year of flooding before returning to baseline levels the second year (Table 4.6). Pore water–surface water exchange decreased through both years of flooding and was a small fraction of baseline levels by the end of the second year of flooding (Tables 4.5 and 4.6).

Litter/Sediment Pool

Litter leaching to surface water generally increased during the first year of flooding and then with the reductions of aboveground litter inputs declined during the second year to levels similar to baseline years (Table 4.5). In contrast, litter leaching to pore water remained the same for organic forms, but declined significantly for inorganic forms during both years of flooding. Aboveground particulate litter loss to the litter/sediment pool declined with flooding. The belowground loss of particulate litter to the litter/sediment pool declined during the flooding years (except for belowground N losses) in response to the overall decline in overall litter-fall rates.

To put some perspective on the size of the litter/sediment pool relative to the other pools and fluxes, the amount of N in the total sediment, including litter, is usually very large, up to 10 times that of the rest of the system combined (Barko et al. 1991). Rough calculations based on analyses for total N by agricultural techniques suggested the pool sizes in the root zone of the sediments in the MERP experimental cells were on the order of 8,250 kg N/ha. Given that approximation, the release of 96 kg/ha from pore water to macrophyte uptake in the cells (Table 4.5) during the baseline years represented a little more than 1% of the sediment pool.

DISCUSSION

Both flooding and draining of the MERP experimental cells were done in late September and October (see Chapter 1). Flooding the cells this late in the year had implications for the postflooding nutrient budgets. Firstly, it flooded the aboveground standing litter of the emergent vegetation produced in the cells during the previous summer (i.e., not live aboveground plant tissues). Secondly, in the spring, the cells were already flooded prior to the onset of emergent vegetation growth in late May and early June. This flooding resulted in a reduction in mean standing crop of macrophytes by >80% during the first year of flooding compared with the baseline year (van der Valk (1994); see Chapter 7). This reduction in aboveground macrophyte maximum standing crop was the result of reduced production, and not due to the death of existing, living aboveground macrophyte tissues at the time of flooding. Consequently, the release of nutrients into the surface-water compartment due to the death of living aboveground plant material was minimal in the MERP study. Nelson et al. (1990) found much greater rate of loss of nutrients by leaching and higher subsequent decomposition rates of green litter compared with naturally senesced litter in the Delta Marsh. If the cells had been flooded in midsummer instead of the late fall, a much greater mass of nutrients would have been released immediately from the macrophyte pools.

As predicted, the deep water levels reduced both the aboveground and belowground N and P macrophyte pools. The aboveground macrophyte pool declined very quickly, whereas the belowground pool declined more slowly due to the initial survival of belowground tissues following flooding (van der Valk 1994). Belowground tissues have stored reserves that allow them to survive through periods of adverse conditions (Linde et al. 1976). The survival of the belowground macrophyte pool meant that this pool remained the largest single pool in the MERP experimental cells during the deep-flooding years (as it was during the baseline years). Consequently, a significant quantity of nutrients remained in this pool and was unavailable to other pools.

Intrasystem nutrient pools and fluxes clearly did respond to the reduction of aboveground emergent macrophytes. Because the major changes in nutrient pool sizes, and consequently fluxes, were primarily the result of changes in macrophyte biomass, the cycles of N and P closely paralleled each other in these marshes. Essentially, intrasystem cycles in these systems were "driven" by uptake by macrophyte biomass in late spring–early summer and losses due to leaching and litter-fall in late summer–fall. The major impact of flooding was to reduce aboveground macrophyte production, resulting in lower macrophyte pool sizes and associated fluxes. Macrophyte uptake from sediments decreased rapidly with the reduction of aboveground macrophyte production, and consequently aboveground fluxes to litter pool and sediment sink also decreased. With the delay in the mortality of the belowground tissues, changes in these fluxes were delayed compared with the aboveground macrophyte pools.

The total mass of N in all measured pools during the baseline years, first year, and second year of deep flooding was 204, 170, and 186 kg/ha, respectively. The total mass of P was 24.8, 15.9, and 21.2 kg/ha, respectively. The combined macrophyte pools lost 68 kg/ha of N and nearly 12 kg/ha of P between the baseline year and the first year of flooding (Figure 4.1). The mass of N and P in metaphyton, invertebrates, surface-water and pore-water pools all increased during the first year of flooding as predicted. The total mass of nutrients in all nonmacrophyte pools combined during the first year of deep flooding, however, increased by only 34 kg/ha of N and 3 kg/ha of P, far short of the total losses from the macrophyte pools during the same period. This difference indicates that a significant amount of both N and P was still tied up in the litter and sediment pool and had not been released to the other pools.

During the second year of deep flooding, there was a slight increase in the aboveground macrophyte nutrient pools over those found during the first year of flooding (as submersed vegetation became established in the experimental cells) although the belowground macrophyte pools continued to decline. The total increases in the mean mass of nutrients in macrophytes and the other nonmacrophyte pools combined during the second year of flooding was ≈ 33 kg/ha for N and 6.7 kg/ha for P, the equivalent of only 50% of the N and

60% of the P reductions in the macrophyte pools between the baseline and the first year of flooding. It appears that a significant amount of the N and P in the plant litter inundated during the initial deep flooding had still not been released through litter decomposition by the end of the second year of deep flooding.

Our prediction that other pools (most notably surface-water, pore-water, algae, and invertebrate pools) would increase during the first year of flooding was supported by the data from the experimental cells. The combined mass of all nonmacrophyte pools during the first year of deep flooding was over 70% and 85% greater for N and P, respectively, than during the baseline years. These nonmacrophyte pools continued to increase during the second year of flooding, which was not predicted. During the second year of deep flooding, the combined mass of all nonmacrophyte pools was 130% and 230% greater for N and P, respectively, than during the baseline years. During this time the nonmacrophyte nutrient pools collectively were larger than the remaining macrophyte pools, although no one of them alone was yet equal in mass to the mass of nutrients in the macrophyte pools. A large portion of the nonmacrophyte pools was associated with the increase in the metaphyton communities during the deep-flooding period. The metaphyton was probably responding to the availability of nutrients and the habitat conditions provided by the flooded litter (Robinson et al. 1997a,b). Metaphyton standing crops, however, during the deep-flooding years never approached the mass of the aboveground macrophyte compartment during the baseline years and were about equal to the remaining aboveground macrophyte pool by the second year of deep flooding.

The continued increase in the nonmacrophyte pools during the second year of flooding suggests that the release of nutrients from flooded emergent litter was relatively slow. Only during the second year of flooding were sufficient nutrients available for the total size of all measured nutrient pools to approach their combined mass during the baseline year. Our prediction that much of the litter would decompose early in flooding and leave little additional nutrients to support production during the second year of flooding was clearly wrong. The total mass of all the measured nutrient pools in the cells during the second year of flooding was still ≈10% for N and 15% for P below that during the baseline years. Concurrent studies of emergent-litter decomposition during the deep-flooding years in the MERP cells (Murkin et al. 1989; van der Valk et al.1991; Wrubleski et al. 1997a,b) indicate that release of nutrients was generally accelerated by flooding, but that significant quantities (often 50% or more) of N and P remained in the litter of the four dominant emergent species after 1 or 2 years of flooding. In fact, some litter types actually increased in N and P during the first year of decomposition under flooded conditions due to microbial colonization of the litter. These results are also consistent with studies of litter decomposition in the Delta Marsh and other prairie marshes that indicate that this is a very slow process for the litter of the dominant emergent species (Davis and van der Valk 1978a, b; Neely and Davis 1985; Nelson et al. 1990). Ultimately, changes in the size of the nonmacrophyte nutrient pools and fluxes during the deep-flooding years are primarily a function of patterns of N and P release from decomposing litter (van der Valk et al. 1991). During the deep-flooding years, both surface water and pore water accumulated dissolved inorganic and organic N and P because of release of nutrients from above- and belowground litter.

In general, the result of deep flooding was the reduction of the macrophyte pools and the increase in the nonmacrophyte pools in the system. However, a portion of the nutrients lost from the macrophytes could not be accounted for in the increase in the non-macrophyte pools. These unaccounted nutrients were probably incorporated in the litter/sediment pool and eventually over time would be buried and lost from the available nutrient pools or decompose slowly and become available to the other pools in the system but over an extended time period. A portion of the N may have been lost through denitrification. Denitrification can be a significant process in the removal of N in many wetland types (Neely and Baker 1989). Because conditions suitable for denitrification were present in the experimental cells (e.g., anaerobic conditions and availability of carbon substrates [emergent litter]), it is likely that some denitrification did occur. The lack of information on denitrification and other microbial processes associated with litter and the sediments in the MERP study and their potential role in the nutrient cycling of prairie wetland systems indicates that future work in these systems must address these processes. In addition, the long-term losses of nutrients through burial (due to sedimentation and litter buildup) require further investigation.

Raising water levels in the MERP complex differed somewhat from what happens during high-water events in prairie marshes in several significant ways. In natural marshes, lateral seepage losses should be much lower than in the MERP cells during high-water years because the entire marsh would have the same water level. There could, however, still be losses of water and nutrients to groundwater and bank storage (Winter 1989). The N and P concentrations in the water entering the marsh due to precipitation, surface flow, or groundwater flow are not likely to be similar to those in the marsh already. The latter was the case during the MERP study because water from the adjacent marsh was pumped into the cells. Surface water normally does not drain out of the marsh at the end of the high-water years. Water levels drop primarily because of evapotranspiration, which leaves nutrients in the surface-water pool in the marsh. Although there may have been a greater input to the MERP experimental cells due to the pumping of Delta Marsh water, there was an offsetting loss of these nutrients through increased seepage and the pumping down of the cells at the end of the deep-flooding experiments. In spite of these differences in nutrient inputs–outputs between the MERP cells during the deep-flooding experiments and natural marshes under natural above-normal flooding conditions, overall nutrient cycling was dominated and controlled by the macrophytes and therefore these input–output differences would be of minor importance to the overall nutrient cycling in these systems. In conclusion, similar results should be expected in natural systems under conditions of natural flooding as were found in the MERP experimental cells during the deep-flooding experiments.

We predicted that by the second year of deep flooding that overall productivity in the experimental cells would decline and the conditions in the cells would approach those characteristic of the lake marsh stage of the prairie wetland wet–dry cycle. However, our results indicated that during the second year of flooding, conditions were somewhat different from those expected during the lake marsh stage. Under natural conditions, the lake marsh stage develops in prairie marshes during periods of continuous flooding often at stable water levels (van der Valk and Davis 1978). The normal development of the lake marsh stage occurs over an extended period of time as the emergent vegetation slowly dies back due to factors related to continuous flooding. Thus, there is a gradual declining input of litter as the emergent vegetation is eliminated from the central portions of the basin. Overall productivity within the water column declines as aboveground nutrient pools and associated fluxes are reduced and pools associated with the sediments (buried litter and pore water) increase. Drawdown conditions are required (through germination of plants and oxidation of the sediments) to reestablish the aboveground pools and associated productivity.

In the MERP deep-flooding experiments, a single, large pulse of litter was added to the system as water levels were raised to 1 m above normal. This pulse of litter resulted in increased sizes of nonmacrophyte pools, including algae and invertebrates, for the 2 years of flooding. So although the emergent macrophyte nutrient pools were much reduced and indicative of lake marsh conditions, the algal and invertebrate pools by the end of the second year of flooding were still at levels above the baseline years (i.e., above the low levels expected during the lake marsh stage). Murkin (1989) reported that invertebrate levels would be very low during the lake marsh stage due to the reduced litter inputs and the reduced algal production during this stage of the wet–dry cycle. Goldsborough and Robinson (1996) in modeling the development of algal communities during the wet–dry cycle suggest that phytoplankton will dominate during lake marsh conditions. Phytoplankton does not attain high biomass levels in prairie marshes (Robinson et al 1997a,b). Metaphyton was still abundant in the MERP cells during the second year of flooding (Hosseini and van der Valk 1989).

Because the nutrient budgets of the experimental cells appear to be "driven" by the pools and fluxes associated with the macrophytes, it would be expected, with continued flooding, and as the macrophyte litter fluxes decreased, that the aboveground pools would eventually decrease and the cells would enter the true lake marsh stage. Interestingly, pore-water inorganic pools increased over the period of flooding, as would be expected under lake marsh conditions with the elimination of the macrophytes and associated fluxes such as plant uptake. The pore water–surface water exchange also declined during the second year of flooding, another indication that the cells were on their way to the lake marsh stage.

Pumping of water from the adjacent Delta Marsh into the MERP cells greatly increased the volume of the surface-water pool and the mass of N and P in this pool. Draining water out of the cells at the end of the deep-flooding years resulted in more nutrients being lost from the cells than were added to them when they were pumped up. This net loss of nutrients from the cells was equivalent to ≈10% of the N and 25% of the P that were present in all measured nutrient pools during the baseline years. Using water-level manipulations to manage wetland vegetation can result in a loss of nutrients and possibly the reduction of the long-term productivity of the wetland. Raising and lowering water levels in an entire marsh by pumping, as was done in the MERP cells, could result in the net loss of dissolved nutrients from the wetland. This is because the mass of dissolved N and P in the surface water that can be removed by pumping is much higher in wetlands when they are deeply flooded than when they are not. In the deep-flooded MERP experimental cells, the mean total mass of N and P in the surface water during the second year of flooding was 330% and 180% greater, respectively, than in the baseline years. When marshes are being drained after a period of deep flooding to kill emergents, there is a real possibility of significant losses of N and P. These losses, however, may only be short term due to the release of nutrients from the large sediment pool over the long term.

ACKNOWLEDGMENTS

This Paper No. 97 of the Marsh Ecology Research Program, a joint project of Ducks Unlimited Canada and the Delta Waterfowl and Wetlands Research Station. Penny Alisauskas and Jonathon Scarth supervised the field crews during the baseline and deep-flooding years.

LITERATURE CITED

Barko, J.W., D. Gunnison, and S.R. Carpenter. 1991. Sediment interactions with submersed macrophyte growth and community dynamics. Aquatic Botany 41:41-65.

Cook, A.H., and C.F. Powers. 1958. Early biochemical changes in the soils and waters of artificially created marshes in New York. New York Fish & Game Journal 5:9-65.

Davis, C.B., and A.G. van der Valk. 1978a. Litter decomposition in prairie glacial marshes. In *Freshwater Wetlands*. (Eds.) R.E. Good, D.F. Whigham, and R.L. Simpson, pp.99-113. New York: Academic Press.

Davis, C.B., and A.G. van der Valk. 1978b. The decomposition of standing and submerged litter of *Typha glauca* and *Scirpus fluviatilis*. Canadian Journal of Botany 56:662-675.

Goldsborough, L.G., and G.G.C. Robinson. 1996. Pattern in wetlands. In *Algal Ecology: Freshwater Benthic Ecosystems*. (Eds.) R.J. Stevenson, M.L. Bothwell, and R.L. Lowe, pp.77-117. New York: Academic Press.

Hecky, R.E., R.W. Newbury, R.A. Bodaly, K. Patals, and D.M. Rosenberg. 1984. Environmental impact prediction and assessment: the Southern Indian Lake experience. Canadian Journal of Fisheries and Aquatic Sciences 41:720-732.

Hosseini, S.M., and A.G. van der Valk. 1989. The impact of prolonged above-normal flooding on metaphyton in a freshwater marsh. In *Freshwater Wetlands and Wildlife*. (Eds.) R.R. Sharitz and J.W. Gibbons, pp.317-324. USDOE Symposium Series No. 61, Oak Ridge: USDOE Office of Scientific and Technical Information.

Kadlec, J.A. 1962. Effect of a drawdown on the ecology of a waterfowl impoundment. Ecology 43:267-281.

Kadlec, J.A. 1983. Water budgets for small diked marshes. Water Resources Bulletin 19:223-229.

Kadlec, J.A. 1986a. Effects of flooding on dissolved and suspended nutrients in small diked marshes. Canadian Journal of Fisheries and Aquatic Sciences 43:1999-2008.

Kadlec, J.A. 1986b. Input-output nutrient budgets for small diked marshes. Canadian Journal of Fisheries and Aquatic Sciences 43:2009-2016.

Kantrud, H.A., J.B. Millar, and A.G. van der Valk. 1989. Vegetation of wetlands of the prairie pothole region. In *Northern Prairie Wetlands*. (Ed.) A.G. van der Valk, pp.132-187. Ames: Iowa State University Press.

Linde, A.F., T. Janisch, and D. Smith. 1976. Cattail—the significance of its growth, phenology, and carbohydrate storage to its control and management. Wisconsin Department of Natural Resources Technical Bulletin 94. 27pp.

Murkin, H.R. 1989. The basis for food chains in prairie wetlands. *In Northern Prairie Wetlands*. (Ed.) A.G. van der Valk, pp.316-339. Ames: Iowa State University Press.

Murkin, H.R., and J.A. Kadlec. 1986. Responses by benthic macroinvertebrates to prolonged flooding of marsh habitat. Canadian Journal of Zoology 64:65-72.

Murkin, H.R., J.A. Kadlec, and E.J. Murkin. 1991. Effects of prolonged flooding on nektonic invertebrates in small diked marshes. Canadian Journal of Fisheries and Aquatic Sciences 48:2355-2364.

Murkin, H.R., A.G. van der Valk, and C.B. Davis. 1989. Decomposition of four dominant macrophytes in the Delta Marsh, Manitoba. Wildlife Society Bulletin. 17:215-221.

Neely, R.K., and J.L. Baker. 1989. Nitrogen and phosphorus dynamics and the fate of agricultural runoff. In *Northern Prairie Wetlands.* (Ed.) A.G. van der Valk, pp.92-131. Ames: Iowa State University Press.

Neely, R.K., and C.B. Davis. 1985. Nitrogen and phosphorous fertilization of *Sparganium eurycarpum* Engelm. and *Typha galuca* Godr. Stands. I. Emergent plant decomposition. Aquatic Botany 23:347-361.

Nelson, J.W., J.A. Kadlec, and H.R. Murkin. 1990. Seasonal comparisons of weight loss for 2 types of *Typha glauca* Godr. leaf litter. Aquatic Botany 37:299-314.

Ott, L. 1988. *An Introduction to Statistical Methods and Data Analysis.* PWS-Kent Publishing, Boston, Massachusetts.

Robinson, G.G.C., S.E. Gurney, and L.G. Goldsborough. 1997a. Response of benthic and planktonic algal biomass to experimental water-level manipulation in a prairie lakeshore wetland. Wetlands 17:167-181.

Robinson, G.G.C., S.E. Gurney, and L.G. Goldsborough. 1997b. The primary productivity of benthic and planktonic algae in a prairie wetland under controlled water-level regimes. Wetlands 17:182-194.

Vallentyne, V.R. 1952. Insect removal of nitrogen and phosphorus compounds from lakes. Ecology 33:573-577.

van der Valk, A.G. 1994. Effects of prolonged flooding on the distribution and biomass of emergent species along a freshwater wetland coenocline. Vegetatio 110:185-196.

van der Valk, A.G., and C.B. Davis. 1978. The role of seed banks in the vegetation dynamics of prairie glacial marshes. Ecology 59:322-335.

van der Valk, A.G., J.M. Rhymer, and H.R. Murkin. 1991. Flooding and the decomposition of litter of 4 emergent plant species in a prairie wetland. Wetlands 11:1-16.

Weller, M.W., and L.H. Fredrickson. 1974. Avian ecology of a managed glacial marsh. Living Bird 12:269-291.

Weller, M.W., and C.E. Spatcher. 1965. Role of habitat in the distribution and abundance of marsh birds. Department of Zoology and Entomology Special Report 43. Agricultural and Home Economics Experiment Station, Iowa State University, Ames, Iowa.

Whitman, W.R. 1974. Impoundments for waterfowl. Canadian Wildlife Service Occasional Paper 22.

Winter, T.C. 1989. Hydrologic studies of wetlands in the northern prairies. In *Northern Prairie Wetlands.* (Ed.) A.G. van der Valk, pp.16-54. Ames: Iowa State University Press.

Wrubleski, D.A., H.R. Murkin, A.G. van der Valk, and C.B. Davis. 1997a. Decomposition of litter of three mudflat annual species in a northern prairie marsh during drawdown. Plant Ecology 129:141-148.

Wrubleski, D.A., H.R. Murkin, A.G. van der Valk, and J.W. Nelson. 1997b. Decomposition of emergent roots and rhizomes in a northern prairie marsh. Aquatic Botany 58:121-134.

5

The Drawdown and Reflooding Years

Arnold G. van der Valk
Henry R. Murkin
John A. Kadlec

ABSTRACT

Pumping the experimental cells to three different levels had a significant impact on the nutrient budgets and fluxes as they proceeded through the wet–dry cycle following flooding. The inputs from pumping to initially establish and then maintain the experimental water levels and resulting outputs due to increased seepage resulted in little net difference in input–output budgets among the water-level treatments. Therefore, the observed differences among treatments were due to the different water levels maintained over the course of the experiment. As during the other MERP experiments, macrophytes dominated the pools and fluxes within the nutrient budgets. During the initial reflooding and regenerating stage, macrophyte pools and fluxes reacted as predicted with an initial increase after flooding, a small decline after 2 or 3 years of flooding (due to the loss of short-lived emergents), and then an increase again as the long-term emergents expanded into all available habitats. The low-treatment cells were still in the regenerating stage at the end of 5 years of flooding and had not reached the levels found in the reference cells (at the same water level) as predicted. During the degenerating stage that occurred in the deeper-flooding treatments (medium and high), aboveground macrophyte pools declined significantly. The decline in the belowground pools was delayed and occurred at a slower rate. The permanent decline in the belowground pool marks the start of the degenerating stage. The degenerating stage did not start sooner in the high-water treatment because the difference in depth between the two treatments was not large enough to result in differential macrophyte survival. As the macrophyte pools declined, other pools and associated fluxes began to increase. It does not appear that any of the treatments moved to the lake marsh stage during the reflooding experiment.

INTRODUCTION

In wet–dry cycles in prairie wetlands, the lake marsh stage is suggested to be the period of lowest productivity (van der Valk and Davis 1978a; van der Valk 1981). During this stage, the macrophyte nutrient pools are at minimum levels with large masses of nutrients stored in sediment/litter and pore-water pools (see Chapter 4). Because emergent macrophyte seeds cannot germinate under flooded conditions, a dry period or managed draw-down free of standing water is required to reestablish macrophytes and associated nutri-ent pools and fluxes within the basin (see Figure 5.1). The newly reflooded stage after the drawdown is termed the regenerating marsh stage during which emergent vegetation that became established during the drawdown expands due to clonal growth (van der Valk and Davis 1978a). This expansion continues until all suitable habitat for emergent vegetation is fully occupied (van der Valk 1994). The eventual decline of emergent populations due to the effects of continued flooding marks the beginning of the degenerating stage. In this chapter, we examine the nutrient pools during the drawdown years and the subsequent changes to the nutrient budgets, pools, and fluxes when the MERP experimental cells were reflooded (Figure 5.2) to the three different water levels (see Tables 1.1 and 1.2 in Chapter 1). We also examine how the different water levels following the drawdown affected the length of the various stages of the wet–dry cycle.

During the wet–dry cycle, the regenerating marsh stage ends when the emergents begin to die back irreversibly (van der Valk and Davis 1978b). A decline in emergents can occur for two reasons. Firstly, some emergent species cannot survive being permanently flood-ed for more than 2 or 3 years and die back. In the Delta Marsh, two dominant emergent species fall into this category, *Scolochloa festucacea* (Neckles et al. 1985; Merendino and Smith 1991; Neill 1992) and *Scirpus lacustris* subsp. *validus* (Macauley 1973; van der Valk and Davis 1978b; Shay and Shay 1986). Although these species usually disappear during the second or third year of continuous flooding, the degenerating marsh stage may not begin at this point in time. Other longer-lived emergent species, such as *Typha* spp. and *Phragmites australis*, invade areas where the short-lived emergents were formerly found. Thus, the regenerating phase may continue because short-lived emergents species are replaced by long-lived species that can tolerate longer-term inundation (de Geus 1987). Secondly, stable water levels during the regenerating stage eventually result in the loss of most emergent species, even those that can survive longer-term inundation, due to their eventual elimination in water depths that exceed their tolerances (Squires and van der Valk 1992), their increasing susceptibility to diseases and insect herbivores when stressed by continuous flooding, or grazing by muskrats whose populations increase in deeper areas of the marsh because of improved winter survival (Weller and Fredrickson 1974; van der Valk and Davis 1978a; van der Valk 1981; Clark and Kroeker 1993). The decline in long-lived emergent populations with continuous flooding is an unequivocal indication of the onset of the degenerating stage.

Alternating periods of flooding and drying during wet–dry cycles increase the overall long-term productivity of prairie wetlands (van der Valk and Davis 1978a) and the avail-ability of nutrients for macrophyte growth (Bayley et al. 1985). During a natural period of decreased precipitation or artificial drawdown, nutrients accumulated in sediment and pore-water pools during the lake marsh stage (see Chapter 4) begin to be released as plants germinate on the dry substrate and uptake from the pore-water pools is reestablished (i.e., the plants "pump" nutrients to aboveground pools). The increased macrophyte pools and resulting litter pools produced during the dry stage add significant amounts of nutrients to surface-water pools through leaching and subsequent litter decomposition when the

Figure 5.1   Development of vegetation on the exposed substrate of a MERP experimental cell during the first year of drawdown (1983).

marshes are reflooded. The reestablishment of these aboveground pools and associated fluxes contributes to the overall productivity of the system. During the early years of the regenerating stage, macrophyte production is expected to increase and should reach maximum pool size during this stage. Algal production is hypothesized to be high during the first several years of the regenerating stage because of the availability of nutrients and light (the emergent plants have not yet begun to shade most of the standing water) (Goldsborough and Robinson 1996; Robinson et al. 1997a,b). Likewise, invertebrate production is hypothesized to be high in the first several years in response to the increased algal and litter pools (Murkin 1989).

In 1985, the MERP cells were reflooded to three different water levels. Four cells were reflooded to 247.5 m above mean sea level (AMSL), the natural levels found in the Delta Marsh since the early 1960s when regulation of Lake Manitoba began (see Chapter 2). This also was the water level during the baseline years in the MERP experimental cells

Figure 5.2   Vegetation changes in a MERP experimental cell in the medium-flooding (30 cm above normal) treatment (1985–1989); 1985 was the first year of flooding after 2 years of drawdown conditions.

(Table 1.1 in Chapter 1). This treatment is designated as the low or normal treatment. Three cells (medium treatment) were reflooded to 30 cm above the level in the low treatment, and three remaining cells (high treatment) were reflooded to 60 cm above the level in the low treatment. All three water-level treatments were maintained for 5 years (1985–1989) by using electric pumps.

Specific predictions made with respect to effects of reflooding at three different water levels on the nutrient budgets of the MERP experimental cells were as follows:

1. Macrophyte, algal, and invertebrate pools should reach their maximum sizes soon after flooding (regenerating stage) and then decline with continued flooding (degenerating stage). A slight decrease in the macrophyte pools would occur when the short-lived emergent species died back after 2 or 3 years; however, they would be replaced by long-lived species and the macrophyte pools would then continue to increase through the regenerating stage.

2. Increased water depths would accelerate the change from regenerating to degenerating stages. The high treatments would move from the regenerating to the degenerating stage most quickly while the low treatment would remain in the regenerating stage for the longest period.

3. Because they were at the same water levels (247.5 m AMSL), nutrient pools and fluxes in the low-treatment cells would eventually approach conditions found in the reference cells.

To test these predictions, the overall input–output nutrient budgets (i.e., nutrient inputs and outputs due to pumping, precipitation, seepage, and evapotranspiration) and their intrasystem cycles (i.e., the mass of nitrogen (N), phosphorus (P), and carbon (C) in the various pools present and fluxes amongst these pools) during the reflooding period (1985–1989) were estimated for the experimental and reference cells by using the nutrient-budget framework and equations outlined in Chapter 3. Because a complete set of data on dissolved inorganic carbon (DIC) and dissolved organic carbon (DOC) pools was available during this period, construction of a C budget as well as N and P budgets was possible. Losses of C due to methanogenesis, however, were not measured, and this flux in the C budgets was treated as losses to the DIC pool and eventually the litter pool. Pool sizes and fluxes among years and flooding treatments were compared by using analysis of variance, and in cases where the assumptions of parametric procedures were violated, Kruskal–Wallis nonparametric tests (Ott 1988) were used.

We begin with the hydrologic input–output nutrient budgets to examine the effects of pumping on the pools and fluxes in the different water-level treatments. Keeping the experimental cells at their designated water levels required substantial pumping especially in the medium and high treatments. Hydrologic inputs can have significant effects on the size of their nutrient pools and seasonal fluxes among pools (Devito et al. 1989). The importance of this input of nutrients can best be examined by comparing overall water and nutrient budgets for the three water-level treatments to those of the reference cells where water levels were not manipulated in any way. We then examine the changes in the mean size of the nutrient pools and fluxes in response to the different water-level treatments during reflooding.

HYDROLOGY

In the low-treatment and reference cells, the water budget was dominated by precipitation and losses due to evapotranspiration (Table 5.1). In the two deeper treatments, however, pumping during the summer equaled or exceeded precipitation. In the high treatment, the volume of water pumped into cells in the summer was nearly double the volume of precipitation inputs. Because of the increased surface area flooded and greater hydrostatic heads in the medium and high treatments, evapotranspiration and seepage were higher compared with that of the low treatment. These losses offset the greater inputs of water due to pumping. In effect, the medium and especially the high treatment were equivalent to flow-through prairie wetland systems (Winter 1989).

Available weather data provided an estimate of precipitation (water equivalent) inputs for the 6-month winter period (Table 5.1). These data probably underestimated winter water inputs into the cells because they trapped drifting snow. With that caveat and the assumption that winter evapotranspiration was near zero (actually very low but probably not zero), winter water budgets were calculated as shown in Table 5.1. Seepage from the medium (+30 cm) and high (+60 cm) treatments over the 6-month winter period were estimated to be approximately half of the amount lost through seepage during the summer. In fact, most of the winter seepage indicated by these analyses probably occurred in late fall before freeze-up and in early spring after the initial thaw.

Table 5.1.   Summary annual water budgets during the summer (May–October) and winter (November–April) for the MERP experimental cells, 1985–1989

Flooding status	Season	Flooded area – ha (SE)	Pumped[b]	Volume – 1,000 m^3/experimental cell (SE)[a]		
				Precipitation	Evapo-transpiration[c]	Net seepage
Low	Summer	3.75(0.10)	−0.38(2.43)	21.6(0.80)	−24.3(1.70)	3.93(1.29)
n = 4	Winter		—	8.96(0.94)	—	−2.50(1.02)
Medium	Summer	4.67(0.08)	19.1(1.99)	21.6(0.56)	−30.1(1.39)	−13.5(1.21)
n = 3	Winter		—	8.97(0.74)	—	−6.52(0.64)
High	Summer	5.90(0.12)	47.3(1.98)	24.1(0.56)	−37.9(1.39)	−29.1(1.21)
n = 3	Winter		—	9.98(0.88)	—	−13.2(1.64)
Reference	Summer	3.28(0.14)	0	20.7(0.69)	−19.3(1.70)	1.60(1.29)[d]
n = 2	Winter		—	8.59(0.88)	—	−4.00(2.36)

[a] The standard error represents the statistical summation of the data and does not reflect the error of measurement inherent in the methods used to obtain the data.

[b] No pumping during winter.

[c] Evapotranspiration assumed to be zero during the winter.

[d] Seepage in reference cells equal to surface-water inflow and outflow.

N AND P BUDGETS

Inputs–Outputs

Pumping added significant quantities of NH_4, dissolved organic nitrogen (DON), soluble reactive phosphorous (SRP), and dissolved organic phosphorous (DOP) to the experimental cells in the medium and high treatments (Table 5.2). For NH_4, losses due to seepage equaled or exceeded the inputs due to pumping in the deeper treatments (net loss). The DON lost through seepage, however, was always lower than that input by pumping (net gain). For SRP and DOP, seepage losses exceeded pumping inputs in the high treatment, but not in the medium treatment where seepage losses were not as high as pumping inputs. Only for DON and DOP did pumping add more nutrients to the cells in the medium and high treatments than precipitation. The result was that, in spite of pumping, net dissolved nutrient inputs to the medium- and high-treatment cells were not much higher than inputs to the reference cells and low treatments where some pumping out was required to maintain the experimental water levels (Table 5.2).

Intrasystem Cycling

Pools

Macrophytes

For a detailed description of changes in the composition and distribution of the vegetation in the three water-level treatments during the reflooding years (1985–1989), see de Swart et al. (1994) and van der Valk (1994). For changes in macrophyte standing crops and primary production see Chapter 8. Above- and belowground macrophyte pools were the only nutrient pools measured during the drawdown years (see Appendix 1). Figure 5.1 shows the development of vegetation during the first year of drawdown in experimental cells; Figure 5.2 shows the changes in vegetation following reflooding (in this case, a medium cell). Aboveground macrophyte N and P pools during the drawdown period were larger than during the reflooding years and were similar in size to the N and P pools in the reference cells (Table 5.3). In contrast, the aboveground macrophyte C pools were about the same size during the drawdown and reflooding years and especially during the early years of reflooding. Mean aboveground macrophyte N and P pools (Table 5.3) did not change significantly during the reflooding years, although the trend was for a slight decline during the second year of flooding followed by an increase during the third year. After the third year of flooding, pool sizes in the medium and high treatments declined over the next 2 years, with a significant decline in the high treatment from the early-flooding years. The C aboveground macrophyte pool increased from 1986 to 1987 and then declined, especially in the medium and high treatments.

Mean belowground N, P, and C pools were smaller during the drawdown years than during the reflooding years and were also much smaller than belowground pools in the reference cells (Table 5.3). The vegetation during the drawdown years was dominated by annual species (Welling et al. 1988) with very small root systems compared with the below-

Table 5.2. Mean annual input–output nutrient budgets (kg/ha/yr) for various forms of nitrogen (N), phosphorus (P), and carbon (C) in surface water of the MERP experimental cells, 1985–1989.

Nutrient	Flooding status	Pumped	Precipitation	Seepage	ΔStorage	SYSF[a]
NH$_4$	Low	−0.12	2.63	0.14	0.54	−2.12
	Medium	0.48	2.65	−0.46	0.81	−1.86
	High	0.62	2.64	−1.11	0.91	−1.23
	Reference	—	2.50	0.15[b]	−0.15	−2.80
Dissolved	Low	−1.17	4.11	5.92	2.46	−6.45
organic N	Medium	11.7	4.07	−7.49	3.34	−4.95
	High	21.8	4.05	−17.2	4.59	−4.24
	Reference	—	4.32	4.29	−0.29	−8.90
Suspended N	Low	0.36	1.69	—[c]	0.04	−1.02
	Medium	2.87	1.67	—	0	−4.53
	High	5.15	1.67	—	0.08	−6.75
	Reference	—	1.75	—	0.66	−1.09
Soluble	Low	−0.09	0.63	0.07	−0.10	−0.71
reactive P	Medium	0.15	0.63	−0.04	−0.07	−0.80
	High	0.28	0.62	−0.46	0.08	−0.36
	Reference	—	0.67	0.46	−0.87	−1.48
Dissolved	Low	−0.04	0.08	0.13	0.14	−0.16
organic P	Medium	0.23	0.07	−0.14	0.05	−0.11
	High	0.45	0.07	−0.50	−0.34	−0.37
	Reference	—	0.06	0.20	−0.24	−0.51
Suspended P	Low	0.05	0.26	—	−0.01	−0.32
	Medium	0.36	0.26	—	0.01	−0.61
	High	0.72	0.26	—	0.01	−0.97
	Reference	—	0.27	—	−0.28	0.56
Dissolved	Low	−132.4	9.05	173.8	128.0	77.6
inorganic C	Medium	399.1	8.78	−52.7	141.3	−213.9
	High	803.4	8.81	−210.7	160.3	−441.4
	Reference	—	10.0	75.6	7.32	−102.1
Dissolved	Low	−3.84	2.81	12.8	10.1	−1.73
organic C	Medium	27.1	2.82	−19.6	11.4	2.02
	High	49.9	2.76	−35.0	11.7	−6.00
	Reference	—	2.96	9.91	0.01	−12.9
Suspended C	Low	0.35	1.67	—	−0.05	−2.09
	Medium	2.70	1.67	—	0.10	−4.26
	High	4.91	1.67	—	0.13	−6.46
	Reference	—	1.82	—	−0.14	−1.98

[a]SYSF = pumped + precipitation − seepage − ΔS.
[b]Seepage for undiked cells is surface inflow and outflow.
[c]No losses of suspended nutrients possible through seepage.

Table 5.3. Mean (SE) macrophyte nutrient pools (kg/ha) (N, nitrogen; P, phosphorus; C, carbon) in the MERP experimental cells by flooding status and year, 1983–1989. During 1983 and 1984, the cells in a given flooding treatment were drawn down, except in the undiked reference cells

Flooding status	Year	Aboveground			Belowground		
		N	P	C	N	P	C
Low = 4	1983(n=2)	70.3(4.25)a[a]	37.6(11.7)a	1,022(62.5)a	5.65(0.80)a	0.40(0.06)a	176(26.1)a
	1984	72.3(4.98)a	10.8(0.45)b	1,353(103.4)a	14.1(0.31)a	1.67(0.03)b	458(121)a
	Mean	71.3(1.00)	23.7(12.9)	1,187(165)	9.89(4.23)	1.04(0.63)	317(141)
	1985	44.0(1.54)b	8.30(0.47)b	1,301.9(57.9)	227.7(4.53)ab	2.70(0.45)bc	884(144)ab
	1986	39.4(2.23)b	5.90(0.31)b	1,485(83.6)ab	45.6(5.59)b	4.81(0.61)c	1,315(159)bc
	1987	49.5(5.98)b	6.15(0.53)b	1,740(260)b	61.2(16.5)b	5.74(1.55)c	1,804(487)c
	1988	36.2(6.04)b	4.73(0.75)b	1,050(183)a	72.8(16.1)b	5.64(1.28)c	2,061(458)c
	1989	41.8(9.03)b	5.53(1.10)b	1,335(315)a	61.5(12.6)b	5.37(1.09)c	2,053(417)c
	Mean	42.2(2.24)	6.12(0.60)	1,382(113)	53.8(7.84)	4.85(0.56)	1,623(229)
Medium = 3	1983	78.4(1.84)a	34.6(3.61)a	1,516(65.3)a	8.29(2.49)a	0.59(0.18)a	257(76.9)a
	1984	73.6(0.20)a	23.5(0.94)a	1,347(100)a	17.6(1.90)a	2.10(0.22)ab	556(60.6)a
	Mean	76.0(2.41)	29.1(5.57)	1,431(84.5)	13.0(4.67)	1.35(0.76)	406(149)
	1985	42.2(5.02)b	8.20(1.13)b	1,499(238)a	33.7(6.68)ab	3.30(0.68)b	948(195)b
	1986	40.0(5.07)b	6.17(0.77)b	1,340(223)a	41.5(1.15)bc	4.28(0.10)b	1,096(29.5)b
	1987	52.1(3.65)b	7.20(0.53)b	1,744(179)a	67.6(17.8)c	7.03(1.93)c	1,835(490)c
	1988	38.9(7.75)b	5.77(1.22)b	1,014(233)b	60.2(17.2)c	5.90(1.21)b	1,782(512)c
	1989	30.3(9.46)b	4.96(1.65)b	831(291)b	44.5(11.2)bc	4.77(1.25)b	1,407(364)bc
	Mean	40.7(3.50)	6.46(0.57)	1,285(164)	49.5(6.26)	5.07(0.65)	1,413(178)
High = 3	1983	69.3(6.40)a	30.9(1.99)a	1,301(136)a	15.2(0.40)a	1.08(0.25)a	474(117)a
	1984	72.9(3.90)a	22.6(5.50)b	1,456(90.2)a	31.4(3.31)a	3.67(0.38)b	1,044(116)b
	Mean	71.1(1.80)	26.8(4.16)	1,378(77.7	23.3(8.12)	2.38(1.30)	759(285)
	1985	49.4(1.70)b	9.47(0.77)c	1,664(60.1)a	33.7(3.42)a	3.17(0.32)b	971(105)b
	1986	37.3(6.36)b	6.53(1.03)c	1,446(256)a	52.4(9.62)ab	5.58(1.09)c	1,520(288)c
	1987	51.8(5.56)ab	7.97(1.58)c	1,923(508)a	61.8(19.9)b	6.90(2.13)c	1,824(585)c
	1988	31.5(9.47)bc	5.50(1.67)cd	926(204)b	54.1(14.3)ab	5.75(1.26)c	1,596(238)c
	1989	24.2(9.00)c	3.33(1.17)d	715(263)b	35.3(14.8)a	3.15(1.45)ab	1,012(303)ab
	Mean	38.8(5.26)	6.56(1.05)	1,334(226)	47.5(5.54)	4.91(0.75)	1,384(168)
Reference = 2	1983	106(4.0)a	28.1(14.9)a	2,667(122)a	94.9(13.8)a	9.42(1.31)a	3,073(452)a
	1984	103(3.0)a	32.3(11.7)a	3,531(359)a	117(12.0)a	19.0(1.90)a	4,571(468)a
	1985	66.4(8.60)a	10.0(0.70)a	2,360(90.0)a	131(17.1)a	14.4(1.90)a	4,451(580)a
	1986	77.7(5.00)a	11.3(1.20)a	2,672(373)a	145(10.6)a	16.0(1.20)a	4,653(313)a
	1987	54.1(7.40)a	7.85(1.35)a	2,526(119)a	141(5.53)a	14.4(0.58)a	4,821(187)a
	1988	49.5(6.50)a	8.10(1.70)a	1,609(393)a	159(14.1)a	11.0(0.77)1a	4,834(430)a
	1989	70.8(12.8)a	11.9(3.25)a	2,448(823)a	158(46.0)a	14.8(4.25)a	5,277(520)a
	Mean	69.8(4.13)	11.3(1.30)	2,401(138)	135.5(8.75)	14.2(1.19)	4,525(261)

[a] Within a column by flooding status, means with different letters are significantly different ($P < 0.05$).

83

ground roots and rhizomes of the emergent vegetation that dominated the cells during the reflooding years and that were dominant in the reference cells during the entire study. Belowground macrophyte pools of N, P, and C appeared to increase from 1985 to 1988 with a decline in 1989 (especially the medium and deep treatments), but these changes were generally not statistically significant (Table 5.3).

None of the macrophyte nutrient pools, above- or belowground, in the reflooded cells ever reached levels comparable to those pools in the reference cells (Table 5.3). From 1985 to 1989, mean aboveground biomass nutrient pools in the low-treatment cells were 50 to 60% of those in the reference cells and, only 35 to 40% of those in the reference cells for belowground pools. Pool sizes in the medium and high treatments were lower still.

Algae and Invertebrates

Algal nutrient pools (Table 5.4) all showed the same general pattern in the low-treatment cells; they were lowest in 1985, increased in 1986, reached their highest levels in 1987 and 1988, and declined in 1989 to levels similar to those in 1986. Mean algal nutrient-pool sizes in the low cells from 1985 to 1989 were comparable to those in the reference cells. In the medium and high treatments, the algal nutrient pools were larger in 1988 and 1989 than from 1985 to 1987 (Table 5.4). In the medium treatment, algal nutrient pools were slightly higher than in the reference cells from 1985 to 1989, whereas in the high treatment they were slightly lower.

In the low treatments, invertebrate nutrient pools (Table 5.4) were not significantly different from 1985 to 1988; however, they dropped by >50% in 1989. Mean invertebrate nutrient pool sizes in the three water-level treatments from 1985 to 1989 had double the mass of those in the reference cells. In the medium and high treatments, invertebrate nutrient pools were generally smaller in 1988 and 1989 than in 1987, and significantly smaller in all cases in 1989. By 1989, invertebrate nutrient pools in the medium and high treatments were comparable in size to those in the reference cells.

Dissolved Nutrients and Suspended Solids

In the low-treatment cells, surface-water nutrient pools generally showed no significant change in mass from 1985 to 1989 with the exception of the NH_4 pool, which was highest in 1988 and 1989 and DIC, which declined after 2 or 3 years of flooding and then increased to previous levels, especially in the low and medium treatments (Table 5.5). The mean mass of the surface-water nutrient pools during the reflooding years in the low treatment and reference cells were comparable, although DON was higher in the low treatment and SRP and SSP considerably higher in the reference cells in most years. In the medium and high treatments, surface-water nutrient pools were generally larger in 1988 and 1989 compared with 1987 (Table 5.5), although suspended N and P were lower late in flooding. Except for SRP, SSP, and SSC, surface-water pools in the medium and high treatments had a mean mass equal to or higher than in the reference cells from 1985 to 1989.

In the low treatment, pore-water nutrient pools had a pattern nearly identical to the surface-water pools. Mean NH_4 and SRP pool masses in the reflooded cells were higher than in the reference cells (Table 5.6), whereas DON and DOP pools were similar across the water-level treatments and reference cells. In the medium and high treatments, pore-water

Table 5.4.   Mean (SE) algae and invertebrate nutrient pools (kg/ha) in the MERP experimental cells by flooding status and year, 1985–1989

Flooding status	Year	Algae			Invertebrates		
		N	P	C	N	P	C
Low	1985	7.74(0.90)a[a]	1.10(0.12)a	109(12.2)a	1.54(0.21)a	0.15(0.02)a	4.45(0.98)a
n = 4	1986	11.2(0.75)b	1.64(0.12)b	161(12.4)b	1.98(0.10)a	0.20(0.01)a	9.24(0.50)a
	1987	14.8(1.15)c	1.91(0.12)c	190(12.5)c	1.70(0.12)a	0.17(0.01)a	7.93(0.55)a
	1988	15.3(0.89)c	2.06(0.10)c	203(10.1)c	1.31(0.20)a	0.13(0.02)a	6.12(0.96)a
	1989	10.7(0.93)b	1.41(0.11)b	139(10.8)b	0.64(0.01)b	0.06(0.01)b	2.99(0.06)b
	Mean	11.9(1.40)	1.62(0.18)	160(17.1)	1.43(0.23)	0.14(0.02)	6.15(1.13)
Medium	1985	10.4(1.30)a	1.50(0.20)a	148(19.9)a	1.69(0.18)a	0.17(0.02)a	4.19(0.27)a
n = 3	1986	11.1(0.54)a	1.60(0.10)a	157(9.13)a	2.21(0.12)b	0.22(0.01)b	10.3(0.57)b
	1987	13.7(1.03)a	1.82(0.12)a	180(12.2)a	1.83(0.18)a	0.18(0.02)a	8.53(0.83)b
	1988	16.6(2.15)a	2.20(0.28)a	217(27.3)a	1.07(0.12)c	0.11(0.01)c	5.00(0.55)a
	1989	14.9(1.99)a	1.98(0.27)a	196(26.3)a	0.43(0.04)d	0.04(0.01)d	2.02(0.22)c
	Mean	13.3(1.16)	1.82(0.13)	180(12.8)	1.45(0.31)	0.14(0.03)	6.01(1.50)
High	1985	5.88(0.33)a	0.78(0.05)a	76.9(4.63)a	1.49(0.04)a	0.15(0.01)a	3.49(0.17)a
n = 3	1986	8.88(0.88)b	1.26(0.14)b	124(13.6)b	1.98(0.03)b	0.20(0.01)b	9.26(0.14)b
	1987	9.39(1.71)b	1.31(0.24)b	129(23.6)b	1.53(0.10)a	0.15(0.01)a	7.12(0.44)c
	1988	11.5(2.02)b	1.56(0.27)b	154(27.1)b	1.37(0.15)a	0.14(0.01)a	6.39(0.68)c
	1989	12.2(2.27)b	1.51(0.24)b	150(24.9)b	0.59(0.04)c	0.06(0.01)c	2.76(0.18)a
	Mean	9.57(1.12)	1.28(0.14)	127(13.9)	1.39(0.23)	0.14(0.02)	5.80(1.19)
Reference	1985	6.30(0.91)a	0.95(0.15)a	93.7(14.3)a	0.49(0.23)a	0.05(0.02)a	1.88(0.56)a
n = 2	1986	12.5(0.78)b	1.83(0.15)b	180(14.2)b	0.61(0.06)a	0.06(0.02)a	2.86(0.86)a
	1987	15.5(1.28)b	1.89(0.13)b	188(13.1)b	0.62(0.17)a	0.06(0.02)a	2.89(0.81)a
	1988	16.0(0.69)b	2.07(0.05)b	205(4.81)b	0.58(0.31)a	0.06(0.03)a	2.71(1.46)a
	1989	11.1(0.90)b	1.43(0.07)b	141(7.55)b	0.48(0.20)a	0.05(0.02)a	2.24(0.91)a
	Mean	12.3(1.76)	1.63(0.20)	161(20.0)	0.56(0.03)	0.06(0.01)	2.52(0.20)

[a] Within a column by flooding status, means with different letters are significantly different ($P < 0.05$).

nutrient pools were generally larger in 1988 and 1989 compared with 1987 (Tables 5.6), although once again these increases were not always significant.

Litter/Sediment Pool
As indicated in Chapter 3, no estimates of total litter pool size were included in the budgets because of the accumulation of litter over many years. Only fluxes to and from the litter pool and to the sediment sink were determined on an annual basis.

Fluxes

Tables 5.7–5.9 summarize information on the fluxes in the major N, P, and C nutrient pools by flooding treatment during the reflooding years 1985–1989. Detailed year-by-year nutrient flux estimates for N, P, and C during the reflooding years can be found in Tables A2.3–A2.5 in Appendix 2.

As during the baseline and deep-flooding years (see Chapter 4), the largest fluxes dur-

Table 5.5. Mean (SE) surface-water nutrient pools (kg/ha) in the MERP experimental cells by flooding status and year, 1985–1989.

Flooding status	Year	NH_4	DON	SSN	SRP	DOP	SSP	DIC	DOC	SSC
Low	1985	0.38(0.07)a[a]	17.7(2.48)a	3.46(0.9)a	0.59(0.44)a	0.45(0.10)a	0.43(0.10)a	932(69.6)a	416(65.0)a	25.4(7.10)a
n = 4	1986	0.30(0.02)a	15.3(0.73)a	3.67(0.8)a	0.19(0.07)a	0.17(0.06)a	0.35(0.1)ab	752(58.5)b	367(45.0)a	24.9(3.63)a
	1987	0.37(0.04)a	10.9(0.38)a	2.26(0.3)ab	0.04(0.01)a	0.15(0.01)a	0.43(0.13)a	438(38.2)c	229(12.0)b	15.7(2.3)ab
	1988	0.82(0.16)b	16.9(1.37)a	2.27(0.3)ab	0.08(0.03)a	0.27(0.02)a	0.61(0.16)a	677(11.5)b	329(19)ab	19.7(6.5)ab
	1989	1.81(0.36)c	16.4(0.59)a	1.73(0.2)ab	0.27(0.07)a	0.28(0.02)a	0.28(0.02)b	693(4.52)b	350(26.0)a	12.5(1.00)b
	Mean	0.74(0.28)	15.4(1.20)	2.68(0.20)	0.23(0.10)	0.26(0.05)	0.42(0.06)	698(79.0)	347(37.0)	19.6(4.82)
Medium	1985	0.32(0.02)a	15.3(0.43)a	1.88(0.41)a	0.16(0.05)a	0.35(0.03)a	0.33(0.05)a	722(55.9)a	325(38.0)a	15.6(3.53)a
n = 3	1986	0.31(0.03)a	15.8(0.56)a	3.68(0.34)b	0.09(0.03)a	0.33(0.05)a	0.40(0.03)a	548(28.8)b	248(21.1)b	22.2(2.01)a
	1987	0.59(0.15)b	14.0(0.45)a	2.10(0.5)ab	0.05(0.01)b	0.19(0.01)b	0.47(0.05)a	534(38.4)b	267(7.7)ab	14.3(1.4)ab
	1988	0.86(0.20)b	20.8(0.54)b	2.80(0.3)ab	0.15(0.06)a	0.32(0.02)a	0.38(0.03)a	687(24.6)a	363(10.0)a	21.6(2.53)a
	1989	2.01(0.65)c	19.6(0.61)b	1.33(0.12)a	0.12(0.02)a	0.32(0.02)a	0.32(0.01)a	678(47.4)a	361(33.0)a	10.3(1.01)b
	Mean	0.82(0.32)	17.1(1.32)	2.36(0.41)	0.11(0.02)	0.30(0.03)	0.38(0.03)	634(38.7)	313(24.2)	16.8(2.31)
High	1985	0.58(0.11)a	22.5(2.05)a	3.95(1.3)ab	0.85(0.17)a	0.55(0.10)a	0.54(0.21)a	712(65.6)a	327(33.2)a	30.6(9.02)a
n = 3	1986	0.39(0.04)a	18.1(0.61)a	6.78(2.32)a	0.39(0.08)b	0.52(0.03)a	0.56(0.15)a	457(36.6)b	192(11.1)b	32.1(9.64)a
	1987	0.79(0.24)a	19.0(1.19)a	4.75(2.2)ab	0.18(0.03)c	0.42(0.04)a	0.61(0.31)a	698(77.8)a	292(14.3)a	27.7(11.0)a
	1988	0.86(0.23)a	22.0(1.66)a	3.55(1.1)ab	0.40(0.11)b	0.49(0.03)a	0.53(0.09)a	731(99.0)a	349(36.3)a	39.7(6.84)a
	1989	2.92(0.95)b	23.8(1.20)a	2.51(0.42)b	0.87(0.16)a	0.56(0.04)a	0.40(0.01)a	725(104)a	393(20.4)a	17.0(2.00)b
	Mean	1.11(0.46)	21.1(1.09)	4.31(0.72)	0.54(0.14)	0.51(0.03)	0.52(0.04)	665(52.0)	311(34.4)	29.4(3.73)
Reference	1985	0.40(0.14)a	7.98(0.55)a	3.85(0.61)a	0.65(0.31)a	0.27(0.05)a	0.80(0.38)a	522(84.0)a	382(26.4)a	27.5(5.42)a
n = 2	1986	0.46(0.08)a	13.5(1.51)a	4.12(1.96)a	0.50(0.21)a	0.13(0.15)a	0.85(0.25)a	550(86.0)a	306(64.3)a	37.3(13)a
	1987	1.37(0.70)a	11.2(1.38)a	5.84(1.05)a	0.76(0.29)a	0.32(0.06)a	1.13(0.40)a	324(6.40)a	195(22.1)b	37.9(8.66)a
	1988	0.77(0.50)a	9.20(1.41)a	4.14(0.50)a	0.97(0.26)a	0.29(0.07)a	0.63(0.40)a	415(4.40)a	281(35.3)a	28.8(3.22)a
	1989	1.07(0.61)a	9.12(0.75)a	3.45(0.52)a	1.30(0.75)a	0.32(0.10)a	0.88(0.12)a	408(16.0)a	223(57)ab	20.2(3.33)a
	Mean	0.81(0.18)	10.2(0.98)	4.28(1.61)	0.83(0.14)	0.27(0.04)	0.85(0.14)	443(117)	277(38.6)	30.4(6.32)

Note: Ammonia (NH_4), dissolved organic nitrogen (DON), suspended solids nitrogen (SSN), soluble reactive phosphorus (SRP), dissolved organic phosphorus (DOP), suspended solids phosphorus (SSP), dissolved inorganic carbon (DIC), dissolved organic carbon (DOC), suspended solids carbon (SSC).
[a] Within a column by flooding status, means with different letters are significantly different ($P < 0.05$).

Table 5.6. Mean (SE) pore-water nutrient pools (kg/ha) in the MERP experimental cells by flooding status and year, 1985–1989

Flooding status	Year	NH$_4$	DON	SRP	DOP	DIC	DOC
Low	1985	6.51(0.98)a[a]	25.1(1.07)a	2.31(0.35)a	0.62(0.12)a	684(11.6)a	308(17.9)a
$n = 4$	1986	9.21(0.58)b	24.8(1.42)a	2.84(0.26)a	0.59(0.05)a	1,074(130)b	469(75.6)a
	1987	15.0(1.34)b	27.6(1.53)a	2.88(0.22)a	1.04(0.17)a	1,122(140)b	474(77.6)a
	1988	17.2(1.31)b	32.2(1.61)a	2.58(0.18)a	0.74(0.14)a	1,256(143)b	487(67.9)a
	1989	17.1(2.05)b	31.9(1.45)a	1.95(0.17)a	0.84(0.11)a	1,237(131)b	499(63.8)
	Mean	13.0(2.18)	28.3(1.60)	2.51(0.17)	0.77(0.08)	1,075(103)	447(35.1)
Medium	1985	7.13(0.39)a	27.6(1.13)a	1.58(0.27)a	0.53(0.06)a	686(65.5)a	340(17.8)a
$n = 3$	1986	10.5(0.62)ab	26.6(1.04)a	2.67(0.24)a	0.6(0.11)8a	1,001(28.7)b	459(7.50)b
	1987	13.8(0.64)b	27.9(0.67)a	2.38(0.17)a	0.72(0.11)a	1,020(95.6)b	494(24.4)b
	1988	19.6(1.75)bc	34.0(1.43)a	3.58(0.26)a	0.64(0.10)a	1,313(57.2)c	493(14.3)b
	1989	22.7(2.08)c	31.3(1.00)a	3.42(0.33)a	0.89(0.10)a	1,124(65.3)b	473(22.1)b
	Mean	14.8(2.87)	29.5(1.38)	2.73(0.36)	0.69(0.06)	1,029(102)	452(28.6)
High	1985	6.61(1.15)a	35.3(1.97)a	1.61(0.25)a	0.89(0.21)a	707(18.0)a	425(62.1)a
$n = 3$	1986	8.42(0.60)a	29.6(1.64)a	3.84(0.19)b	0.6(0.06)5a	966(37.0)b	464(43.4)a
	1987	13.0(1.90)b	30.5(1.79)a	3.87(0.31)b	0.65(0.08)a	1,004(76.0)b	459(62.8)a
	1988	13.6(0.92)b	31.9(1.76)a	3.14(0.19)b	0.96(0.26)a	922(106)b	369(42.7)a
	1989	14.5(0.72)b	29.1(2.35)a	2.98(0.18)b	0.81(0.17)a	814(75.3)ab	387(70.8)a
	Mean	11.2(1.56)	31.3(1.11)	3.09(0.41)	0.79(0.06)	883(54.0)	421(19.0)
Reference	1985	4.99(1.23)a	22.2(1.85)a	1.03(0.21)a	0.38(0.04)a	692(79.5)a	302(63.3)a
$n = 2$	1986	4.47(0.77)a	23.6(2.64)a	1.76(0.42)a	0.50(0.08)a	821(133)a	384(124)a
	1987	5.52(1.87)a	25.0(2.47)a	1.37(0.28)a	0.42(0.08)a	739(118)a	392(110)a
	1988	5.34(1.31)a	28.2(3.92)a	1.40(0.44)a	0.65(0.14)a	760(41.1)a	422(141)a
	1989	10.5(2.22)a	29.5(4.65)a	2.25(0.65)a	0.63(0.17)a	715(9.77)a	509(174)a
	Mean	6.16(1.10)	25.7(1.38)	1.56(0.21)	0.52(0.05)	746(22.0)	402(33.4)

Note: Ammonia (NH$_4$), dissolved organic nitrogen (DON), soluble reactive phosphorus (SRP), dissolved organic phosphorus (DOP), dissolved inorganic carbon (DIC), dissolved organic carbon (DOC).
[a] Within a column by flooding status, means with different letters are significantly different ($P < 0.05$).

ing the reflooding years were dominated by those associated with macrophytes (uptake, translocation, leaching, and litter-fall), leaching from litter, and movement of litter to the sediment sink. There were relatively few statistically significant differences recorded in these fluxes over the 5 years of the experiment (Tables A2.3–A2.5 in Appendix 2), although a number of trends are discussed in the following sections.

Macrophytes

Mean aboveground macrophyte nutrient fluxes did not change significantly over time or among treatments during the reflooding years (Tables A2.3–A2.5 in Appendix 2), however, some trends were apparent across treatments and years. The uptake and translocation of N generally increased through the third year of flooding and declined thereafter, reflecting the patterns seen in the macrophyte nutrient pools. Uptake and translocation of P

Table 5.7.   Mean (SE) (1 Oct. – 30 Sept.) nitrogen (N) fluxes (kg/ha/yr) in the three flooding treatments in the diked experimental cells and the reference cells from 1985 to 1989

Flux	Low (n = 5)	Medium (n = 5)	High (n = 5)	Reference (n = 5)
Macrophytes				
Uptake	52.9(5.31)a[a]	47.9(5.25)a	54.3(5.32)a	97.8(15.5)b
Translocation	23.1(2.35)a	25.2(1.68)	28.2(1.49)a	44.9(2.53)b
Aboveground leaching	10.7(0.98)a	11.6(0.71)a	13.0(0.62)a	20.7(1.08)b
Belowground leaching	4.65(0.96)a	5.02(0.74)a	4.20(0.83)a	8.16(1.29)b
Aboveground litter-fall	12.6(1.21)a	13.9(0.64)a	15.3(0.77)a	23.7(1.23)b
Belowground litter-fall	19.4(5.78)	a19.7(3.04)a	20.3(6.03)a	32.7(5.18)b
Algae and invertebrates				
Algal uptake	15.5(2.73)a	9.09(2.50)b	7.86(0.86)b	5.71(0.78)b
Algal litter-fall	9.74(3.76)a	4.62(0.94)b	4.11(1.31)b	3.96(1.35)b
Invertebrate uptake	0.71(0.34)a	0.82(0.38)a	0.65(0.35)a	0.40(0.12)a
Invertebrate loss	0.57(0.15)a	0.71(0.21)a	0.57(0.112)a	0.29(0.07)a
Dissolved nutrients				
Mineralization SWDO to SWDI	5.81(1.85)a	4.13(1.05)a	6.04(2.28)a	9.09(2.43)a
Mineralization PWDO to PWDI	4.99(1.14)a	6.64(1.65)a	6.40(2.00)a	8.04(1.72)a
Pore–surface water exchange	–8.03(3.90)a	–11.4(3.61)a	–15.5(2.27)a	–24.61(2.76)b
Litter/sediment pool				
Litter leaching				
Dissolved inorganic to SW	4.21(0.90)a	1.82(0.90)ab	2.52(1.63)ab	0.01(0.01)b
Dissolved organic to SW	5.72(2.64)a	3.62(1.51)a	7.25(4.18)a	2.63(1.21)a
Dissolved organic to PW	5.80(0.36)a	6.59(0.98)a	4.88(0.87)a	10.1(2.70)b
Dissolved inorganic to PW	40.3(3.37)a	27.7(5.10)a	27.9(6.24)a	57.5(5.68)b
Aboveground litter to sediments	21.8(1.76)a	26.5(1.46)a	23.3(1.29)a	30.8(0.85)b
Belowground litter to sediments	22.7(1.22)a	21.9(0.96)a	20.1(1.45)a	34.6(1.86)b

Note: Surface water (SW); pore water (PW); dissolved organic (DO), dissolved inorganic (DI).
[a] Within a row means with different letters are significantly different (P < 0.05)

behaved in a similar manner except for an initial peak during the first year of flooding. Although highly variable, aboveground leaching and litter-fall showed general trends coincident with the changes in the aboveground macrophyte pools. The belowground N litter-fall increased significantly toward the end of the reflooding years, whereas the belowground P litter-fall decreased after 1985 and then remained constant throughout the remainder of the flooding period (Tables A2.3 and A2.4 in Appendix 2). All fluxes involving macrophytes in the three water-depth treatments were consistently lower than those in the reference cells over the 5 years of reflooding (Tables 5.7–5.9), probably a consequence of smaller macrophyte nutrient pools in the reflooded cells.

Algae and Invertebrates
Algal uptake did change over time within treatments, however, this change was only significant for the P fluxes (Tables A2.3–A2.5 in Appendix 2). Generally, uptake was initially higher in the low and medium treatments than in the high treatment, but declined over time in the low and medium treatments and increased in the high treatment (Tables

Table 5.8. Mean (SE) (Oct. 1 – Sept. 30) phosphorus fluxes (kg/ha/yr) in the three flooding treatments in the diked experimental cells and the reference cells from 1985 to 1989.

Flux	Low (n = 5)	Medium (n = 5)	High (n = 5)	Reference (n = 5)
Macrophytes				
Uptake	11.1(1.32)a[a]	12.3(0.91)a	14.4(1.02)a	22.1(0.84)b
Translocation	4.01(0.45)a	4.37(0.35)a	5.01(0.40)a	6.50(0.60)b
Aboveground leaching	1.78(0.19)a	2.00(0.13)a	2.30(0.15)a	2.96(0.25)b
Belowground leaching	1.46(0.16)a	1.61(0.10)ab	1.92(0.09)b	3.09((0.13)c
Aboveground litter-fall	2.41(0.28)a	2.54(0.21)a	2.91(0.23)ab	3.38(0.25)b
Belowground litter-fall	5.51(0.70)a	6.42(0.39)ab	7.67(0.38)b	12.4(0.50)c
Algae and invertebrates				
Algal uptake	1.87(0.33)a	1.12(0.35)b	0.97(0.07)b	0.52(0.12)b
Algal litter-fall	0.99(0.27)a	0.59(0.10)ab	0.56(0.17)ab	0.48(0.10)b
Invertebrate uptake	0.03(0.02)a	0.04(0.02)a	0.03(0.02)a	0.02(0.01)a
Invertebrate loss	0.06(0.02)a	0.07(0.02)a	0.06(0.12)a	0.03(0.01)a
Dissolved nutrients				
Mineralization SWDO to SWDI	0.16(0.03)a	0.16(0.04)a	0.44(0.28)ab	0.63(0.22)b
Mineralization PWDO to PWDI	0.66(0.17)a	0.44(0.08)ab	0.52(0.10)ab	0.29(0.07)b
Pore–surface water exchange	−1.56(0.26)a	−2.55(0.17)b	−2.86(0.55)b	−4.33(0.22)c
Litter/sediment pool				
Litter leaching				
Dissolved inorganic to SW	0.42(0.11)a	0.14(0.08)bc	0.22(0.09)b	0(0)c
Dissolved organic to SW	0.15(0.12)a	0.14(0.06)a	0.32(0.17)a	0.29(0.10)a
Dissolved organic to PW	0.70(0.13)a	0.55(0.06)a	0.61(0.14)a	0.40(0.09)
Dissolved inorganic to PW	6.92(1.61)a	8.21(1.10)a	9.53(0.87)a	14.9(0.86)b
Aboveground litter to sediments	3.19(0.10)a	3.92(0.15)b	4.38(0.50)b	4.24(0.16)b
Belowground litter to sediments	4.15(0.16)a	4.42(0.15)a	3.89(0.36)a	5.24(0.53)b

Note: Surface water (SW); pore water (PW); dissolved organic (DO), dissolved inorganic (DI).
[a] Within a row means with different letters are significantly different ($P < 0.05$).

A2.3–A2.5). By the end of the study (1989) the difference in algal uptake among treatments was negligible; however, the initial differences resulted in higher overall algal uptake in the low than the medium and high treatments when averaged over the 5 years of reflooding (Tables 5.7–5.9). Algal litter-fall generally increased in all treatments over the 5 years of study for N and P. In the low treatment, algal litter-fall was significantly higher in 1989 than in the previous 4 years (Tables A2.3–A2.5). Over the reflooding years, as with algal uptake, mean algal N, P, and C litter-fall was lower in the medium and high treatments than the low treatment (Table 5.7–5.9). Although mean algal nutrient pool sizes from 1985 to 1989 were comparable to those in the reference cells, algal fluxes were larger in the experimental cells than in the reference cells.

Mean invertebrate uptake was initially at similar levels in all three treatments soon after flooding and then declined over time to low levels by the end of 5 years of flooding (Tables A2.3–A2.5). The result was no overall treatment effect on invertebrate uptake (Tables 5.7–5.9). Invertebrate loss increased after 1986 and then declined in 1989 with little difference among treatments (Tables A2.3–A2.5). Although mean invertebrate nutrient pool

Table 5.9. Mean (SE) (1 Oct. – 30 Sept.) carbon (C) fluxes (kg/ha/yr) in the three flooding treatments in the diked experimental cells and the reference cells from 1985 to 1989

Flux	Low (n = 5)	Medium (n = 5)	High (n = 5)	Reference (n = 5)
Macrophytes				
C fixation	1,681(78.8)a[a]	1,513(208)a	1,834(138)a	3,335(375)b
Translocation	774(58.9)a	664(103)a	741(88.8)a	1,341(159)b
Aboveground leaching	361(33.7)a	374(43.6)a	447(31.5)a	878(72.5)b
Belowground leaching	135(28.5)a	131(22.8)a	138(29.1)a	242(26.0)b
Aboveground litter fall	427(41.7)a	484(41.2)a	540(35.4)a	1,052(76.8)b
Belowground litter fall	560(101)a	542(74.1)a	633(186)a	951(101)b
Algae and invertebrates				
Algal uptake	196(34.7)a	117(37.8)b	101(7.59)b	69.4(10.8)b
Algal litter fall	127(38.7)a	63.0(10.0)b	59.5(17.5)b	55.1(9.02)b
Invertebrate uptake	3.32(1.60)a	3.84(1.77)a	3.23(1.56)a	2.04(0.51)a
Invertebrate loss	2.67(0.72)a	3.33(0.94)a	2.64(0.55)a	1.38(0.32)a
Dissolved nutrients				
Mineralization SWDO to SWDI	61.3(14.7)a	34.6(12.3)a	48.2(9.69)a	150(30.0)b
Mineralization PWDO to PWDI	50.8(16.1)a	82.6(33.7)a	82.5(49.5)a	111(33.7)a
Pore–surface water exchange	−176(87.0)a	−226(95.4)a	−332(86.0)a	−829(53.8)b
Litter/sediment pool				
Litter leaching				
Dissolved inorganic to SW	114(15.2)a	153(34.8)a	133(40.7)a	0(0)b
Dissolved organic to SW	150(48.5)a	151(28.9)a	177(79.5)a	78.7(13.0)a
Dissolved organic to PW	116(24.8)a	128(20.7)a	94.7(17.3)a	163(57.7)a
Dissolved inorganic to PW	212(196)a	268(214)a	463(262)a	1,017(132)b
Aboveground litter to sediments	415(24.0)a	423(54.9)a	459(61.9)a	818(33.2)b
Belowground litter to sediments	174(30.1)a	150(42.5)a	157(52.7)a	394(82.8)b

Note: Surface water (SW); pore water (PW); dissolved organic (DO), dissolved inorganic (DI).
[a] Within a row means with different letters are significantly different ($P < 0.05$).

sizes in the reflooding treatments from 1985 to 1989 were double the mass of those in the reference cells (Table 5.4), fluxes associated with invertebrates were comparable in magnitude (Table 5.7–5.9).

Dissolved Nutrients and Suspended Solids
Mineralization of dissolved organic forms to inorganic forms of N, P, and C in surface water generally was low during the first year of flooding, increased for 1 or 2 years, and then declined (Tables A2.3–A2.5 in Appendix 2). Generally surface-water mineralization rates differed little among the flooding treatments when averaged across the 5 years of the study (Tables 5.7–5.9), except that P mineralization in the deep treatment was higher than in the two shallower treatments (Table 5.8). Generally, surface-water mineralization rates were higher in the reference cells than in the reflooded cells. Pore-water mineralization rates were variable over time (Tables A2.3–A2.5) and across treatments (Tables 5.7–5.9). Thus, little difference was detected among treatments. Pore water–surface water exchange

also varied widely from year to year; however, when averaged over the 5 years of the study, rates increased with depth of flooding (Table 5.7–5.9). The overall highest rates, however, were found in the reference cells.

Litter/Sediment Pool

Litter leaching to surface and pore water changed little over time; however, where significant differences did occur, leaching generally declined over time (Tables A2.3–A2.5 in Appendix 2). The flooding treatments show few differences. Surface-water fluxes related to litter leaching were lower in the reference cells, whereas pore-water litter-leaching fluxes in the reference cells were generally higher than in the reflooded cells (Tables 5.7–5.9).

Particulate loss of above- and belowground particulate litter to the sediment sink varied little except for a suggestion of increasing rates over time in the deeper treatments (Tables A2.3–A2.5 in Appendix 2). When averaged over the 5 years of the study, the treatments were similar but rates in the reference cells were generally higher than in the reflooded cells (Tables 5.7–5.9).

DISCUSSION

During the reflooding years, no significant changes occurred in the size of any nutrient pools or fluxes in the reference cells. Thus, changes seen in the three water-level treatments were not due to ongoing changes in the Delta Marsh or to unusual weather conditions during one or more of the reflooding years. As in the earlier deep-flooding experiment (see Chapter 4), the largest nutrient pools in all water-level treatments during all five reflooding years were the macrophyte pools, primarily the belowground macrophyte pool. Changes in the mass of macrophytes were primarily responsible for the fluctuations in all other pool sizes and fluxes in the MERP cells during the reflooding years. Studies have noted that macrophytes dominate the nutrient cycling in those shallow-water areas where they are abundant, most notably shallow littoral zones of lakes (Graneli and Solander 1988; Pieczynska 1993), freshwater marshes (Bayley et al. 1985), and tidal marshes (Bowden et al. 1991).

Although substantial amounts of nutrients were pumped into the experimental cells to maintain the water-level treatments, the net input to individual cells varied very little due to increased seepage in those treatments (medium and high) with elevated water levels. Thus, the changes in pools and fluxes over the 5 years of flooding were due to the different water levels among the treatments rather than to pumping input effects. Within the experimental cells, changes over the regenerating stage and eventually the degenerating stage in some cells were due to the differences in water levels rather than to differences in nutrient inputs or outputs over the same period.

Regenerating Stage

For the sake of simplicity, our discussion of the regenerating stage is restricted primarily to the low-treatment cells. During the first 3 years, the cells in the medium and high treatments were in the regenerating state and they did not differ much in either the pool sizes or fluxes among pools from the cells in the low treatment.

Specifically, we hypothesized that above- and belowground macrophyte pools and associated fluxes in the low treatment would increase in size from 1985 to 1986 and then decline for a short period (due to the decline of short-lived emergents), and finally increase again to their size during the baseline years and in the reference cells. Algal and invertebrate pools and fluxes, however, were predicted to be largest during the first few years after reflooding and then to decline to sizes similar to those found in the cells during the baseline years and in the reference cells. Actual changes in the size of the macrophyte pools in the low-treatment cells were consistent with those predicted for the regenerating stage. By 1989, however, they had not reached the size of the reference cells. Changes in the algal pools in the low cells were not similar to those predicted to occur during the regenerating stage. The algal pool size was not larger in the low treatment than in the reference cells but increased steadily from 1985 to 1989 especially in the medium and high treatments. Changes in the size of the invertebrate nutrient pools were as predicted for the regenerating stage: they were higher than in the reference cells and declined from 1986 to 1989, although only the decline in 1989 was significant.

Our results confirm that during the regenerating stage there can be two peaks in aboveground standing crop. The first occurs early during reflooding, when both short-lived and long-lived emergents are present in the marsh. The second occurs when populations of long-lived emergents have completed their expansion into areas formerly occupied by the short-lived emergents. The latter had not been completed in 1989 in the low-treatment cells because populations of the long-lived emergents *Typha glauca* and *Phragmites australis* were still expanding (see Chapter 8). These two peaks in aboveground macrophyte production also resulted in similar peaks in uptake and translocation, especially for P.

Throughout the 5 years of flooding, the belowground biomass in the low cells was far below the belowground biomass during the baseline years in these cells. The different macrophyte pools evidently change at different rates during the regenerating stage. The belowground pool takes longer to reach its maximum than the aboveground pool. The regenerating stage has usually been described on the basis of changes in aboveground biomass (van der Valk and Davis 1978a,b), but changes in belowground biomass and associated nutrient pools seem more reliable indicators of this stage. The end of the regenerating stage would be marked by drop in belowground macrophyte nutrient pools.

These results indicate that the cells in the low treatment were still in the regenerating stage at the end of the reflooding years (1989). The mean total N (170 kg/ha) and P (16.9 kg/ha) mass in the major compartments in the low treatment from 1985 to 1989 were only 63% and 48%, respectively, of masses in the reference cells during the same period. Although the low-treatment cells had begun to resemble the reference cells more closely by 1989, they still were significantly lower in the size of their macrophyte compartments. For example, by 1989 in the low treatment, the total macrophyte N pool was only 103 kg/ha, whereas in the reference cells it was 230 kg/ha. Most of this difference was due to the much lower belowground mass of N in the low-treatment cells; in 1989, it was only 40% of that in the reference cells. Aboveground N mass, in contrast, in the low treatment in 1989 was ≈60% of that in the reference cells.

Total N in surface- and pore-water compartments in 1989 was 80% higher in the low-treatment cells than in the reference cells. This finding would indicate that nutrients were still available within the surface- and pore-water pools to support further production (e.g., plant and algal uptake), another indication that these cells were still in the regenerating stage. Bayley et al.(1985) also recorded an increase in available nutrients in pore water in response to flooding. These nutrients are then reduced as plant uptake reduces N and P concentrations in the pore water (Dean and Biesboer 1985; Chen and Barko 1988). The still-developing emergent plant communities in the MERP experimental cells had not reduced these available pools (through plant uptake) in the time available during the reflooding experiment.

Algal and invertebrate N and P mass were similar in the low treatment and reference cells by 1989. Algal pools increased over time as expected during the regenerating stage and then began to decline as the emergent vegetation stands increased and probably began to shade the water column, another predicted outcome during the regenerating stage. Interestingly, the dissolved nutrients in the water column (especially DIC) declined as algal production increased, and then increased later in flooding as algal uptake declined. Invertebrate pools in the low cells were much higher that in the reference cells, probably in response to the increased algal and litter inputs (Murkin 1989). Although invertebrates declined somewhat during the last year of flooding they were still higher than the reference cells, indicating that regenerating-stage conditions were persisting even after 5 years of flooding at the low-treatment levels.

Aboveground macrophyte production during the regenerating stage was predicted to be higher than in any other time during the wet–dry cycle (van der Valk and Davis 1978b). It appears the populations of emergents in the low treatment cells were still expanding in 1989 and the maximum aboveground macrophyte pools in these cells had not yet been reached; thus, the regenerating stage was not over and macrophyte pools comparable to that in the reference cells would not occur for another year or more. Table 7.4 in Chapter 7 summarizes changes in the mean aboveground biomass of each dominant emergent species in the low cells during the baseline year and from 1985 to 1989. These data indicate that emergent populations of *Typha glauca* and *Phragmites australis* were still expanding in 1989 and total aboveground biomass was still increasing (i.e., the low cells were still in the regenerating stage of the wet–dry cycle and therefore had not attained their maximum macrophyte pool sizes).

Degenerating Stage

Cells in the medium and high treatments were predicted to enter the degenerating stage because of loss of emergent species in areas where water levels following reflooding were too deep for them to survive for the duration of the study. The onset of the degenerating stage would be marked by a decrease in the size of the macrophyte nutrient pools as the plants die back and an increase in the size of the surface-water, pore-water, algal, and invertebrate pools as nutrients become available from the declining macrophyte pools. In the medium and high cells, actual changes in the macrophyte, surface- and pore-water, algal, and invertebrate nutrient pools during the last couple of years of flooding were generally consistent with what was predicted to occur during the first few years of the degenerating stage. The degenerating stage, however, did not begin earlier in the high treatment than in the medium treatment as predicted.

In both the medium and high treatments, the degenerating stage began in 1988 (i.e., during the fourth year of flooding) with a significant drop in aboveground macrophyte biomass (Figure 5.2) and associated N, P, and C pools. Belowground macrophyte biomass and nutrient pools also declined in 1988, but not as much as the aboveground macrophyte pools. In 1989, the total mass of N in the macrophyte pools in the medium and high treatments was ≈70% and 60%, respectively, of that in the low treatment. In both of these deeper treatments, the reduction in the size of the above- and belowground macrophyte compartments in 1988 resulted in an increase of total N and P in either surface- or pore-water pools, or both, and an increase in the algal pools as shading by emergents was eliminated and the algae were able to take advantage of the available dissolved nutrients in the water column. This increase in algae may be responsible for the lower levels of SRP observed in the medium and high treatments later in flooding when the vegetation was eliminated from these cells. With this decline in vegetation and subsequent litter input to the water column in these deeper treatments over time, suspended nutrients in the water column declined as would be expected later in the degenerating stage of the wet–dry cycle.

In 1989, the total mass of N in the major compartments (macrophytes, surface and pore water, algae, and invertebrates) was 180, 170, and 140 kg/ha in the low, medium, and high treatments, respectively. These amounts represent 57, 45, and 42% of the total N mass, respectively. Total P was 16, 17, and 13 kg/ha, respectively, or 69, 54, and 49% of the total mass of P, respectively. These results suggest that in the short term, there is storage or loss of nutrients to the litter/sediment pool, especially in the deeper treatments (as would be predicted during the degenerating stage of the wet–dry cycle).

Why did the degenerating stage not begin earlier in the high-water treatment? The primary reason for this was the behavior of the short-lived emergents. These species seem to have enough belowground reserves after the drawdown years to survive for 2 to 3 years of flooding, regardless of water depth. When the belowground reserves were exhausted, these species were extirpated in all three treatments simultaneously (see Chapter 8). The reason for the decline of aboveground biomass in 1988 in the medium and high treatments was a decline in the size of populations of long-lived emergents *Typha glauca* and *Phragmites australis* (de Swart et al. 1994; van der Valk 1994). This delayed decline in the emergent populations is consistent with experimental studies of the water-depth tolerances of these species, which indicate that growth is possible for several years in water depths that cannot be tolerated over the long term (McKee et al. 1989; Squires and van der Valk 1992). These species seem able to sustain growth by using carbohydrates stored in their rhizomes (Linde et al. 1976; Grace 1989). Consequently, it is not possible to significantly shorten the regenerating stage within the depth ranges used in this part of the MERP study. (It is possible to eliminate much of the regenerating and degenerating stages entirely by raising water levels so high that emergents plants cannot grow, as was done during the MERP deep-flooding years [see Chapter 4]).

In most cases with the species mix of the Delta Marsh region, the degenerating stage should begin in the third year after the onset of the regenerating stage and should be detectable during the fourth year. Once the degenerating stage began in the MERP cells, water-depth differences between the medium and deep treatments were not large enough to influence the rate of change in decline of the macrophyte pools. However, levels well above normal, as took place during the MERP deep-flooding (1 m above normal) experiment, can affect this rate of change (i.e., accelerated progression to the lake marsh stage) (see Chapter 4).

The degenerating marsh stage ends with the effective elimination of the emergent vegetation and the onset of the lake stage (van der Valk and Davis 1978a). Our data do not allow prediction of when this would happen because the experimental cells were still in the regenerating stage (low) or degenerating stage (medium and high) at the end of the experiment. In prairie marshes, water levels may continue to rise during the entire regenerating stage as the basin refills. Thus, with time, the rate of loss of emergents should increase. Changes in muskrat populations also occur and potentially affect macrophyte production during these periods (Weller and Spatcher 1965; Weller and Fredrickson 1974). In the medium and high treatments during the reflooding years, muskrats invaded the cells early (Clark and Kroeker 1993), however, their feeding and lodge-building activities appear to have had only a minor impact on macrophyte standing crops (Clark 1994; see Chapter 11). In the MERP cells with their constant water levels, the degenerating stage could end with only a partial extirpation of the emergent communities, the amount of emergent vegetation left being determined by water level. In the medium and high treatments, an arrested degenerating stage would probably have developed if the reflooding treatments had been continued longer with stable populations of emergents interspersed among areas of open water, too deep for emergent long-term survival.

ACKNOWLEDGMENTS

Elaine Murkin and Lisette Ross supervised the field crews during the drawdown and reflooding years of MERP. This is Paper No. 98 of the Marsh Ecology Research Program, a joint project of Ducks Unlimited Canada and the Delta Waterfowl and Wetlands Research Station.

LITERATURE CITED

Bayley, S.E., J. Zoltek, Jr., A.J. Hermann, T.J. Dolan, and L. Tortora. 1985. Experimental manipulation of nutrients and water in a freshwater marsh: Effects on biomass, decomposition, and nutrient accumulation. Limnology and Oceanography 30:500-512.

Bowden, W.B., C.J. Vorosmarty, J.T. Morris, B.J. Peterson, J.E. Hobbie, P.A. Steudler, and B.I. Moore. 1991. Transport and processing of nitrogen in a tidal freshwater marsh. Water Resources Research 27:389-408.

Chen, R.L., and J.W. Barko. 1988. Effects of freshwater macrophytes on sediment chemistry. Journal of Freshwater Ecology 4:279-289.

Clark, W.R. 1994. Habitat selection by muskrats in experimental marshes undergoing succession. Canadian Journal of Zoology 72:675-680.

Clark, W.R., and D.W. Kroeker. 1993. Population dynamics of muskrats in managed marshes at Delta, Manitoba. Canadian Journal of Zoology 71:1620-1628.

Dean, J.V., and D.D. Biesboer. 1985. Loss and uptake of ^{15}N-ammonium in submerged soils of a cattail marsh. American Journal of Botany 72:1197-1203.

deGeus, P.M. 1987. Vegetation change in the Delta Marsh, Manitoba between 1948-1980. M.S. thesis, University of Manitoba, Winnipeg, Manitoba.

de Swart, E.O.A.M., A.G. van der Valk, K.J. Koehler, and A. Barendregt. 1994. Experimental evaluation of realized niche models for predicting responses of plant species to a change in environmental conditions. Journal of Vegetation Science 5:541-552.

Devito, K.J., P.J. Dillon, and B.D. Lazerte. 1989. Phosphorus and nitrogen retention in five Precambrian shield lakes. Biogeochemistry 8:185-204.

Goldsborough, L.G., and G.G.C. Robinson. 1996. Pattern in wetlands. In *Algal Ecology: Freshwater Benthic Ecosystems*. (Eds.) R.J. Stevenson, M.L. Bothwell, and R.L. Lowe, pp.77-117. New York: Academic Press.

Grace, J.B. 1989. Effects of water depth on *Typha latifolia* and *Typha domingensis*. American Journal of Botany 76:762-768.

Graneli, W., and D. Solander. 1988. Influence of aquatic macrophytes on phosphorus cycling in lakes. Hydrobiologia 170:245-266.

Linde, A.F., T. Janisch, and D. Smith. 1976. Cattail—the significance of its growth, phenology, and carbohydrate storage to its control and management. Wisconsin Department of Natural Resources Technical Bulletin 94.

Macauley, A. J. 1973. Taxonomic and ecological relationships of *Scirpus acutus* Muhl. and *S. validus* Vahl. (Cyperacea) in southern Manitoba. Ph.D. dissertation, University of Manitoba, Winnipeg.

McKee, K.L., I.A. Mendelssohn, and D.M. Burdick. 1989. Effect of long-term flooding on root metabolism response in five freshwater marsh plant species. Canadian Journal of Botany 67:3446-3452.

Merendino, M.T., and L.M. Smith. 1991. Influence of drawdown date and reflood depth on wetland vegetation establishment. Wildlife Society Bulletin 19:143-150.

Murkin, H.R. 1989. The basis for food chains in prairie wetlands. *In Northern Prairie Wetlands*. (Ed.) A.G. van der Valk, pp.316-339. Ames: Iowa State University Press.

Neckles, H.A., J.A. Nelson, and R.L. Pederson. 1985. Management of whitetop (*Scholochloa festucacea*) marshes for livestock forage and wildlife. Delta Waterfowl and Wetlands Research Station Technical Bulletin 1.

Neill, C. 1992. Life history and population dynamics of whitetop (*Scolochloa festucacea*) shoots under different levels of flooding and nitrogen supply. Aquatic Botany 42:241-252.

Ott, L. 1988. *An Introduction to Statistical Methods and Data Analysis*. PWS-Kent Publishing, Boston, Massachusetts.

Pieczynska, E. 1993. Detritus and nitrogen dynamics in the shore zone of lakes: a review. Hydrobiologia 251:49-58.

Robinson, G.G.C., S.E. Gurney, and L.G. Goldsborough. 1997a. Response of benthic and planktonic algal biomass to experimental water-level manipulation in a prairie lakeshore wetland. Wetlands 17:167-181.

Robinson, G.G.C., S.E. Gurney, and L.G. Goldsborough. 1997b. The primary productivity of benthic and planktonic algae in a prairie wetland under controlled water-level regimes. Wetlands 17:182-194.

Shay, J.M., and C.T. Shay. 1986. Prairie marshes in western Canada, with specific reference to the ecology of five emergent macrophytes. Canadian Journal of Botany 64:443-454.

Squires, L., and A.G. van der Valk. 1992. Water depth tolerances of the dominant emergent macrophytes of the Delta Marsh, Manitoba. Canadian Journal of Botany 70:1860-1867.

van der Valk, A.G. 1981. Succession in wetlands: a Gleasonian approach. Ecology 62:688-696.

van der Valk, A.G. 1994. Effects of prolonged flooding on the distribution and biomass of emergent species along a freshwater wetland coenocline. Vegetatio 110:185-196.

van der Valk, A.G., and C.B. Davis. 1978a. The role of seed banks in the vegetation dynamics of prairie glacial marshes. Ecology 59:322-335.

van der Valk, A.G., and C.B. Davis. 1978b. Primary production of prairie glacial marshes. In *Freshwater Wetlands: Ecological Processes and Management Potential*. (Eds.) R.E. Good, D.F. Whigham, and R.L. Simpson, pp.21-38. New York: Academic Press.

Weller, M.W., and L.H. Fredrickson. 1974. Avian ecology of a managed glacial marsh. Living Bird 12:269-291.

Weller, M. W., and C. E. Spatcher. 1965. Role of habitat in the distribution and abundance of marsh birds. Department of Zoology and Entomology Special Report #43. Agricultural and Home Economics Experiment Station, Iowa State University, Ames, Iowa.

Welling, C.H., R.L. Pederson, and A.G. van der Valk. 1988. Temporal patterns in recruitment from the seed bank during drawdowns in a prairie wetland. Journal of Applied Ecology 25:999-1007.

Winter, T.C. 1989. Hydrologic studies of wetlands in the northern prairies. In *Northern Prairie Wetlands*. (Eds.) A.G. van der Valk, pp.16-54. Ames: Iowa State University Press.

6

Nutrient Budgets and the Wet–Dry Cycle of Prairie Wetlands

Henry R. Murkin
Arnold G. van der Valk
John A. Kadlec

ABSTRACT

The Marsh Ecology Research Program (MERP) nutrient budgets, described in detail in Chapters 4 and 5 and summarized in Figure 6.1 and Table 6.1, show that pools and fluxes associated with the macrophytes dominate the overall nutrient budgets within the experimental cells during all stages of the wet–dry cycle, except possibly during the lake marsh stage when pore-water pools are among the largest nutrient pools in the system. Macrophyte uptake is normally the largest annual flux followed by leaching, translocation, and litter-fall. All other pools and fluxes were minor by comparison, except again during the lake stage. During the dry marsh, regenerating, and degenerating stages, nutrients cycle quickly as the macrophytes take up nutrients from the sediment pools and incorporate them into aboveground tissues, which eventually decompose and return nutrients to the belowground pools. Nutrients released by the macrophytes through a variety of fluxes are incorporated into other pools (e.g., algae, invertebrates, vertebrates), and thereby support a diversity of producers and consumers. As macrophyte production is eliminated by deep or continuous flooding, plant uptake is reduced during the degenerating and lake stages, and nutrients become "locked" in the sediment pools. Only the reestablishment of the macrophytes during the dry marsh stage "unlocks" these nutrients and reestablishs the diversity of nutrient pools and fluxes characteristic of productive prairie wetlands.

The increase in productivity following the reflooding of prairie wetlands after a dry period is generally considered to be due to the aeration of the sediments resulting in increased decomposition and release of nutrients from the sediments. The MERP results indicate that although aeration of the sediments is probably important to meet the nutrient requirements of the renewed plant uptake, the pulse of productivity associated with the water column is due to the leaching of nutrients from aboveground litter soon after flooding, the presence of algae to take advantage of these nutrients, and the abundant, submersed litter providing surfaces for algal growth and invertebrate food and habitat.

Microbial processes were not addressed directly by MERP, however, they probably play a major role in prairie-wetland nutrient cycling. Research on the wide range of microbial and biogeochemical processes in the water and sediments of prairie wetlands is required before definitive conclusions can be made regarding the factors controlling nutrient cycling in these dynamic systems.

INTRODUCTION

Studies of nutrient cycling and budgets in wetlands from a total ecosystem perspective are relatively recent (Davis and van der Valk 1978; Klopatek 1978). Stimuli for much of the work to date have been interest in the nutrient-removal functions of wetland systems (van der Valk et al. 1978; Kadlec and Kadlec 1979; Mitsch et al. 1995; Kadlec and Knight 1996), management of wetlands and shallow ponds for waterfowl (Kadlec 1962; Whitman 1974) and fish (Neess 1946), and wetland rice culture (Harter 1966; Mahapatra and Patrick 1969). Although advances in our understanding provided by this previous work have been considerable (Howard-Williams 1985; Bowden et al. 1991; Kadlec and Knight 1996; inter alia), there has been little published research focused directly on prairie wetlands (Neely and Baker 1989). The dynamic nature of the prairie environment and the consequent changes in water levels and plant communities ensure that nutrient budgets in prairie wetlands vary significantly during the wet–dry cycle as described by van der Valk and Davis (1978a).

The overall objective of MERP was to assess the effects of water-level changes on the nutrient budgets during a simulated wet–dry cycle in the MERP experimental cells. Previous chapters in Part II of this book summarized the nutrient budgets during the various phases of the MERP study (see Table 1.1 in Chapter 1). The deep-flooding years (see Chapter 4) were intended to simulate the lake marsh stage of the wet–dry cycle, whereas the subsequent drawdown and reflooding (see Chapter 5) were intended to simulate the dry marsh, regenerating, and degenerating stages. In general, the magnitude and duration of the MERP water-level treatments fall within the ranges found in prairie wetlands during natural wet–dry cycles (see Chapter 13). In this chapter, our objective is to extrapolate the nutrient data and budgets from the various phases of MERP to characterize the nutrient budgets (pools and fluxes) during each stage of an idealized wet–dry cycle as described in van der Valk and Davis (1978b). We briefly consider the effects of diking on the primary production and nutrient budgets of the MERP experimental cells. We close with a general discussion of nutrient cycling in prairie marshes and the Delta Marsh in particular.

DIKED AND UNDIKED MERP CELLS

One of the striking results of the Marsh Ecology Research Program (MERP) nutrient-budget determinations was the lower mass of the nitrogen (N), phosphorus (P), and carbon (C) macrophyte pools in the diked cells during the various phases of the study compared with those same pools in the undiked reference cells during the same time periods (see Table 5.3 in Chapter 5). The macrophyte pools in the diked experimental cells during the baseline year in 1980 (i.e., soon after the dikes were constructed and before any water-level manipulations began) were similar in size to the macrophyte pools in the undiked reference cells that same year (see Chapter 4). The main issue in question is whether diking the cells had a negative impact on their primary production or whether there simply had not been adequate time during the various flooding experiments for macrophyte biomass and associated nutrient pools in the diked cells to return to baseline levels. If the former were true, extrapolating the MERP results to prairie wetland systems in general would be difficult.

Macrophyte production and the associated N, P, and C pools in the diked cells were expected to decrease during the MERP deep-flooding years (1981–1983) (see Chapter 4). However, they were expected to recover during the reflooding years (1985–1989) following drawdown (1983–1984), at least in the low treatment (cells maintained at the same water level as baseline year and undiked reference cells), to levels comparable to those found during the baseline year (1980) and in the reference cells. The MERP nutrient budgets indicated that 5 years of reflooding was evidently not long enough for this recovery to baseline levels to occur. In fact, because the short-lived emergents were not extirpated until 1987 in the low treatment (i.e., third year of flooding following drawdown) (see Chapters 5 and 7), the nutrient pools associated with the longer-lived emergents that replaced these short-lived species had only 2 years to develop before the study ended in 1989. In comparison, the experimental cells during the baseline year (1980) and the undiked reference cells had experienced stable water levels and sustained macrophyte growth for >15 years prior to the study (see Chapter 2). Consequently, there is no reason to suspect that diking per se had an adverse effect on the primary production of the MERP cells or their nutrient budgets. It just appears that the 5-year time frame of the reflooding experiments (1985–1989) was not long enough for the macrophyte pools in the diked cells (and specifically the low treatment) to attain the levels found during the baseline year (1980) and in the undiked reference cells.

NUTRIENT BUDGETS DURING THE WET–DRY CYCLE

The changes in the nutrient budgets of prairie wetlands during the various stages of an idealized wet–dry cycle as extrapolated from the MERP results are summarized in Table 6.1 and Figure 6.1. Table 6.1 gives a qualitative summary of changes in the macrophyte, algal, invertebrate, and surface- and pore-water pools for the four stages of the cycle. Figure 6.1 provides a quantitative estimate of pool sizes during the various stages by using appropriate data from the diked MERP experimental cells and undiked reference cells (see paragraph following). There was little difference among the budgets for the three nutrients (N, P, and C) during the MERP experiments except that the uptake of C was by aboveground plant tissues, whereas N and P uptake was by belowground roots and rhizomes. The only

Table 6.1. Relative sizes and changes in nutrient pools during the wet–dry cycle in prairie marshes

Pool	Dry	Regenerating	Degenerating	Lake marsh
		Stage of the wet–dry cycle		
Macrophytes				
Aboveground	Highest for nitrogen (N) and phosphorus (P), mostly annual species with developing emergent vegetation, carbon (C) pool lower than reflooding years (due to lower C concentrations in annual species)	Low initially, N and P increases rapidly as emergent vegetation expands through vegetative growth, C increases to levels higher than drawdown period	All pools (N,P,C) begin to decrease as emergent vegetation eliminated from deeper portions of basin, some submersed vegetation present	Very low as emergent vegetation is restricted to shallow fringe of basin, some submersed vegetation present in shallow, sheltered areas of the basin
Belowground	Low, mostly annual species with little belowground biomass	Increases (N, P, C) through vegetative growth, reaches highest pool size near the end of this stage	Begins to decline in size (N,P,C), not as quickly as aboveground pools	Low, some survival for a year or two after aboveground pool reduced, but eliminated shortly thereafter
Algae	Lowest, some epipelon biomass associated with moist surfaces	Increases initially in response to available nutrients in surface water and availability of light to highest levels during cycle, epiphyton and metaphyton dominate, however begin to decline as emergent stands develop (shading)	Increases initially in response to available nutrient from leaching litter and availability of light as emergent stands begin to die back, eventually begins to decline as community switches to phytoplankton	Low, phytoplankton predominates, some metaphyton on shallow sheltered areas
Invertebrates	No aquatic invertebrate pool, some terrestrial and semi-terrestrial species present	Increases in response to flooded litter and increased algal pools, attain highest levels of cycle, only a small fraction of macrophyte and algal pools	Remains high early in response to litter availability and increased algal pools, begins to decline rapidly as macrophytes eliminated and phytoplankton dominates open water	Declines to lowest levels of cycle, however some persistence of benthic species in sediments of shallow open areas

Table 6.1. (*Continued*)

Pool	Dry	Regenerating	Degenerating	Lake marsh
		Stage of the wet-dry cycle		
Surface water				
Organic	No surface water	Begins high, due to leaching of submersed litter, declines quickly to lowest levels seen during cycle, then increases to second peak during death of short-lived macrophytes	Increases to fairly high levels due to decomposition of emergent vegetation	Maintained at moderate levels due to rapid turnover of algae
Inorganic	No surface water	High initially due to leaching of submersed litter, N increases while P declines to very low levels before returning to levels found early in flooding	Levels increase throughout period due to leaching and decomposition of macrophyte and algal litter	Maintained at fairly high levels without many sources of uptake in the water column
Suspended solids	No surface water	Low initially, some increase over time with macrophyte and algal decomposition	Increase initially with macrophyte and algal litter, declines as these sources are eliminated	Reduced as emergent vegetation and algae eliminated, maybe episodic peaks due to wind effects on the sediments
Pore water				
Organic	Some increase with annual and macrophyte development	Slight increase with production and decomposition of below-ground tissues	Slight increase with decomposition of below ground tissues	Stays at levels found in earlier stages (slight increase in P), input from litter decomposition
Inorganic	Decreases over time due to uptake associated with high annual plant and macrophyte production	Increase over time, in spite of elevated uptake over this period	Levels continue to increase with belowground decomposition and reduction in plant uptake	Levels increase to maximum levels with elimination of plant uptake and inputs through belowground decomposition and mineralization

103

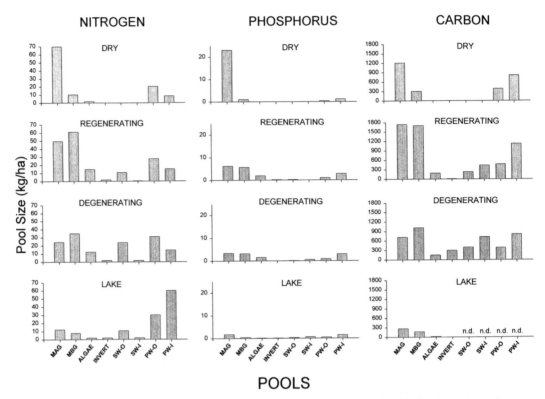

POOLS

Figure 6.1   Relative nutrient (N, P, C) pool sizes during the various stages of an idealized wet–dry cycle. (MAG, aboveground macrophytes; MBG, belowground macrophytes; SWO, surface water organic; SWI, surface water inorganic; PWO, pore water organic; PWI, pore water inorganic). All data are derived from various parts of the MERP experiments: dry stage (drawdown period 1983–1984), regenerating stage (reflooding experiment 1985–1989), degenerating stage (deep-flooding experiment 1981–1983 and reflooding experiment 1985–1989), and lake marsh stage (open-water areas of undiked reference cells (1980–1989). (n.d., no data).

prior attempts to estimate changes of the functional attributes of prairie wetlands during a wet–dry cycle were made by van der Valk and Davis (1978b) for macrophytes, Voigts (1976) for invertebrates, and most recently by Goldsborough and Robinson (1996) and Robarts et al. (1995) for the various algae communities in prairie wetlands.

We begin with the dry marsh stage, which was simulated by the MERP drawdown period (1983–1984) (see Table 1.1 in Chapter 1). We follow with a consideration of the regenerating, degenerating, and lake marsh stages. For reasons already noted in Chapters 4 and 5, the extrapolation of the MERP data from the reflooding years (1985–1989) and deep-flooding years (1981–1983) to the lake marsh stage is not straightforward. The deep-flooding phase of MERP following the baseline year (1980) was intended to move the cells to the lake marsh stage. By the end of the second year of deep flooding, aboveground macrophyte pools were much reduced, as would be expected during the lake marsh stage, but the belowground pools were still very large (see Table 4.3 in Chapter 4). There was also a substantial amount of standing litter remaining (van der Valk 1986) and this litter supported a large community of attached algae and invertebrates, much larger than would be expected to occur during the lake stage in a prairie wetland (Hosseini and van der Valk 1989a, b; Murkin and Kadlec 1986). Consequently, algal, invertebrate, and surface-water pool sizes for the lake marsh stage in Figure 6.1 are based on data collected from open-water areas in the undiked reference cells. The open-water areas of the Delta Marsh are considered to be in the lake marsh stage of the wet–dry cycle due to the long-term stabilization of the marsh water levels (see Chapter 2).

Dry Marsh Stage

The dry marsh stage in an idealized wet–dry cycle is preceded by the lake marsh stage where much of the emergent vegetation has been eliminated by flooding and the remaining vegetation within the basin is dominated by submersed aquatic species (see Figure 1.3 in Chapter 1). As in other studies of the vegetation response to dry conditions in prairie wetlands (van der Valk and Davis 1978a), a combination of mudflat annuals and newly recruited emergents dominated the vegetation during the MERP drawdown period (Welling et al. 1988; see Chapter 7). Major fluxes associated with macrophytes, most notably nutrient uptake, leaching, and litter-fall, were reestablished as important components of the nutrient budgets during this dry stage (see Chapter 5). Translocation remained low because the vegetation was dominated primarily by annual species that do not translocate nutrients between above and belowground tissues to any great degree. The aboveground biomass of annuals during the dry marsh stage can equal and often exceed that of emergent vegetation during the regenerating stage of the wet–dry cycle (van der Valk and Davis 1978a). This outcome was certainly realized with the vegetation response to the MERP drawdown (see Table 7.2 in Chapter 7). An important difference between annuals and emergents is the much lower belowground production in annual species. Although emergent macrophytes became established during the MERP dry stage, development of belowground tissues was relatively slow (see Table 7.2 in Chapter 7). Thus, nutrient pools associated with belowground roots and rhizomes remained much lower than the aboveground biomass pools (annuals and emergent combined) during the dry marsh stage (see Table 5.3 in Chapter 5).

 Perhaps the most important development during the dry marsh stage is the reestablishment of the plant uptake flux to move N and P from the sediments and pore water to aboveground pools and fluxes. Nutrients "locked" in the sediment and pore-water pools during the lake marsh stage (see paragraphs following and Chapter 4) are rapidly incorporated into plant biomass during the dry marsh stage. Several studies have documented increases in pore-water concentrations during drawdowns (Kadlec 1962; Klopatek 1978); however, during the MERP studies, concentrations within the inorganic pools declined during the drawdown (Kadlec 1989), presumably in response to the increased plant uptake by annuals and emergents germinating on the newly exposed substrates. Chen and Barko (1988) also recorded reductions in sediment N and P concentrations due to uptake by macrophytes.

A commonly cited reason for the increase in productivity during and following drawdowns is the aeration of the sediments resulting in more efficient aerobic microbial decomposition of the litter/sediment pools (Kadlec 1989). During MERP, the uptake of N and P by the vegetation during the early dry marsh stage exceeded the mass of dissolved inorganic nutrients in the pore-water pool (see Chapter 5). This would indicate increased and rapid inputs to the pore-water inorganic pool from the other sediment pools, most probably from the mineralization of organic compounds in the dissolved-organic pore-water pool. These results support increased aerobic decomposition during the drawdown stage. Verhoeven and Arts (1987) also noted rapid mineralization in the sediments to meet the uptake requirements of emergent vegetation in small mesotrophic fens.

During the dry marsh stage, other pools dependent on standing water, such as algae and aquatic invertebrates, are much reduced. Algae are restricted to epipelon associated with damp substrates (see Chapter 8), whereas aquatic invertebrates are present only as dormant stages (see Chapter 9), although some terrestrial invertebrates may colonize the dry basins. Surface-water and associated nutrient pools are eliminated as the water is lost from the basin. An important consideration during the dry marsh stage is how the water was lost from the system. Under natural drying conditions (e.g., during prolonged prairie drought), most of the water is lost through evapotranspiration and therefore the nutrients remain within the basin, particularly in basins where groundwater outflow is not a factor (e.g., groundwater discharge basins [Winter 1989]). In wetlands where there is groundwater outflow (e.g., groundwater recharge and flow-through wetlands), some nutrients are lost from the system. In basins that are actively pumped down as part of a drawdown management scheme, the pools associated with the surface water would be lost, and at certain times of the year this could represent a significant amount of nutrients (see Chapter 4; Cook and Powers 1958; Kadlec 1979). For example, the MERP experimental cells lost 10% and 25% of their N and P respectfully, when they were drawndown after the 2 years of deep flooding (not including N and P in the large litter/sediment pools). The impact of these losses on future productivity is unknown, however, it is generally thought to be minor because of the large stores of nutrients in the litter/sediment pools (Bonetto et al. 1988; Jordan et al. 1989; see Chapter 12).

Regenerating Stage

The return of standing water to the basin marks the onset of the regenerating stage (Figure 1.3 in Chapter 1). Immediately following reflooding in the MERP experimental cells, the annual mudflat plant species were killed and the associated biomass entered the aboveground litter pool (see Chapter 7). Leaching from this aboveground litter resulted in large amounts of nutrients, particularly N and P, entering the water column at the time of flooding (Wrubleski et al. 1997). Interestingly, the monthly schedule of the MERP surface water sampling program did not detect large increases in surface water N and P during the early part of the reflooding experiments (Kadlec, unpublished data). However, Robinson et al. (1997a,b) noted that algal production, particularly metaphyton, in the experimental cells was very high soon after flooding, and thus probably played a significant role in reducing surface-water nutrient concentrations. They noted that algal production began soon after ice-out and therefore was well underway by the first MERP water-sampling period in May. Robarts and Waiser (1998) noted that the importance of algae in the primary production of prairie wetlands is due in part to their presence and continual production throughout the ice-free period. The MERP nutrient-budget data showed that the algal uptake flux during the early years of reflooding in the MERP experimental cells was similar to the leaching losses from the aboveground litter during the same period (Tables A2.3–A2.5 in Appendix 2).

A pulse of productivity during the reflooding of wetlands after a dry or drawdown period is well-known (Kadlec 1962; Whitman 1974; van der Valk and Davis 1978b). As noted, the aeration of the sediments resulting in increased decomposition and release of nutrients from the sediments is generally considered as the reason for this increased productivity (Kadlec 1962). The MERP results suggested that the initial productivity increase was also due in large part to the leaching of nutrients from the aboveground litter (see Chapter 5), the presence of algae to take advantage of the released nutrients (see Chapter 8), and the abundant submersed litter providing surfaces for algal colonization as well as invertebrate food and habitat (see Chapter 9).

As macrophyte productivity is maintained through the early years of reflooding, overall productivity of the system continues to be high (van der Valk and Davis 1978b). During the MERP reflooding experiments (1985–1989), the macrophyte pools and associated fluxes remained the dominant components of the nutrient budgets (see Chapter 5) and, in turn, continued to support other sources of productivity such as algae, invertebrates, and other components of the wetland food chains. The ongoing macrophyte litter production continued to add nutrients to the water column and sediments. Eventually, the various decomposition and mineralization processes returned these nutrients to the dissolved inorganic pools where they were available for plant uptake once again. The importance of plants in recycling nutrients and maintaining overall productivity in aquatic systems has been noted by other researchers, including Mitsch and Gosselink (1993) for other freshwater wetlands and Wetzel (1989) for shallow lake systems.

The regenerating stage is maintained until the long-lived emergent vegetation species begin to die back due to continued flooding or a subsequent increase in water levels beyond the tolerance levels of the species present in the basin (van der Valk and Davis 1978b). During MERP, the short-lived emergent species died early in the regenerating stage, however, even the species adapted to continuous flooding eventually began to die back due to stress, disease, and other factors (see Chapter 7). As reported in Chapter 5, the aboveground pools began to decline first, followed eventually by the belowground pool. If water levels are reduced before the belowground tissues die, it is likely that the emergent vegetation would survive and the regenerating stage could be prolonged (see Chapter 7). However, once the belowground tissues of the long-lived emergent species die and the belowground pool begins to decline, the wetland enters the degenerating stage of the wet–dry cycle (see Chapter 5).

Degenerating Stage

As noted previously, the onset of the degenerating stage (see Figure 1.3 in Chapter 1) begins with the permanent decline in the abundance of long-lived emergent species within the basin. The switch from the regenerating stage to the degenerating stage is often considered to be the most productive period for consumers during the wet–dry cycle in prairie wetlands (Murkin 1989). During the degenerating stage in the MERP experimental cells, the remaining, although declining, macrophyte production and the litter availability combined to support a broad range of primary and secondary consumers within the system (see Chapters 7–9). The MERP algal monitoring indicated that early in the degenerating stage, there was sufficient litter production and associated leaching to support significant algal production within the water column (Hosseini and van der Valk 1989a, b; see Chapter 8). The remaining emergent vegetation and litter also provided substrate for algal growth, whereas the thinning of the stands allowed light penetration for photosynthesis within the water column. This resulted in a period of high algal production as the algal communities responded to the available nutrients and habitat conditions (Robinson et al. 1997a, b). In their model of prairie wetland algal dynamics, Goldsborough and Robinson (1996) referred to this stage as the sheltered state where the shelter of the existing vegetation and the increased light availability allowed dense growth of epiphytic algae that eventually breaks free and results in mats of metaphyton forming in open-water areas of the wetland. They suggested that algal production eventually declines as the submersed habitat provided by the emergent stands and associated nutrient fluxes to the water column are eliminated. During the MERP studies, the algal pools remained fairly high even after 5 years of flooding, although a slight decline was noted in some treatments late in the study (see Table 5.4 in Chapter 5). However, the low algal levels recorded in the open-water sites of the reference cells suggested that algal abundance in the experimental cells would eventually decline with continued flooding. Invertebrate production in the MERP experimental cells also declined as the macrophytes and the algae were eliminated from the basins (see Chapter 9).

With the loss of the major aboveground and belowground macrophyte pools in the experimental cells, the dominant fluxes associated with the macrophytes (uptake, translocation, leaching, and litter-fall) were reduced over time (see Tables A2.4 and A2.5 in Appendix 2). The nutrients once associated with macrophytes accumulated in other pools (Figure 6.1) as macrophyte pools were reduced and fluxes to the macrophyte pools eliminated. The pools with inputs from macrophytes and outputs to pools other than macrophytes decreased in size as well. These pools included surface water, algae, and invertebrates that during the late degenerating and lake marsh stages would be expected to reach low levels compared with other stages of the wet–dry cycle. During the MERP reflooding experiments, these pools did decline during the later years of flooding (see Chapter 5) and were very low in the open-water areas of the reference cells and Delta Marsh throughout the period of study (see Chapters 7–9).

With the decline of macrophyte uptake during the MERP deep-flooding experiment and continued inputs to the pore-water inorganic pools through mineralization, there was an overall increase in the pore-water inorganic nutrient pools during this period (see Tables 4.3 and 4.4 in Chapter 4). Similar conditions would be expected during the degenerating stage of the wet–dry cycle. During this stage, inputs to the sediment and pore water continue as the remaining emergent litter falls and decomposes. With the elimination of plant uptake as the major flux of dissolved inorganic nutrients from the pore water and continued microbial activity within the sediments (mineralization), this pool would increase over time and eventually with continued flooding become one of the largest pools in the system (Figure 6.1). Nutrients remained "locked" in the pore water and sediment until emergent vegetation and associated plant uptake are reestablished to move the nutrients from these belowground pools.

The rate at which a wetland proceeds to and through the degenerating stage is determined in part by the water depth within the basin. During the MERP reflooding experiments in 1985–1989, the medium and deep treatments proceeded more quickly to degenerating conditions than the low treatment, due to the deeper water in these treatments. The aboveground macrophyte pools and associated fluxes were reduced first in the deeper cells (see Table 5.3 in Chapter 5) and the pore-water pools began to increase sooner, indicating that these cells had proceeded to the degenerating stage sooner than the more shallowly flooded treatment.

Lake Marsh Stage

The lake marsh stage is generally considered the period of lowest overall productivity as the emergent vegetation is reduced to a vestigial band around the periphery of the basin (van der Valk and Davis 1978a). During this stage, the standing water column is often turbid (as a result of wind-induced wave action due to the loss of the sheltering effects of the emergent vegetation) with phytoplankton as the dominant algal community (Goldsborough and Robinson 1996). Phytoplankton are generally considered to be poor competitors for nutrients and can only flourish when other competitors such as epiphyton, metaphyton, and macrophytes are absent. Phytoplankton communities during the lake marsh stage do not reach the biomass levels of the various algal communities that dominate other stages of the wet–dry cycle (Goldsborough and Robinson 1996). In prairie wetlands subjected to nutrient inputs from surrounding land uses (e.g., fertilization in agricultural operations), an elevated surface-water pool may be maintained which, in turn, may "fuel" some algal and associated invertebrate production throughout the lake marsh stage in these wetlands (Neely and Baker 1989; Robarts and Waiser 1998).

The low primary productivity of the lake marsh stage is because the nutrient cycling within the system is reduced to the large pool within the sediments with few fluxes available to move nutrients among other potential pools. Overall productivity is related to the number and diversity of fluxes within the system (see Chapter 5). With reduced numbers of fluxes during the lake marsh stage, productivity is low and remains so until additional fluxes are reestablished (e.g., plant germination and associated plant uptake) during the dry marsh stage.

DISCUSSION

Nutrient pool sizes and fluxes among pools vary significantly during the four stages of a wet–dry cycle, primarily because of differences in the abundance of emergent vegetation. It must be emphasized, however, that actual wet–dry cycles are highly variable both in the duration of the various stages and in water depths and rates of change of water depths during flooded stages (see Chapter 13). The MERP water-level treatments were applied in the form of a step function with very rapid transitions from one depth to another. During natural wet–dry cycles, water levels change as a more-or-less continuous function, although fairly abrupt changes are possible, particularly the transition from the dry marsh stage to the regenerating stage. It is also possible for water-level changes during a cycle to affect only part of the marsh (e.g., shallow areas or peripheral zones of the basin) (van der Valk and Davis 1978b).

Algal research on the MERP experimental cells and elsewhere in the Delta Marsh (Goldsborough and Robinson 1996; Robinson et al. 1997a, b) indicated that algae play an important role during various stages of the wet–dry cycle. Although algal biomass may not reach the levels found in macrophytes at any one point in time, the high production and rapid turnover of algal cells and tissues result in annual algal production that may equal or exceed macrophyte production (see Chapter 9). Algae also may affect nutrient budgets and cycling through oxygenation of the water column, production of dissolved organic material, mediation of the flux of nutrients from the sediments to the overlying water (primarily epipelon), and sediment formation (Goldsborough and Robinson 1996). This varied role of algae in the nutrient cycling of wetlands, and prairie wetlands in particular, requires further investigation (see Chapter 9).

To put the MERP study in perspective, we compare the MERP results with those from comparable studies on nutrient dynamics of freshwater wetlands. Specifically, we examine plant uptake and nutrient retention in the MERP cells with those in other freshwater wetlands. We also compare P pools and fluxes in the MERP cells with those reported for a comparable study of an arctic flooded meadow.

Plant Uptake

Annual N and P uptake by macrophytes in the MERP cells was within the range of literature values for freshwater wetlands (Table 6.2). In common with several other studies (Chapin et al. 1978; Barko and Smart 1980), the MERP nutrient budgets (see Chapters 4 and 5; Figure 6.1) indicated that vascular plant pools of N and P were many times larger than any other pool, except during the lake marsh stage. However, estimates of total sediment N and P pools are very large in relation to macrophyte uptake, so that mobilization of only a very small percentage (0.01–5.0%) of this pool is required to meet plant needs over the course of the growing season. For example, Bowden et al. (1991) estimated that net microbial mobilization of 18 g N/m², or ≈2–3% of total N in the top 30 cm of peat, was adequate to meet the annual plant uptake requirements for a mostly emergent freshwater tidal marsh. For the Delta Marsh, mobilization of N from the sediments sufficient

Table 6.2.   Estimates of annual vascular plant uptake of nitrogen (N) and phosphorus (P) (kg/ha/yr) in freshwater wetlands

	N	P
Delta Marsh (MERP)	48.6	9.63
95% CI	(41.44–55.96)	(7.97–11.29)
Tidal Fresh Marsh, New England (Bowden et al. 1991)	155	—
Thoreau's Bog, Massachusetts (Hemond 1983)	30	—
Wet meadow tundra, Alaska (Chapin et al. 1978)	—	1.55
Freshwater marsh, Louisiana (DeLaune et al. 1986)	88.7[a]	—
Peatland, Michigan (Richardson et al. 1978)	30	1.7
Cattail marsh, Wisconsin (Prentki et al. 1978)	—	31
River bulrush, Wisconsin (Klopatek 1978)	207.5	53
Sawgrass, Florida (Steward and Ornes 1975)	—	1.8
Fens, Netherlands[b] (Koerselman et al. 1990)	54–101	2.6–10.9
Phragmites, Czechoslovakia[b] (Ulehlova et al. 1973)	453	130

[a]Aboveground only.
[b]Standing crop, perhaps a rough estimate of uptake.

to meet plant uptake requirements would range from ≈1% during the dry marsh stage to only 0.1% during the lake marsh stage. For P, ≈1.4% mobilization of the sediment pool would be required during the regenerating stage and only 0.4% during the lake marsh stage.

Nutrient Retention

Nutrient retention is often estimated as the difference between inputs and outputs on a seasonal or annual basis (Devito and Dillon 1993). On average, the MERP cells retained 3.04 kg N/ha/yr and 0.69 kg P/ha/yr of the nutrients entering the cells through pumping, seepage, or precipitation (Table 6.3). This retention represented ≈6 or 7% of vascular plant uptake within the cells and therefore involved pools other than macrophytes, most likely the sediments. The retention of N and P was variable among cells and years, as were other aspects of the nutrient pools and fluxes (see Chapters 4 and 5). Retention of N tended to be especially high during deep flooding, perhaps reflecting denitrification rather than retention in sediments (see Table 4.2 in Chapter 4). Seasonal MERP nutrient budgets

Table 6.3.   Estimated nitrogen (N) and phosphorus (P) retention and accumulation in lake and wetland sediments (kg/ha/yr)

	N	P
Delta Marsh (MERP)	3.04	0.69
Lakes, Ontario (Dillon and Evans 1993)	—	0.2–1.5
Arctic lake (Whalen & Cornwell 1985)	2.94	0.5
Gulf Coast salt marsh (DeLaune et al. 1990)	170	—
Louisiana freshwater marsh (DeLaune et al. 1986)	120	—
Northern peatlands (Nichols 1983)	1–47	1–2
Everglades, Florida (Davis 1991)	75	1.8
Gulf Coast brackish marsh (DeLaune and Patrick 1990)	72	—
Gulf Coast freshwater lake (DeLaune et al. 1990)	180	12
Ohio riparian wetland (Niswander and Mitsch 1995)	—	29

showed P retention consistent across seasons, whereas N retention was higher in the spring–early summer (Kadlec, unpublished data). Annual nutrient budgets calculated for the MERP cells were consistent with other studies of nutrient retention in freshwater wetlands (van der Valk et al. 1978; MacCrimmon 1980; Phipps and Crumpton 1994; Kadlec and Knight 1996) . Prairie wetlands during flooded stages of the wet–dry cycle generally retain nutrients (Neely and Baker 1989). However, given the seasonal variability in the nutrient budgets discussed previously, prairie wetlands may occasionally function as sources rather than sinks for N and P for short periods of time (see Chapters 4 and 5; Neely and Baker 1989).

P in a Flooded Arctic Meadow and the Delta Marsh

The P budgets derived for the MERP cells are compared with those in a flooded arctic meadow at Barrow, Alaska (Chapin et al. 1978) in Table 6.4. Both studies are based on the quantification of storage of P in various pools in the ecosystem and derived fluxes from a variety of assumptions about processes that result in changes in pool sizes. The approaches were sufficiently similar to permit direct comparisons of major pools and fluxes (Table 6.4). Arctic tundra vegetation is, as expected, much smaller in stature and less productive than that of the MERP cells. Thus, most fluxes and pool sizes were smaller at Barrow, but for the most part were more or less proportional to the pools and fluxes in the MERP experimental cells (i.e., almost all the pools and fluxes at Barrow tended to be

Table 6.4.   Comparison of phosphorus pools and fluxes in an arctic flooded meadow (Chapin et al. 1978) and the MERP experimental cells (kg/ha or kg/ha/yr)

	Pool		Annual flux	
	Meadow	MERP	Meadow	MERP
Macrophyte: Aboveground	1.16	9.34	—	—
Belowground	5.96	8.52	—	—
Total	7.12	17.86	—	—
Macrophyte uptake	—	—	1.55	9.63
Nonvascular plants	1.12[a]	1.50[a]	—	—
Nonvascular plant uptake	—	—	0.62	1.02[b]
Total litter fall	—	—	1.53	7.29
Pore water DOP	0.22	0.70	—	—
Total sediment P	542	447	—	—
Hydrologic input–output	—	—	.01	0.69
Invertebrates	0.20	0.14	—	—
Dissolved inorganic P	0.01	2.6	—	—
Sediment to pore water	—	—	2.15[c]	8.73[c]

[a] Mosses at Barrow, algae in MERP, that have very different turnover rates.
[b] Annual uptake of P by algae in the MERP cells was probably 3–5 times this amount when turnover is taken into account.
[c] In both locations, some fluxes were combined to make sums as comparable as possible.

about 0.2 to 0.4 times the corresponding mean pool or flux over the entire study in the MERP cells, with a few conspicuous exceptions). Hydrologic inputs at Barrow were much smaller than those in the MERP cells, and dissolved inorganic P was very much lower, probably reflecting differences in substrates and perhaps limiting nutrients. Surprisingly, invertebrates were more abundant in the tundra than in the MERP cells. In both ecosystems, total sediment P was by far the greatest pool of P, so much so that all other fluxes and pools were small by comparison.

What Controls Prairie Wetland Nutrient Cycling?

Chapin et al. (1978) hypothesized that decomposition was the "major control point" in P cycling in the flooded meadow at Barrow, Alaska. In the MERP study, as well as others (Barko and Smart 1980), macrophyte uptake exceeded quantities of available N and P in the inorganic pore-water pools at any point in time. Consequently, nutrients lost by plant uptake from these pore-water pools must be quickly replaced by inputs from organic pools in the sediments. For P, that means the breakdown of organic or inorganic compounds, whereas for N, decomposition processes could be supplemented by N-fixation. Decomposition of macrophyte standing litter (above and below ground) is a logical candidate process, but most decomposition studies, including those done during MERP (Murkin et al. 1989; Nelson et al. 1990; van der Valk et al. 1991; Wrubleski et al. 1997), do not show the seasonal patterns and levels of released N or P to supply the macrophyte uptake in the spring. However, if a relatively small percentage of the large sediment pools of N and P were mineralized, the macrophyte nutrient needs would easily be met.

The uptake of N and P by plants is highly seasonal. Monthly measurements of NH_4^+ in pore water during MERP did not show consistent seasonal patterns that might imply accumulation and depletion of this nutrient. Although the previous arguments suggest the pool size may not be the appropriate measure (i.e., fluxes are more important), it seems clear that the processes involved must be operating at short time scales. The inorganic nutrients taken up by macrophytes are rapidly replaced in the pore-water pools. However, there is a mismatch in the timing of macrophyte uptake and postulated microbial mineralization of organic matter to supply the inorganic nutrients needed by macrophytes. Uptake occurs in spring and early summer when substrate temperatures are still low, whereas one would expect microbial activity to peak later when temperatures are higher (Robarts et al. 1994).

Chapin et al. (1978), considering these problems, postulated that the main source of P for macrophytes in early spring was P released from bacterial cells that lyse during the freeze–thaw cycle. They cite evidence that bacteria are active and abundant in the litter and sediments even at the low temperatures found in arctic tundra wet meadows. Although anaerobic metabolic pathways are generally less efficient than aerobic processes (Hoge 1994), recent studies in lakes suggest that the role of anaerobic bacteria may be much greater than suspected. Cole and Pace (1995) found that anaerobic bacterial populations in cold hypolimnia were in general larger, and in sum, more productive than those in warmer, aerobic environments. Gachter and Meyer (1993) reviewed the role of bacteria in P cycling at the sediment–water interface and suggested that, in eutrophic lakes, sediment bacteria contain as much P as settles with organic detritus in any one year. Both Davelaar (1993) and Eckerrot and Pettersson (1993) suggest bacteria may regulate the supply of P in lakes. It would not seem unreasonable to suggest that the role of anaerobic sediment bacteria in wetland P cycling may be even greater than in lakes. If sediment bacteria are a major pool of P, are they also a major pool of N? Martinova (1993) suggests sediment bacteria contain >50% of total organic N and >70% of total organic P in lakes and reservoirs. Mayer and Rice (1992) found bacteria contained 5–31% of sediment protein. Most studies of bacteria with respect to N have been concerned with fixation, denitrification or mineralization (Bowden 1984; Reddy and Graetz 1988; Groffman 1994), and although these processes require further investigation, the overall role of bacteria in wetland nutrient cycling needs to be understood in more detail.

If, as seems likely, bacterial processes and populations are a key to more fully understanding wetland nutrient cycling, the issue of substrate limitation of bacterial growth needs to be considered. Even though most wetland sediments are highly organic, much of that organic matter may not be suitable as a substrate for some microbial activities. For example, Bowden (1984) concluded that substrate quality and availability were very important in a study of ammonium consumption and production by microbial processes in sediment from a tidal freshwater marsh. Several other studies have suggested substrate amount or quality limits bacterial processes in spite of apparently abundant organic matter (Kerner 1993, Sander and Kalff 1993). Waiser and Robarts (1997) showed that only a small percentage of the available DOC in Redberry Lake was suitable to bacteria as an energy and nutrient source. It seems likely that the issue of quality of organic matter as substrate for bacteria is critical in anaerobic environments (Bridgham and Richardson 1992; Moore et al. 1992; Schallenberg and Kalff 1993); however, this topic requires more research in prairie wetland systems.

Microbial cells themselves may be a source of labile organic matter and, indeed, particulate organic matter may be much less important than dissolved organic matter in these systems. Robarts and Waiser (1998) discussed the importance of DOC to the biogeochemical processes of shallow prairie lakes. Besides "fueling" microbial food chains, DOC plays a major role in several photochemical reactions and flocculates other chemical substances, including P. Waiser and Robarts (1995) showed that little of the SRP in Redberry Lake in Saskatchewan was available for microbial processes due to the high concentrations of DOC. In wetland studies, concentrations of dissolved organic nutrients are not measured or reported as often as other forms of nutrients, but the MERP study suggests that large quantities are present during various stages of the wet–dry cycle. Although not all dissolved organic nutrients are labile, that is, easily decomposed by bacteria (Moran and Hodson 1990, 1992), they may play a variety of roles in the nutrient and energy budgets and overall function of prairie wetlands (Robarts and Waiser 1998). This, too, is an important area for further investigation in prairie wetlands.

In short, it is not clear what controls the nutrient cycles in prairie wetlands. Although plant uptake plays a major role in restoring nutrients to aboveground pools and fluxes, the respective roles of litter decomposition and microbial processes in the water column and pore water need to be elucidated in future research. There is relatively little known about in-water biogeochemical processes and the factors that regulate them (Robarts and Waiser 1998). Considering their potential importance to nutrient cycling in prairie wetlands, significant new effort is required to improve our understanding of these important processes.

Salinity and Nutrient Cycling in Prairie Wetlands

The MERP study did not address the issue of salinity and its effects on nutrient cycling in prairie wetlands. However, many prairie wetlands are saline, and as they proceed through the wet–dry cycle, large changes in salinity occur (Curtis and Adams 1995; LaBaugh et al. 1998). Besides affecting the macrophyte species composition (see Chapter 7), salinity affects algae (see Chapter 8), and invertebrates (see Chapter 9) and their potential roles in wetland nutrient cycling. In addition, salinity has effects on dissolved nutrients within the system (Robarts and Waiser 1998). Waiser and Robarts (1997) suggest that salinity affects the molecular composition of DOC and thus can have significant impacts on C cycling in these systems. With the potential influence of salinity in many other wetlands, its impact on overall system function, and particularly nutrient cycling, requires further attention.

Nutrient Cycling in the Delta Marsh

The Delta Marsh is generally considered to be in the lake marsh stage of the wet–dry cycle (see Chapter 2); however, the long-term, stable water levels imposed on the marsh since the early 1960s have had different effects in different areas of the marsh. The deeper bays have progressed to the classic lake marsh stage described by van der Valk and Davis (1978a). These bays are now open-water areas fringed by a narrow band of emergent vegetation. The nutrient budgets of these open-bay areas would be as described for the lake marsh stage previously in this chapter.

In the shallow fringe areas of the marsh, the long-term flooding depths are less than the depth tolerance of the emergent vegetation (primarily *Typha* spp.), and therefore these areas have become overgrown with continuous unbroken stands of vegetation (deGeus 1987). All open-water areas have been eliminated as emergent vegetation has occupied all available habitats. Over time, these areas have developed large aboveground and below-ground biomass and associated nutrient pools. Algae populations (particularly epiphyton and metaphyton) in these densely vegetated areas of the Delta Marsh were low due to the dense crowding of the emergent vegetation and the resulting lack of light within the water column (Hosseini and van der Valk 1989a,b). Invertebrate densities were lower in the densely vegetated habitats of the reference cells than in the diked experimental cells following the baseline year in 1980 (Murkin and Ross 1999). It appears the nutrient budgets of the shallow, densely vegetated areas of the Delta Marsh are very simple and dominated by the macrophyte pools and associated decomposition fluxes. Because the stabilized water levels are within the tolerance range of the plants and there are no significant disturbance factors present (e.g., muskrat population levels are low in these areas of the Delta Marsh because the water is shallow enough that it freezes to the bottom each winter, thereby eliminating the muskrats), the plants continue to survive and over the long term have developed large pools of nutrients associated with living tissue and litter. Without some type of disturbance to kill the emergent vegetation (see Chapter 2), this simplified nutrient budget and cycling will continue in these shallow areas of the Delta Marsh.

ACKNOWLEDGMENTS

Ducks Unlimited Canada provided the resources to make completion of this entire volume possible, but they deserve special recognition for providing Henry Murkin with the time and resources required to revise the Part II chapters and complete this final chapter on nutrient budgets. We thank the Florida Center for Environmental Studies for providing office facilities during the initial drafting of this chapter. Suzanne Bayley, Bill Clark, and Richard Robarts reviewed earlier drafts and provided helpful input to this final product. This is Paper No. 99 of the Marsh Ecology Research Program, a joint project of Ducks Unlimited Canada and the Delta Waterfowl and Wetlands Research Station.

LITERATURE CITED

Barko, J.W., and R.M. Smart. 1980. Mobilization of sediment phosphorus by submersed freshwater macrophytes. Freshwater Biology 10:229-238.

Bonetto, C., F. Minzoni, and H.L. Golterman. 1988. The nitrogen cycle in shallow water sediment systems of rice fields: Part II - Fractionation and bio-availability of organic nitrogen compounds. Hydrobiologia 159:203-210.

Bowden, W.B. 1984. Nitrogen-15 isotope dilution study of ammonium production and consumption in a marsh sediment. Limnology and Oceanography 29:1004-1015.

Bowden, W.B., C.J. Vorosmarty, J.T. Morris, B.J. Peterson, J.E. Hobbie, P.A. Steudler, and B. Moore. 1991. Transport and processing of nitrogen in a tidal freshwater wetland. Water Resources Research 27:389-408.

Bridgham, S.D., and C.J. Richardson. 1992. Mechanisms controlling soil respiration (CO_2 and CH_4) in southern peatlands. Soil Biology and Biochemistry 24:1089-1099.

Chapin, F.S., R.J. Barsdate, and D. Barel. 1978. Phosphorus cycling in Alaskan coastal tundra: a hypothesis for the regulation of nutrient cycling. Oikos 31:189-199.

Chen, R.L., and J.W. Barko. 1988. Effects of freshwater macrophytes on sediment chemistry. Journal of Freshwater Ecology 4:279-289.

Cole, J.J., and M.L. Pace. 1995. Bacterial secondary production in oxic and anoxic freshwaters. Limnology and Oceanography 40:1019-1027.

Cook, A.H., and C.F. Powers. 1958. Early biochemical changes in the soils and waters of artificially created marshes in New York. New York Fish and Game Journal 5:9-65.

Curtis, P.J., and H.E. Adams. 1995. Dissolved organic matter quantity and quality from freshwater and saltwater lakes in east-central Alberta. Biogeochemistry 30:59-76.

Davelaar, D. 1993. Ecological significance of bacterial polyphosphate metabolism in sediments. Hydrobiologia 253:179-192.

Davis, C.B., and A.G. van der Valk. 1978. Decomposition of standing and fallen litter of *Typha glauca* and *Scirpus fluviatilis*. Canadian Journal of Botany 56:662-675.

Davis, S.M. 1991. Growth, decomposition, and nutrient retention of *Cladium jamaicense* Crantz and *Typha domingensis* Pers in the Florida everglades. Aquatic Botany 40:203-224.

deGeus, P.M. 1987. Vegetation changes in the Delta Marsh, Manitoba between 1948–1980. M.S. thesis, University of Manitoba, Winnipeg, Manitoba.

DeLaune, R.D., and W.H. Patrick, Jr. 1990. Nitrogen cycling in Louisiana gulf coast brackish marshes. Hydrobiologia 199:73-79.

DeLaune, R.D., C.W. Lindau, R.S. Knox, and C.J. Smith. 1990. Fate of nitrogen and phosphorus entering a Gulf Coast freshwater lake: a case study. Water Resources Bulletin 26:621-631.

DeLaune, R.D., C.J. Smith, and M.N. Sarafyan. 1986. Nitrogen cycling in a freshwater marsh of *Panicum hemitomon* on the deltaic plain of the Mississippi River. Journal of Ecology 74:249-256.

Devito, K.J., and P.J. Dillon. 1993. Importance of runoff and winter anoxia to the P and N dynamics of a beaver pond. Canadian Journal of Fisheries and Aquatic Sciences 50:2222-2234.

Dillon, P.J., and H.E. Evans. 1993. A comparison of phosphorus retention in lakes determined from mass balance and sediment core calculations. Water Research 27:659-668.

Eckerrot, A., and K. Pettersson. 1993. Pore water phosphorus and iron concentrations in a shallow, eutrophic lake - indications of bacterial regulation. Hydrobiologia 253:165-177.

Gachter, R., and J.S. Meyer. 1993. The role of microorganisms in mobilization and fixation of phosphorus in sediments. Hydrobiologia 253:103-121.

Goldsborough, L.G., and G.G.C. Robinson. 1996. Pattern in wetlands. In *Algal Ecology: Freshwater Benthic Ecosystems*. (Eds.) R.J. Stevenson, M.L. Bothwell, and R.L. Lowe, pp.77-117. New York: Academic Press.

Groffman, P.M. 1994. Denitrification in freshwater wetlands. Current Topics in Wetland Biogeochemistry 1:15-35.

Harter, R.D. 1966. Effect of water levels on soil chemistry and plant growth of the Magee Marsh Wildlife Area. Ohio Game Monographs No. 2. Ohio Division of Wildlife, Columbus, Ohio.

Hemond, H.F. 1983. Nitrogen budget of Thoreau's Bog. Ecology 64:99-109.

Hoge, B.E. 1994. Wetland ecology and paleoecology: relationships between biogeochemistry and preservable taxa. Current Topics in Wetland Biogeochemistry 1:48-67.

Hosseini, S.M., and A.G. van der Valk. 1989a. The impact of prolonged, above-normal flooding on metaphyton in a freshwater marsh. In *Freshwater Wetlands and Wildlife.* (Eds.) R.R. Sharitz and J.W. Gibbons, pp.317-324. USDOE Symposium Series 61, Oak Ridge: USDOE Office of Scientific and Technical Information.

Hosseini, S.M., and A.G. van der Valk. 1989b. Primary productivity and biomass of periphyton and phytoplankton in flooded freshwater marshes. In *Freshwater Wetlands and Wildlife.* (Eds.) R.R. Sharitz and J.W. Gibbons, pp. 303-315. USDOE Symposium Series 61, Oak Ridge: USDOE Office of Scientific and Technical Information.

Howard-Williams, C. 1985. Cycling and retention of nitrogen and phosphorus in wetlands: a theoretical and applied perspective. Freshwater Biology 15:391-431.

Jordan, T.E., D.F. Whigham, and D.L. Correll. 1989. The role of litter in nutrient cycling in a brackish tidal marsh. Ecology 70:1906-1915.

Kadlec, J.A. 1962. Effect of a drawdown on the ecology of a waterfowl impoundment. Ecology 43:267-281.

Kadlec, J.A. 1979. Nitrogen and phosphorus dynamics in inland freshwater wetlands. In *Waterfowl and Wetlands: An Integrated Review.* (Ed.) T.A. Bookhout, pp.17-41. Madison: The Wildlife Society.

Kadlec, J.A. 1989. Effects of deep flooding and drawdown on freshwater marsh sediments. In *Freshwater Wetlands and Wildlife.* (Eds.) R.R. Sharitz and J.W. Gibbons, pp.127-143. USDOE Symposium Series 61, Oak Ridge: USDOE Office of Scientific and Technical Information.

Kadlec, R.H., and J.A. Kadlec. 1979. Wetlands and water quality. In *Wetland Functions and Values: The State of Our Understanding.* (Eds.) P.E. Greeson, J.R. Clark, and J.E. Clark, pp.436-456. American Water Resources Association Technical Publication TPS 79-2.

Kadlec, R.H., and R.L. Knight. 1996. Treatment Wetlands. Boca Raton: Lewis Publishers.

Kerner, M. 1993. Coupling of microbial fermentation and respiration processes in an intertidal mudflat of the Elbe Estuary. Limnology and Oceanography 38:314-330.

Klopatek, J.M. 1978. Nutrient dynamics of freshwater riverine marshes and the role of emergent macrophytes. In *Freshwater Wetlands: Ecological Processes and Management Potential.* (Eds.) R.E. Good, D.F. Whigham, and R.L. Simpson, pp.195-216. New York: Academic Press.

Koerselman, W., S.A. Bakker, and M. Blom. 1990. Nitrogen, phosphorus and potassium budgets for two small fens surrounded by heavily fertilized pastures. Journal of Ecology 78:428-442.

LaBaugh, J.W., T.C. Winter, and D.O. Rosenberry. 1998. Hydrologic functions of prairie wetlands. Great Plains Research 8:17-37.

MacCrimmon, H.R. 1980. Nutrient and sediment retention in a temperate marsh ecosystem. Internationale Revue der Gesamten Hydrobiologie 65:719-744.

Mahapatra, I.C., and W.H. Patrick, Jr. 1969. Inorganic phosphate transformation in waterlogged soils. Soil Science 107:281-288.

Martinova, M.V. 1993. Nitrogen and phosphorus compounds in bottom sediments: mechanisms of accumulation, transformation and release. Hydrobiologia 252:1-22.

Mayer, L.M., and D.L. Rice. 1992. Early diagenesis of protein - a seasonal study. Limnology and Oceanography 37:280-295.

Mitsch, W.J., and J.G. Gosselink. 1993. Wetlands, 2nd edition. New York: Van Nostrand Reinhold.

Mitsch, W.J., J.K. Cronk, X.Y. Wu, R.W. Nairn, and D.L. Hey. 1995. Phosphorus retention in constructed freshwater riparian marshes. Ecological Applications 5:830-845.

Moore, P.A., K.R. Reddy, and D.A. Graetz. 1992. Nutrient transformations in sediments as influenced by oxygen supply. Journal of Environmental Quality 21:387-393.

Moran, M.A., and R.E. Hodson. 1990. Bacterial production on humic and nonhumic components of dissolved organic carbon. Limnology and Oceanography 35:1744-1756.

Moran, M.A., and R.E. Hodson. 1992. Contributions of 3 subsystems of a freshwater marsh to total bacterial secondary productivity. Microbial Ecology 24:161-170.

Murkin, H.R. 1989. The basis for food chains in prairie wetlands In. *Northern Prairie Wetlands*. (Ed.) A.G. van der Valk, pp.316-339. Ames: Iowa State University Press.

Murkin, H.R., and J.A. Kadlec. 1986. Responses by benthic macroinvertebrates to prolonged flooding of marsh habitat. Canadian Journal of Zoology 64:65-72.

Murkin, H.R., and L.C.M. Ross. 1999. Macroinvertebrate response to a simulated wet-dry cycle in the Delta Marsh, Manitoba. In *Invertebrates in Freshwater Wetlands of North America: Ecology and Management*. (Eds.) D.P. Batzer, R.B. Rader, and S.A. Wissinger, pp.543-569. New York: Van Nostrand Reinhold.

Murkin, H.R., A.G. van der Valk, and C.B. Davis. 1989. Decomposition of four dominant macrophytes in the Delta Marsh, Manitoba. Wildlife Society Bulletin 17:215-221.

Neely, R.K., and J.L. Baker. 1989. Nitrogen and phosphorus dynamics and the fate of agricultural runoff. In *Northern Prairie Wetlands*. (Ed.) A.G. van der Valk, pp.92-131. Ames: Iowa State University Press.

Neess, J.C. 1946. Development and status of pond fertilization in central Europe. Transactions of the American Fisheries Society 76:335-358.

Nelson, J.W., J.A. Kadlec, and H.R. Murkin. 1990. Seasonal comparisons of weight loss for 2 types of *Typha glauca* Godr leaf litter. Aquatic Botany 37:299-314.

Nichols, D.S. 1983. Capacity of natural wetlands to remove nutrients from wastewater. Journal of the Water Pollution Control Federation 55:495-505.

Niswander, S.F., and W.J. Mitsch. 1995. Functional analysis of a two-year-old created in-stream wetland: hydrology, phosphorus retention, and vegetation survival and growth. Wetlands 15:212-225.

Phipps, R.G., and W.G. Crumpton. 1994. Factors affecting nitrogen loss in experimental wetlands with different hydrologic loads. Ecological Engineering 3:399-408.

Prentki, R.T., T.D. Gustafson, and M.S. Adams. 1978. Nutrient movements in lakeshore marshes. In *Freshwater Wetlands: Ecological Processes and Management Potential*. (Eds.) R.E. Good, D.F. Whigham, and R.L. Simpson, pp.169-194. New York: Academic Press.

Reddy, K.R., and D.A. Graetz. 1988. Carbon and nitrogen dynamics in wetland soils. In *The Ecology and Management of Wetlands*. (Eds.) D.D. Hook, W.H. McKee, H.K. Smith, J. Gregory, V.G. Burrell, M.R. DeVoe, R.E. Sojka, S. Gilbert, R. Banks, L.H. Stolzy, C. Brooks, T.D. Matthews, and T.H. Shear, pp.307-318. Portland: Timber Press.

Richardson, C.J., D.L. Tilton, J.A. Kadlec, J.P.M. Chamie, and W.A. Wentz. 1978. Nutrient dynamics in northern wetland ecosystems. In *Freshwater Wetlands: Ecological Processes and Management Potential*. (Eds.) R.E. Good, D.F. Whigham, and R.L. Simpson, pp.217-241. New York: Academic Press.

Robarts, R.D., and M.J. Waiser. 1998. Effects of atmosphere change and agriculture on the biogeochemistry and microbial ecology of prairie wetlands. Great Plains Research 8:113-136.

Robarts, R.D., M.T. Arts, M.S. Evans, and M.J. Waiser. 1994. The coupling of heterotrophic bacterial and phytoplankton production in a hyper-eutrophic, shallow lake. Canadian Journal of Fisheries and Aquatic Sciences 51:2210-2226.

Robarts, R.D., D.B. Donald, and M.T. Arts. 1995. Phytoplankton primary production of three temporary northern prairie lakes. Canadian Journal of Fisheries and Aquatic Sciences 52:897-902.

Robinson, G.G.C., S.E. Gurney, and L.G. Goldsborough. 1997a. Response of benthic and planktonic algal biomass to experimental water-level manipulation in a prairie lakeshore wetland. Wetlands 17:167-181.

Robinson, G.G.C., S.E. Gurney, and L.G. Goldsborough. 1997b. The primary productivity of benthic and planktonic algae in a prairie wetland under controlled water-level regimes. Wetlands 17:182-194.

Sander, B.C., and J. Kalff. 1993. Factors controlling bacterial production in marine and freshwater sediments. Microbial Ecology 26:79-99.

Schallenberg, M., and J. Kalff. 1993. The ecology of sediment bacteria in lakes and comparisons with other aquatic ecosystems. Ecology 74:919-934.

Steward, K.K., and W.H. Ornes. 1975. Autecology of sawgrass in the Florida Everglades. Ecology 56:162-171.

Ulehlova, B., S. Husak, and J. Dvorak. 1973. Mineral cycles in reed stands of Nesyt Fishpond in southern Moravia. Polski Archiwum Hydrobiologii 20:121-129.

van der Valk, A.G. 1986. The impact of litter and annual plants on recruitment from the seed bank of a lacustrine wetland. Aquatic Botany 24:13-26.

van der Valk, A.G., and C.B. Davis. 1978a. Primary production of prairie glacial marshes. In *Freshwater Wetlands: Ecological Processes and Management Potential*. (Eds.) R.E. Good, D.F. Whigham, and R.L. Simpson, pp.21-37. New York: Academic Press.

van der Valk, A.G., and C.B. Davis. 1978b. Role of seed banks in the vegetation dynamics of prairie glacial marshes. Ecology 59:322-335.

van der Valk, A.G., C.B. Davis, J.L. Baker, and C.E. Beer. 1978. Natural fresh water wetlands as nitrogen and phosphorus traps for land runoff. In *Wetland Functions and Values: The State of Our Understanding*. (Eds.) P.E. Greeson, J.R. Clark, and J.E. Clark, pp.457-467. American Water Resources Association Technical Publication TPS79-2.

van der Valk, A.G., J.M. Rhymer, and H.R. Murkin. 1991. Flooding and the decomposition of litter of 4 emergent plant species in a prairie wetland. Wetlands 11:1-16.

Verhoeven, J.T.A., and H.H.M. Arts. 1987. Nutrient dynamics in small mesotrophic fens surrounded by cultivated land. II. N and P accumulation in plant biomass in relation to the release of inorganic N and P in the peat soil. Oecologia (Berl.) 72:557-561.

Voigts, D.K. 1976. Aquatic invertebrate abundance in relation to changing marsh vegetation. American Midland Naturalist 95:313-322.

Waiser, M.J., and R.D. Robarts. 1995. Microbial nutrient limitation in prairie saline lakes with high sulfate concentration. Limnology and Oceanography 40:566-574.

Waiser, M.J., and R.D. Robarts. 1997. Impacts of a herbicide and fertilizers on the microbial community of a saline lake. Canadian Journal of Fisheries and Aquatic Sciences 54:320-329.

Welling, C.H., R.L. Pederson, and A.G. van der Valk. 1988. Recruitment from the seed bank and the development of zonation of emergent vegetation during drawdown in a prairie marsh. Journal of Ecology 76:483-496.

Wetzel R.G. 1989. Wetland and littoral interfaces of lakes: productivity and nutrient regulation in the Lawrence Lake ecosystem. *In Freshwater Wetlands and Wildlife*. (Ed.) R.R. Sharitz and J.W. Gibbons, pp.283-302. USDOE Symposium Series 61, Oak Ridge: USDOE Office of Scientific and Technical Information.

Whalen, S.C., and J.C. Cornwell. 1985. Nitrogen, phosphorus, and organic carbon cycling in an arctic lake. Canadian Journal of Fisheries and Aquatic Sciences 42:797-808.

Whitman, W.R. 1974. Impoundments for waterfowl. Canadian Wildlife Service Occasional Paper 22.

Winter, T.C. 1989. Hydrologic studies of wetlands in the northern prairie. In *Northern Prairie Wetlands*. (Ed.) A.G. van der Valk, pp.16-54. Ames: Iowa State University Press.

Wrubleski, D.A., H.R. Murkin, A.G. van der Valk, and C.B. Davis. 1997. Decomposition of litter of three mudflat annual species in a northern prairie marsh during drawdown. Plant Ecology 129:141-148.

Part III

Prairie Wetland Ecology

7

Vegetation Dynamics and Models

Arnold G. van der Valk

ABSTRACT

Changes in the composition and net primary production of the Marsh Ecology Research Program (MERP) cells over the 10 years of the study are summarized with emphasis on changes in the abundance of the dominant emergent species. The mean standing crop of emergents was used to estimate their abundance in a cell from year to year. Prior to MERP, very little quantitative information about changes in the composition and primary production of prairie wetland vegetation due to water-level changes was available. Most of this chapter, however, deals with three questions: (1) How did the vegetation of the cells in cells flooded 30 and 60 cm higher than normal respond to an increase in mean water level during the reflooding years? (2) How did zonation patterns develop in the cells during the drawdown and reflooding years? and (3) What have been the contributions of MERP to the development of vegetation dynamics models for prairie wetlands?

INTRODUCTION

Cyclical changes in the abundance and composition of vegetation due to water-level changes were first documented in prairie potholes by Weller and Spatcher (1965) and Weller and Fredrickson (1974) in Iowa and by Millar (1973) in Saskatchewan. Walker (1959, 1965), working at the same time in the Delta Marsh, Manitoba, demonstrated that such changes were not confined to prairie potholes, and that they were also characteristic of large lacustrine marshes in the northern prairie region. Although such cycles were recognized as being highly variable, van der Valk and Davis (1978a) divided an idealized wet–dry cycle into four stages or phases: lake stage when high water has eliminated most or all of the emergents; dry marsh stage when there is no standing water in most or all of the marsh; regenerating stage when the marsh has reflooded and the emergent populations are spreading clonally; and degenerating stage when the emergent populations begin to decline (see Figure 1.3 in Chapter 1).

These early studies raised many questions about the nature of these cycles and how the vegetation of prairie wetlands responds to both increases and decreases in water levels. For example, why are emergent species eliminated during years with high-water levels? Is this due primarily to muskrat "eat-outs" when expanding muskrat populations eliminate most of the emergent vegetation through their feeding and lodge-building activities? Or,

125

is this primarily due to the emergent populations being eliminated directly by high water levels that they are not able to tolerate for prolonged periods? Nearly all of the early studies of vegetation cycles in prairie potholes were primarily concerned with how animal populations, primarily waterfowl, were affected. Consequently, other frequently asked questions concerned primary and secondary production and food chains in prairie potholes (Voigts 1976; Murkin 1989; Neill and Cornwell 1992). How is annual primary production of prairie wetlands affected by water-level changes? What is the most productive stage or phase of the cycle? All these and many more questions took on greater significance as concern about the fate of prairie wetlands due to global climatic change began to mount in the 1980s (Poiani and Johnson 1991). Concern about global climatic change also raised a new set of questions about the effects of permanent changes in the range of water-level fluctuations and duration of a wet–dry cycle on the flora and fauna and distribution of these wetlands.

The MERP vegetation studies were not designed to answer specific questions about how to better manage prairie wetlands for waterfowl or how to predict the effects of global climatic change on them. They were designed to improve our understanding of how the vegetation in prairie wetlands is affected by water-level changes. It was believed that improving our understanding of what controls the establishment, size, and extirpation of populations of wetland species would improve our ability to predict the effect of water-level changes on these wetlands, regardless of the reason for these changes. Although models of the vegetation dynamics of prairie wetlands were developed by the onset of MERP (Weller and Spatcher 1965; van der Valk and Davis 1978a; van der Valk 1981), another major objective of the MERP vegetation studies was to try to improve these models.

The MERP vegetation studies fall into three classes: (1) field studies of changes in the composition and primary production in the MERP cells; (2) ancillary studies, mostly experimental, on seed germination, water-depth tolerances, responses to nutrient additions, etc.; and (3) development of vegetation dynamics models. The first and last classes are emphasized (i.e., observed changes in the composition and primary production and modeling) in this chapter. The ancillary studies are invoked only as needed to help interpret the field and modeling results.

In this chapter, the major findings of the MERP field studies are presented. These MERP studies are then used to address three questions:

1. How does the vegetation of these wetlands respond to a long-term increase in mean water level?
2. How do zonation patterns develop in prairie wetlands? and
3. What advances have been made in developing wetland vegetation dynamics models?

Throughout this chapter, changes in the abundance and distribution of the four dominant emergents species (*Scirpus lacustris, Typha glauca, Scolochloa festucacea,* and *Phragmites australis*) are emphasized. These four species comprised most of the emergent vegetation of the MERP cells when the study began (Table 7.1) and remained the dominant emergents during the entire study. Information on the initial vegetation in the MERP

Table 7.1. Percentage of the area within an elevation range dominated by *Scirpus lacustris*. *Typha glauca, Scolochloa festucacea*, or *Phragmites australis* in the experimental cells in 1980

Elevation (m)	S. lacustris	T. glauca	S. festucacea	P. australis
248.0	0	0	0	46
247.9	0	0	1	66
247.8	0	2	9	60
247.7	0	13	26	44
247.6	2	34	24	38
247.5	20	54	5	18
247.4	45	43	0	7
247.3	42	24	0	1
247.2	19	6	0	1
247.1	3	0	0	0

Source: From van der Valk and Welling (1988).

cells can be found in Chapter 1 and in Welling et al. (1988a,b), van der Valk and Welling (1988), and van der Valk et al. (1989). Information on the Delta Marsh and its vegetation can be found in Chapter 2 and in Löve and Löve (1954), Walker (1959, 1965), Anderson and Jones (1976), de Geus (1987), and Waters and Shay (1990). For a review of vegetation studies in northern prairie wetlands prior to MERP, see Kantrud et al. (1989a,b). Information on the methods used to sample the vegetation during MERP can be found in Murkin and Murkin (1989) and Appendix 1.

CHANGES IN VEGETATION COMPOSITION AND PRIMARY PRODUCTION

Although previous studies of wet–dry cycles in prairie wetlands (Weller and Fredrickson 1974; Walker 1959, 1965) contain information about the abundance of species during all or most of a wet–dry cycle, MERP provided an opportunity to document in detail the response of each dominant species to changes in water levels during a simulated cycle. Several studies have described changes in the composition of the vegetation during some phases of MERP, especially van der Valk (1994), who dealt with the deep-flooding years; Welling et al. (1988a, b) and van der Valk and Welling (1988) and van der Valk et al. (1989), who dealt with the drawdown years; and van der Valk et al. (1994), who dealt with the reflooding years. In this section, changes in the composition and distribution of the vegetation over the simulated wet–dry cycle are summarized using vegetation maps (Figures 7.1–7.3) derived from low-level aerial photographs and MERP biomass data (van der Valk and Squires 1992; van der Valk 1994). Only data in the normal treatment during the reflooding years are included in this section. Data from the medium- and high-water years are presented in the next section on increasing water levels.

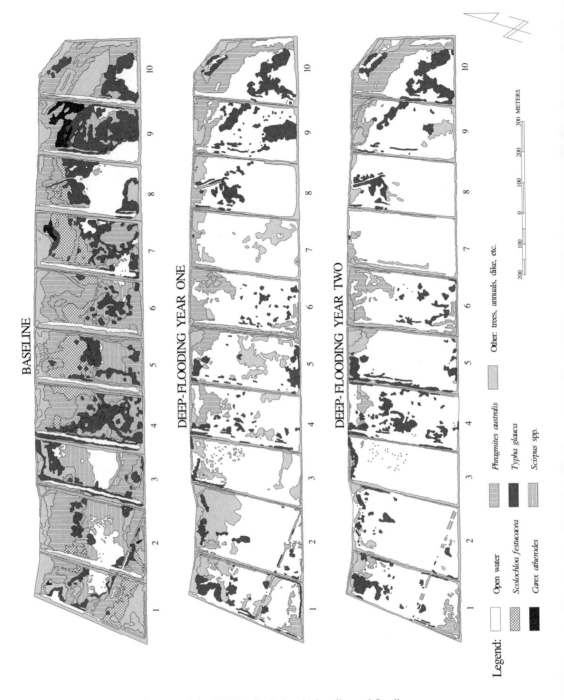

Figure 7.1 Vegetation map of the MERP cells during the baseline and flooding years.

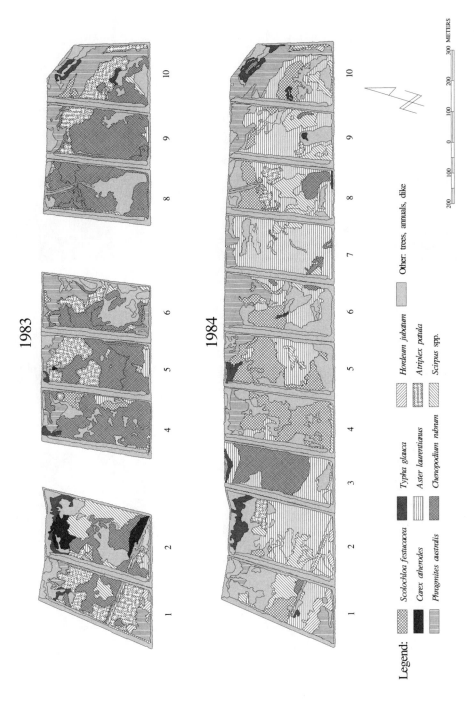

Figure 7.2 Vegetation map of the MERP cells during the drawdown years.

129

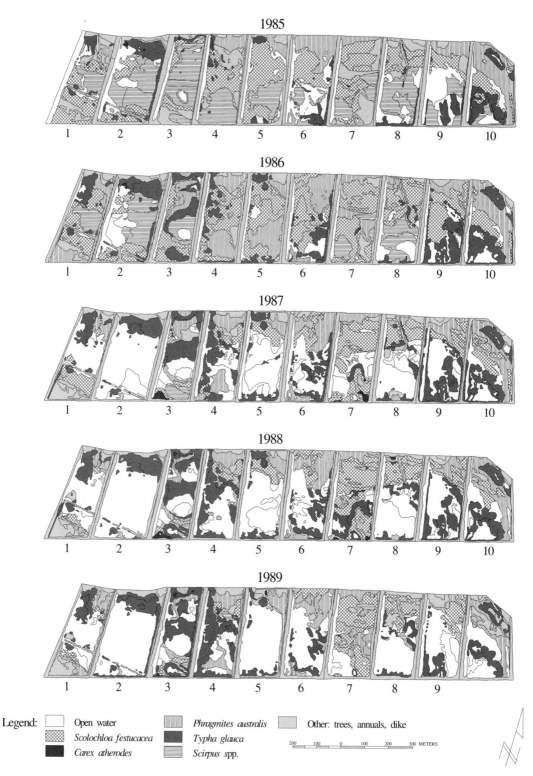

Figure 7.3   Vegetation map of the MERP cells during the reflooding years (1985–1989).

The most reliable and consistent measure of the abundance of emergent species and guilds of other species (annuals, submersed aquatics) over the entire 10 years of MERP were the aboveground biomass data. Aboveground biomass was estimated by harvesting quadrats in late July and early August each year and separating each sample into dominant emergent species as well as other species guilds. Although other measures of the abundance (cover, shoot density) of individual species also were collected (van der Valk 1994), these gave estimates of the fate of species and guilds that are virtually identical to the biomass data. Belowground biomass data, which was collected each spring (June) and each fall before the marsh froze over (September and early October), was not separated into species or species guilds. Nevertheless, belowground biomass data provide the best indicator of the overall state of the emergent vegetation in a cell because they provide a measure of the size of stored energy reserves and shoot meristems located in their rhizomes.

The lake stage of the wet–dry cycle was initiated by raising water levels in the cells by 1 m in the *Typha* zone for 2 years (Figure 7.1). As expected mean aboveground biomass (Table 7.2) for all four dominant emergent species declined significantly between the baseline (400 g/m^2) and first year of flooding (75 g/m^2), but was not significantly different between the first and second year (77 g/m^2) of flooding (van der Valk 1994). *Typha glauca* and *Phragmites australis* were the only emergents that had any appreciable biomass during the two flooding years. Their combined biomass, however, was reduced from 245 g/m^2 during the baseline years to 64 g/m^2 (85% of the total) in the first year to only 27 g/m^2 (35% of the total) in the second year of flooding. Combined *Scirpus lacustris* and *Scolochloa festucacea* biomass was reduced to only about 2 g/m^2 in the first year of flooding. Submersed vegetation had almost no biomass during the first, but increased significantly during the second year of flooding, accounting for about 33% of the total biomass during the second year. Mats of metaphyton were common during the second year of flooding and accounted for about 16% of the total biomass.

The mean total belowground biomass declined significantly between the baseline year and first year or flooding by about 20% (Table 7.2). Mean belowground biomass did not decline significantly from the first to the second year of flooding. After two years of flooding, i.e., by the spring of the first drawdown year, the mean total belowground biomass declined dramatically and was only 38 g/m^2 compared with 460 g/m^2 the previous fall. Most of the roots and rhizomes of *Phragmites australis, Scirpus lacustris,* and *Typha glauca* survived nearly 2 years of deep flooding while those of *Scolochloa festucacea* died during the first year of flooding. Although most of aboveground emergent vegetation in the MERP cells appeared to have been eliminated during the first year of the deep flooding, this was misleading. During the first and most of the second year of deep flooding, most emergent species in deep water were dormant, not dead. They persisted belowground. It took 2 years before their belowground carbon reserves were exhausted and they were extirpated from deep water areas.

In 1983, 8 cells were drawn down for 2 years and, in 1984, two additional cells were drawn down for 1 year (Figure 7.2). All four dominant emergents became re-established in the cells during the drawdown as did many mudflat annuals, particularly *Atriplex patu-*

Table 7.2. Mean aboveground maximum biomass (g/m²) for dominant emergent species and other macrophyte guilds and mean below-ground spring and fall biomass (g/m²) over the entire simulated wet–dry cycle during MERP

	Baseline year	Deep flooding		Drawdown		Reflooding				
		First	Second	First	Second	1985	1986	1987	1988	1989
Aboveground biomass										
Algae	0	0	12	0	0	8	1	2	1	0
Submersed aquatics	2	1	26	0	6	0	17	39	14	11
Scirpus lacustris	61	1	1	23	14	100	47	46	18	21
Typha glauca	160	45	17	5	6	7	18	87	74	160
Scolochloa festucacea	29	1	1	4	16	66	86	150	83	56
Phragmites australis	85	19	10	17	25	18	44	47	37	42
Annuals	11	0	1	200	160	11	3	2	1	2
Total	400	75	77	330	320	310	250	400	240	300
Belowground biomass										
Spring	840	680	510	38	110	130	200	360	320	420
Fall	870	570	460	100	240	250	390	450	670	490
Production	+30	–110	–50	+62	+130	120	190	90	370	70

Note: Data from all 10 cells are given for the baseline, deep-flooding and drawdown treatments. Only biomass data from the normal water-level treatment cells are given for the reflooding years.

la, Aster laurentianus and *Chenopodium rubrum.* Most of the recruitment of emergents during the drawdown occurred in the first couple of months. There were significant differences in recruitment between the cells drawn down in 1983 and 1984. *Typha glauca* and *Phragmites australis* seedlings were more common in 1984 than in 1983 as were seedlings of the annual *Aster laurentianus* (Welling et al. 1988b).

Total aboveground biomass ($\approx$325 g/m^2) during the drawdown years was only $\approx$20% lower than during the baseline year (Table 7.2). Annuals in both years contributed 50% or more of the total aboveground biomass. *P. australis* (5% of the total) and *S. lacustris* (7%) contributed most of the emergent biomass in the first year. During the second year, total emergent biomass increased 25% over that during the first year. The biomass of only three emergent species (*Typha, Scolochloa,* and *Phragmites*) actually increased in the second year of the drawdown, whereas that of *Scirpus* decreased by 40%. In the spring of the first year of the drawdown, total belowground biomass (Table 7.2) was $\approx$40 g/m^2. It increased to 110 g/m^2 by the fall. By the fall of the second year of the drawdown, belowground biomass had reached 240 g/m^2. Emergent species during the drawdown years were allocating most of their energy to roots and rhizomes. Belowground biomass at the end of the 2-year drawdown was still far below the fall belowground biomass (870 g/m^2) during the baseline period.

During the reflooding years, cells drawn down for 1 and 2 years did not behave differently (van der Valk 1994), and their biomass data have been combined in Table 7.2. Figure 7.3 shows changes in the vegetation during the reflooding years for all three water-level treatments. Changes in the aboveground biomass of dominant emergents during the reflooding years in the normal water-level treatment showed three different patterns (Table 7.2). *S. lacustris* biomass declined steadily all 5 years. *S. lacustris* actually is made up of two subspecies (Scoggan 1978–1979; Shay and Shay 1986): *S. lacustris* subsp. *validus* (= *S. validus* Wahl.) and *S. lacustis* subsp. *glaucus* (= *S. acutus* Muhl.). *S. lacustris* subsp. *validus* is intolerant of long-term flooding. Most of the *S. lacustris* present in the MERP cells in 1985 was *S. lacustris* subsp. *validus.* This subspecies began to decline when the cells were reflooded and was gone for all practical purposes after 1987, leaving only the much less common *S. lacustris* subsp. *glaucus,* which is tolerant of flooding and grows in deep water. *S. festucacea* and *P. australis* both had increasing biomass during the first 3 years and a decline in biomass in 1988 and 1989 over that in 1987. *T. glauca* biomass increased steadily, except for a small decline in 1988.

During the reflooding years (from 1985 to 1989), *Typha, Phragmites,* and *Scolochloa* spread clonally into areas with suitable water depths (Figure 7.3). In areas with water depths that they were not able to tolerate, they persisted only through 1987. This explains why the highest total aboveground emergent biomass occurred in 1987, and why there was a sharp drop in 1988 and subsequent increase in 1989. Total aboveground biomass in 1989, however, was still $\approx$40% lower than in the baseline period (Table 7.2).

Unlike aboveground biomass, belowground biomass increased fairly steadily over the 5 years of the reflooding period. This increase was primarily due to *T. glauca* populations with their massive rhizome systems continuing to expand all through this period, and this more than offset losses of belowground biomass due to the decline of *S. festucacea* and *S.*

lacustris. Belowground biomass in the spring of 1985 was ≈130 g/m² (Table 7.2). It reached 360 g/m² in the spring of 1987 and like aboveground biomass it dropped in the spring of 1988, but only slightly. By the spring of 1989, it was ≈420 g/m². Belowground biomass did not exceed aboveground biomass until 1987. Emergent populations were still spreading clonally in 1989, and their lower biomass in 1989 than in 1980 reflects that they did not cover all of available area within the cell and that their populations were not as dense as in 1980.

Although changes in the vegetation of the MERP cells during the experimentally simulated wet–dry cycle generally were similar to those recorded in prairie potholes and previously in the Delta Marsh, total aboveground macrophyte biomass did not vary as much over a cycle as estimated by van der Valk and Davis (1978b) for prairie potholes in Iowa. The lowest aboveground biomass, ≈75 g/m², was found during the second year of the deep flooding. The lake stage, which was being simulated during the high-water years, is expected to have the lowest macrophyte biomass. Maximum aboveground biomass, 400 g/m², occurred in 1987 during the reflooding years (regenerating marsh stage) as expected. Over the simulated wet–dry cycle, however, total aboveground biomass varied only about fivefold, not the 20-fold previously predicted by van der Valk and Davis (1978b). Emergent aboveground production, however, varied more than 11-fold.

Total aboveground and belowground biomass in the Delta Marsh are at the low end of the range reported for prairie wetlands (Murkin 1989). Why the primary production of the Delta Marsh is so low is unknown. There are several possible reasons: (1) short, cool growing season, (2) brackish water, and (3) low fertility. There is a well-known general pattern in the reduction of primary production from the tropics to the arctic. Nevertheless, this seems to be only partly responsible for the Delta Marsh's low production. Salinity is also implicated, especially in areas only seasonally or rarely flooded. Experimental studies by Neil (1993b) in the Delta Marsh have shown that soil salinity significantly reduces annual production of *S. festucacea*. Neil (1990b, 1992, 1993a) also has demonstrated that fertilizing *S. festucacea* and other species in the Delta Marsh can significantly increase their production and that there is an interaction between water depth and fertilizer application rates that is species dependent. In short, several interacting factors seem to be responsible for the low primary production of the Delta Marsh. Additional research is needed to determine the relative importance of each of these limiting factors.

The decline of emergent populations seen in the MERP cells during the third year of reflooding is consistent with an experimental study of the water-depth tolerance of emergent species and field observations. The five dominant emergent species in the Delta Marsh fall into two fairly distinct groups concerning their water-depth tolerances (Squires and van der Valk 1992): species whose peak growth was when flooded 45 cm (*S. lacustris* subsp. *glaucus* and *T. glauca*) and species whose peak growth was when flooded 20 cm (*S. festucacea*, *P. australis*, and *S. lacustris* subsp. *validus*). Unfortunately, this water-depth study ran for only 2 years, whereas it took 3 years for most species to die at water depths that they could not tolerate in the MERP cells (Table 7. 2). Although *S. lacustris* subsp. *validus* and *S. festucacea* survived 2 years of continuous deep flooding in some experimental treatments, their biomass was severely reduced when they were flooded to >45 cm (Squires and van der Valk 1992).

Other investigators have reported that *S. lacustris* subsp. *validus* and *S. festucacea* are intolerant of prolonged flooding (Walker 1965; Macaulay 1973; van der Valk and Davis 1980; Neckles 1984; Neckles et al. 1985; Neill 1990a). An experimental study by Meredino and Smith (1991) established that *S. festucacea* began to die after only 3 months when flooded to depths of 30 and 50 cm, but survived when flooded only 15 cm. In the same study, *S. lacustris* populations declined significantly or were extirpated when flooded 50 cm. For both species, the younger the plants, the greater the negative impacts of flooding. Neil (1993a) also reported that, in a pot experiment, the growth of *S. festucacea* from the Delta Marsh was adversely affected by water depths exceeding 15 cm. Shay and Shay (1986) and van der Valk and Davis (1978a) reported that *S. lacustris* subsp. *validus* clones live a maximum of 3 years when flooded.

Field studies also have reported that 3 years or more of flooding is necessary to eliminate other emergent species in deep water (Walker 1965; Kadlec 1962; Dabbs 1971; Millar 1973; van der Valk and Davis 1980; Sjöberg and Danell 1983; Neckles 1984; Bukata et al. 1988; Wallsten and Forsgren 1989). These field observations are consistent with the results of experimental studies of water-depth tolerance of emergent species (Thomas and Stewart 1969; Lieffers and Shay 1981; Yamasaki and Tange 1981; Stevenson and Lee 1987; Pip and Stepaniuk 1988; Grace 1989; Yamasaki 1990; Waters and Shay 1990). Unfortunately, most experimental studies were much too short, rarely lasting >1 year.

Before and during the MERP study, *Typha* populations were expanding in the Delta Marsh. An examination of aerial photographs of 41 selected areas in the Delta Marsh by de Geus (1987) indicated that *T. glauca* has been replacing *P. australis* stands and invading shallow-water areas since water levels in the Delta Marsh became regulated in the early 1960s. She hypothesized that, in the absence of water-level fluctuations, particularly periods of high water that would periodically destroy *Typha* stands, the Delta Marsh would eventually be taken over by *Typha*. If de Geus' hypothesis is correct, *T. glauca* should have become less common in the MERP cells in 1989 at the end of the study than it had been in 1980 at the beginning because water-level fluctuations had been reestablished. Our data do not support the de Geus hypothesis that reestablishing fluctuating water levels will set back *Typha* expansion. Even killing most of the *Typha* in the MERP cells during the deep-flooding years did not result in other emergent species being able to displace or replace *Typha* during the drawdown and regenerating phases of the cycle.

During the drawdown years, there were significant differences in recruitment of species from the seed bank due to differences in environmental conditions between the 2 years. Factors responsible for these differences are discussed in more detail in the section on vegetation dynamics models in this chapter.

INCREASING WATER LEVELS

The effect of a permanent increase in mean water level was investigated by flooding three MERP cells during the reflooding years to 30 cm (medium treatment) and another 3 to 60 cm (high treatment) above the normal water level of 247.5 m at which the remaining four cells were reflooded (Figure 7.3). Increasing the water level in a cell was hypothesized to hasten the onset of the degenerating marsh stage (van der Valk and Davis 1978a). Thus, emergent populations in the cells in the high treatment should begin to decline first, followed by those in the medium treatment. This decline should be reflected by a drop in the macrophyte biomass of the entire cell.

It also was hypothesized that the emergent populations that survived in the medium and high treatments would begin to migrate upslope in response to the increased water levels (de Swart et al. 1994). Thus, the mean water depth of an emergent species in the medium treatment in 1985 should be 30 cm greater than that in the normal treatment, but by 1989 it should be the same because this species should have migrated 30 cm upslope in the intervening years. Similarly, the mean water depth of individual emergent species in the high treatment should be the same as in the normal treatment by 1989.

Mean total aboveground biomass in the medium treatment (Table 7.3) reached its peak in 1987 at 400 g/m^2, which was ≈10% higher than it had been during the baseline period in the same cells. By 1989, however, it had dropped to only 240 g/m^2, which was 30% below baseline levels in the same cells. *P. australis* and *T. glauca* populations showed an increase in biomass from 1985 to 1987 and then a decline in 1988 and 1989. *S. lacustris*, which began to decline in 1986, was virtually absent in 1987 and in subsequent years. With the decline of emergent populations in 1987, submersed biomass increased significantly, but it was still never >25% of the total aboveground biomass. Belowground spring and fall biomass showed the same pattern as total aboveground biomass and peaked in 1987, but belowground biomass during this period never reached more than 50% of the belowground biomass during the baseline period.

Table 7.3. Mean aboveground maximum biomass (g/m^2) for dominant emergent species and macrophyte guilds and mean belowground spring and fall biomass (g/m^2) during the reflooding years and the baseline year 1980 in the MERP experimental cells in the MEDIUM (30 cm) water-level treatment

	Year					
	1985	1986	1987	1988	1989	Baseline
Aboveground biomass						
Algae	5	2	0	4	0	0
Submersed aquatics	0	37	58	47	61	0
Scirpus lacustris	120	52	3	0	0	55
Typha glauca	27	46	130	100	100	150
Scolochloa festuacea	110	110	87	42	33	29
Phragmites australis	12	17	85	36	23	66
Annuals	7	1	1	0	1	0
Total	340	290	400	240	240	360
Belowground biomass						
Spring	170	190	350	290	320	980
Fall	260	320	490	470	320	910
Production	+90	+130	+140	+180	+0	−70

In the high treatment, the general patterns of aboveground biomass for the dominant emergents and submersed aquatics from 1985 to 1989 were very similar to those in the medium treatment (Table 7.4). Mean total aboveground biomass peaked in 1987 at 440 g/m^2 and dropped steadily to 310 g/m^2 by 1989. Belowground biomass peaked in 1988 and then dropped significantly in 1989. Again, belowground biomass never reached the levels seen in the baseline period at any time. In 1989, belowground biomass was ≈65% below baseline levels.

From 1985 to 1989, ≈35 to 45% of the emergent vegetation in the medium and high treatments was eliminated in the MERP cells (van der Valk and Squires 1992). Although some emergent losses (≈5%) were due to herbivory by muskrats (Clark and Kroeker 1993; see Chapter 11), most was due to the death of two species (*S. festucacea* and *S. lacustris*). An examination of the frequency of occurrence of emergent species in the cells in the three water-level treatments indicates that each species behaved differently (Squires 1991). *S. lacustris* was essentially eliminated from all the cells in all three treatments after 1987 with only some small populations of the deep-water subspecies surviving in a few cells. *S. festucacea* populations were greatly reduced in the medium- and high-water treatments after 1987, but showed a much smaller decline in the normal treatment. It survived in parts of the cells not subject to permanent flooding. *P. australis* was most abundant in 1986 in both treatments and then declined slightly because it was eliminated from areas with water depths beyond its tolerance. At the same time but to a lesser extent, it was expanding simultaneously into other areas with shallower water depths. *T. glauca* behaved generally like *P. australis*.

Table 7.4. Mean aboveground maximum biomass (g/m^2) for dominant emergent species and major macrophyte guilds and mean belowground total spring and fall biomass (g/m^2) during the reflooding years and the baseline year 1980 in the MERP experimental cells in the HIGH (60 cm) water-level treatment

	Year					
	1985	1986	1987	1988	1989	Baseline
Aboveground biomass						
Algae	6	1	17	1	0	0
Submersed aquatics	7	30	41	43	66	0
Scirpus lacustris	120	27	0	3	3	49
Typha glauca	47	80	175	100	150	170
Scolochloa festucacea	87	84	37	25	1	33
Phragmites australis	75	85	160	96	76	120
Annuals	2	0	0	0	0	0
Total	380	320	440	280	310	460
Belowground biomass						
Spring	190	220	380	490	300	840
Fall	230	490	430	620	320	940
Production	+40	+270	+50	+130	+20	+100

An examination of the mean water depths at which *T. glauca*, *S. festucacea*, and *P. australis* occurred from 1985 to 1989 (Table 7.5) indicates these emergent species did not migrate upslope in the medium and high treatment. Mean water depths did not change significantly in the three water-level treatments from 1985 to 1989 for *T. glauca* or *P. australis*. They declined for *S. festucacea* in the high treatment, but only because this species was eliminated in deep water. *S. lacustris* also did not migrate upslope. It occurred too infrequently in most cells to get a reliable estimate of its mean water depth. Field observations indicate, however, that its populations were in the same locations in 1984, 1985, and 1986. It was extirpated after 1986 in many of the cells in the medium and high treatments, except for the deep-water form.

Raising water levels above the long-term mean did not have any detectable effect on total above- or belowground macrophyte biomass until after 1987 in all three treatments (i.e., after the third year of reflooding). Raising water levels eventually resulted in the loss of more emergent vegetation in the medium and high treatments than in the normal treatment. The decline in total aboveground and belowground biomass in the medium and high treatments began in 1988. Thus, it took just as long for the degenerating stage of the wet–dry cycle to begin in the medium as in the high treatments.

The degenerating stage begins with a decline in the emergent populations (van der Valk and Davis 1978a). The MERP results suggest the onset of the degenerating stage in prairie

Table 7.5.   Mean water depth (cm) of *Scolochloa festucacea*, *Phragmites australis*, and *Typha glauca* in the MERP normal, medium and high water-level treatments from 1985 to 1989. Negative numbers indicate elevations above the water line

	Year					
Treatment	1985	1986	1987	1988	1989	Mean
S. festucacea						
Normal	−0.02	0.01	−0.05	−0.08	−0.06	−0.04
Medium	0.15	0.23	0.08	0.04	0.06	0.11
High	0.49	0.45	0.20	0.09	0.19	0.33
Mean	0.18	0.21	0.05	0.00	0.02	0.10
P. australis						
Normal	−0.12	−0.05	0.01	−0.10	−0.11	−0.08
Medium	−0.22	0.02	−0.09	−0.12	−0.08	−0.10
High	0.22	0.06	0.23	0.20	0.22	0.18
Mean	−0.05	0.00	0.04	−0.01	0.00	−0.01
T. glauca						
Normal	0.13	0.10	0.07	0.07	0.10	0.09
Medium	0.25	0.34	0.27	0.25	0.25	0.27
High	0.44	0.57	0.52	0.51	0.42	0.49
Mean	0.27	0.31	0.26	0.26	0.24	0.27

wetlands does not begin until the fourth year of reflooding after a drawdown. By that time, short-lived emergent species have died out as well as subpopulations of other emergent species growing in water too deep for them to tolerate over the long term. Unless water levels continue to increase or herbivores or pathogens become a problem, the degenerating stage can be arrested indefinitely and will not automatically progress to the lake stage. This situation occurred in the shallow, peripheral areas of the Delta Marsh at the onset of MERP (see Chapters 2 and 6).

A comparison of the above- and belowground biomass in the normal, medium, and high treatments in 1989 (Tables 7.2, 7.3, and 7.8) revealed that in the medium and high treatments raising water levels reduced aboveground biomass 33 and 45%, respectively, in 1989 compared with the baseline year, whereas in the normal treatment biomass was down only 20%. Belowground biomass in the fall of 1989 was down 65% in the medium and high treatments from baseline levels compared with 45% in the normal treatment. The ratio of belowground to aboveground biomass in the baseline year 1980 was ≈2.2. In 1989, for the normal, medium, and high treatments, it was 1.6, 1.3, and 1.0, respectively. The lower belowground to aboveground biomass ratios in all the medium and high water-level treatments suggest that the emergent populations in these treatments were still under greater stress in 1989 than in the normal treatment. Studies by McKee et al. (1989), who examined the effects of flooding on emergents in the MERP cells when they were first flooded, and Squires and van der Valk (1992), who studied the water-depth tolerances of the dominant emergents in the Delta Marsh, suggest that belowground biomass is generally reduced relative to aboveground biomass when species are growing at suboptimal water depths.

The hypothesis that emergent species migrate upslope when water levels are raised permanently in a wetland was not supported by the data. In the MERP cells, water levels were raised quickly (i.e., within days or weeks). Raising water levels rapidly means that all emergent species were growing in deeper water than they had been growing during the previous years. This change can have an adverse effect on their growth and survival, especially if they were already growing at water depths approaching their depth limit. Consequently, emergent populations after a rapid increase in water level are often growing under stressed conditions and do not seem to have sufficient energy needed for clonal growth into more optimal water levels. Another complication is that adjacent, shallower water was probably already occupied by other emergent species. Most expansion of surviving emergent populations in the MERP cells in 1988 and 1989 occurred into areas where upslope emergent populations had been killed. In addition to these impediments to upslope migration, in most cases, emergent species cannot colonize flooded areas because of seed dispersal limitations or because their seeds cannot germinate underwater. In the short term, a rapid, permanent increase in water level results in the loss of emergent populations in deep water because they cannot migrate upslope. In the long term, establishment of emergent populations along the new shoreline from seed and their migration downslope may reestablish zonation patterns comparable to those that were present prior to the increase in water levels. This adjustment, however, could take many years.

Although it is often assumed that emergent species can migrate quickly upslope or downslope in response to permanent increases or decreases in water level (Bukata et al. 1988), this is not the case. In fact, lowering and raising water levels does not have the same consequences for emergent species. When mean water level is permanently lowered, emergents can tolerate low water levels for some period and recruitment of new populations from the seed bank at lower levels is possible (van der Valk and Davis 1980; van der Valk 1981). The establishment of emergents during the drawdown years in the MERP cells after they had been eliminated during the deep-flooding years testifies to their ability to become established when water levels are lowered. Even species without seeds in the seed bank should be able to adjust to a drop in water level by migrating downslope through clonal growth or seed dispersal onto newly exposed mudflats. Lowering water levels has been used for many years as a management tool to restore wet-meadow and emergent vegetation that has been eliminated by overgrazing, disease, or high water levels (Weller 1981). Parenthetically, although emergents are adversely affected by an increase in water level, submersed aquatics are not. Submersed aquatics can easily migrate into newly flooded areas because their seeds can germinate underwater. Submersed aquatics, however, are adversely affected by a decrease in water level.

DEVELOPMENT OF ZONATION PATTERNS

Bands or zones of vegetation (coenoclines) at a variety of scales are common features of most wetlands (Spence 1982), including prairie wetlands (Johnson et al. 1987). Wetland ecologists have spent a great deal of time describing and investigating these zones to determine why and how they develop (van der Valk and Davis 1976b; Spence 1982; Snow and Vince 1984; Keddy 1985; Bertness and Ellison 1987; Shipley and Keddy 1991). The MERP study provided a unique opportunity to examine how emergent vegetation zones developed in a prairie wetland during a drawdown and subsequent reflooding.

The various pathways (Figure 7.4) that lead to the establishment of vegetation zones (i.e., the differential distribution of dominant species along an elevation–water depth gradient in a wetland) were outlined in van der Valk and Welling (1988). They identified four major mechanisms: (1) the differential distribution of seeds along the elevation gradient; (2) the differential recruitment of species along the elevation gradient; (3) the differential survival of seedlings and adults along the elevation gradient during the drawdown; and (4) the differential survival of adults after reflooding along the elevation (now water depth) gradient. Several factors can affect both seedling survival and adult survival (e.g., disease, herbivory, competition, and extreme environmental conditions). What role did each of these four mechanisms play in the final distribution of species in the cells at the end of the reflooding period? Data from the cells in the MERP normal treatment will be used to examine how emergent zones developed in these MERP cells.

Seed dispersal patterns per se were not responsible for the development of vegetation zones. Windrowing of seeds at the drift line had resulted in most species having a more or less similar distribution initially along the elevation gradient (Table 7.6). The maximum densities of emergent seeds, regardless of primary dispersal syndrome, were at an elevation of ≈247.5 m (Pederson 1981, 1983; Pederson and van der Valk 1985). Nevertheless,

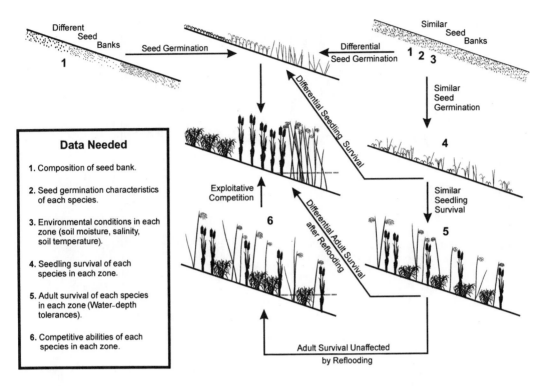

Figure 7.4   Potential pathways of emergent coenocline development in a prairie wetland (van der Valk and Welling 1988).

there were minor differences in the distribution of emergent seed that had some effect on the development of the new coenocline: (1) seeds of *S. lacustris* were more dense at lower elevations and absent at higher elevations; (2) seeds of *S. festucacea* were mostly restricted to elevations above the drift line; and (3) seeds of *P. australis* had a bimodal distribution with one peak at the drift line and a second peak at higher elevations, where this species was most widespread. In other prairie wetlands with normally fluctuating water levels, seeds of all species tend to be more evenly dispersed across elevation gradients (van der Valk and Davis 1976a, 1978a; Poiani and Johnson 1989).

All emergent species had higher seed germination percentages along some sections of the elevation gradient than others (Table 7.6). *T. glauca* and *S. lacustris* seedling densities were highest at slightly lower elevations than that at which their seeds were most abundant. Differential seed germination shifted the distribution of these species down the elevation gradient. Differential seed germination also resulted in *S. festucacea* populations having their maximum seedling density at a higher elevation than that of its seeds. Only *P. australis* seeds at the lower of its two seed-density peaks germinated, shifting its distribution down the elevation gradient significantly.

Seed germination along the elevation gradient influenced the development of the new coenocline for all species. Its most profound impact was on the distribution of *P. australis*

Table 7.6. Mean seed density in the seed bank in 1979 and shoot density (per 10 m^2) in permanent quadrats during the first (1983) and second (1984) years of a drawdown of the dominant emergent species at different elevations in the 8 MERP cells drawdown for 2 years.

Elevation (m)	Scirpus lacustris			Typha glauca			Scolochloa festucacea			Phragmites australis		
		Shoot			Shoot			Shoot			Shoot	
	Seed	1983	1984	Seed	1983	1984	Seed	1983	1984	Seed	1983	1984
248.0	0	0	0	720	0	0	170	70	70	250	0	0
247.9	0	0	30	490	0	0	70	13	12	260	0	0
247.8	600	800	0	1,500	0	0	100	1,400	50	120	8	0
247.7	1,500	1,600	20	1,100	40	4	270	3,400	450	100	5	30
247.6	5,300	150	70	3,100	8	10	510	1,700	400	140	8	1
247.5	21,000	2,000	190	3,600	160	30	690	2,300	400	250	15	4
247.4	8,000	16,000	1,100	6,100	120	40	200	230	330	80	20	90
247.3	3,600	3,100	1,300	1,400	1,300	70	0	0	6	30	10	20
247.2	2,900	500	2,300	180	70	370	0	0	50	10	0	5
247.1	1,300	20	320	320	0	25	0	0	5	40	0	0
Mean	4,400	2,400	530	1,900	170	55	200	910	180	130	7	15

Source: Adapted from Welling et al. (1988b).

142

seedlings. *P. australis* was eliminated at higher elevations, where it had formerly been abundant, because its seeds failed to germinate at these elevations. Although it is not clear from the field data which environmental factors are responsible for the differences in germination observed for a given species along the elevation gradient, subsequent experimental studies indicated that it was primarily a combination of soil moisture, temperature, and salinity (Galinato 1985; Galinato and van der Valk 1986; van der Valk and Pederson 1989) and, locally, litter accumulation, which restricted seed germination (van der Valk 1986; Welling 1987).

Significant mortality during the drawdown (Table 7.6) did occur, but it had little impact on the overall distribution pattern of *S. festucacea* or *P. australis*. The distributions of *S. lacustris* and *T. glauca* were shifted down the elevation gradient. Declining soil moisture at the surface of the marsh seems to have been responsible for most of the seedling mortality observed (Welling et al. 1988b).

After the reflooding of the cells following the drawdown, differential survival resulted in changes in the mean elevation at which the maximum densities of some species occurred in the next 5 years (Table 7.7). In 1986, *S. lacustris* had its mean elevation at a higher level in the normal treatment. In contrast, the mean elevation of maximum shoot density for *P. australis* and *S. festucacea* was not affected and did not differ in 1984 and 1989. *T. glauca* shoot density maxima occurred both at higher and lower elevations. Overall, the reflooding of the cells had only a minor effect on the distribution of emergent species along the elevation gradient.

During coenoclinal development, different combinations of factors influence the final position of a species along an environmental gradient, and no one factor (e.g., dispersal, recruitment, or competition) is responsible for the final distribution of all species. Although the distribution of three of the species in 1989 was centered at or about the same elevations that it had been in 1980, each species reached its 1989 distribution due to a different combination of factors. For example, differential seed germination is primarily responsible for the distribution of *P. australis* along the new coenocline, whereas a combination of seed dispersal, differential seed germination, and seedling and adult mortality is responsible for the position of *S. festucacea*.

Because environmental conditions can differ from year to year and place to place, seed germination, seedling survival, and adult survival patterns that collectively determine the position of a species along an environmental gradient can result in a species being found at noncongruent positions along different coenoclines. Thus, the effects of various environmental filters that control the establishment, growth, and mortality of species along a gradient are not constant. This implies that a given species will not necessarily be found along exactly the same portion of the water-depth gradient in different wetlands or in different places within the same wetland. Unpublished transect data from the Delta Marsh do show that the elevation range over which a species is found varies from place to place. Whether subsequent interactions among the species (i.e., competition) eventually will decrease or eliminate these establishment differences is also unknown. This possibility is implied in the competition literature (Grace and Wetzel 1981), but the significance of competition among emergent species in determining their position along water-depth gradients is still not fully understood (Shipley and Keddy 1991).

Table 7.7.   Mean shoot density per square meter in permanent quadrats in the MERP normal treatment cells from 1984 to 1989 of four dominant emergent species at different elevations

Elevation (m)	n	Drawn down 1984	Flooded				
			1985	1986	1987	1988	1989
Scirpus lacustris							
247.7	4	8	0	0.5	0	0.5	0.5
247.6	7	99	0	0	0	0	0
247.5	6	106	5	8	16	14	23
247.4	12	461	67	53	31	25	30
247.3	4	106	41	41	28	4	0
247.2	2	733	71	76	25	6	4
247.1	3	43	16	21	11	0.3	2
Typha glauca							
247.7	4	5	0	0.3	1	2	2
247.6	7	15	0	0	6	6	6
247.5	6	498	0.8	2	17	20	30
247.4	12	444	2	3	5	11	16
247.3	4	6	5	16	26	47	40
247.2	2	42	6	11	18	22	34
247.1	3	2	0	0	0	0	0
Scolochloa festucacea							
247.8	1	1	1	3	19	63	31
247.7	4	19	11	37	140	242	309
247.6	7	169	51	109	167	388	277
247.5	6	79	96	35	35	22	6
247.4	12	69	62	30	15	2	0.3
247.3	4	0.5	5	5	2	0	0
247.2	2	0	5	0	0.5	0	0
247.1	3	0.3	6	15	1	0	0
Phragmites australis							
247.7	4	0	0	0	3	9	9
247.6	7	0	0	0.1	0.9	0.4	1
247.5	6	33	16	17	13	32	19
247.4	12	36	20	18	13	32	16
247.3	4	0.3	0.5	2	4	6	2
247.2	2	0	5	6	4	1	1

Source: Adapted from Squires (1991).
Note: n = number of permanent quadrats.

VEGETATION DYNAMICS MODELS

Several models have been developed to predict the composition and distribution of vegetation in prairie wetlands during different stages of the wet–dry cycle. They vary from the descriptive models of Weller and Spatcher (1965) to the sophisticated computer simulation models of Poiani and Johnson (1993) and Seabloom (1997) that predict the future status of the vegetation due to changes in environmental conditions, primarily water-level changes. In this section, the life-history model of van der Valk (1981) and two newer models (logistic regression and spatially explicit) that have been used to predict the composition of the vegetation in MERP cells at different water levels are examined and evaluated.

Life-History Model

The life-history model proposed by van der Valk (1981) can be used to predict when species will become established and when they will be extirpated during wet–dry cycles. The model is based on two fundamental concepts: environmental filters and critical life-history traits of wetland plant species. An environmental filter (e.g., water level) determines which species becomes established, persist, or are extirpated at any given time. Critical life-history traits are characteristics of species that determine when they become established, grow, and die. Kautsky (1988) and Boutin and Keddy (1993) also developed comparable functional classifications of aquatic and wetland species. Shipley et al. (1989) did a comparative study of the characteristics of seeds and seedlings and adults of emergent macrophytes in relation to habitat conditions. Stockey and Hunt (1994), Weiher and Keddy (1995), and Seabloom et al. (1998) carried out experimental studies to examine how environmental filters and ecological traits of species combine to determine the composition of wetland communities.

In the life-history model (van der Valk 1981), only three life-history features were used: expected or effective life span, propagule longevity, and propagule establishment requirements. The three life-span types recognized were annuals (A-species), perennials with a fixed life span (P), and perennials with an indefinite life span (V). The two seed-dispersal types were species with long-lived seed (S-species) whose seeds are found in the seed bank at all times, and species with short-lived seeds whose reestablishment is dependent on seed dispersal (D). The two propagule-establishment types were species whose seeds can germinate when there is no standing water, designated type-I species, and species whose seed can germinate under water, type-II species. In the model, all species in a wetland are classified into one of 12 life-history types, each type has a specific set of environmental conditions under which species of that type can become established and extirpated.

The life-history model ignored the distribution of species within the wetland. To add a spatial dimension to the model, the water-depth tolerances of each species were added as a critical life-history feature (van der Valk 1991). Four water-depth types are recognized: annuals, wet-meadow species, emergents, and submersed aquatics. Mudflat species do not tolerate flooding and are only found on exposed sediments during drawdowns. Wet-meadow species are perennials that can tolerate short periods of standing water of a year or so, but not long-term inundation (Neckles 1984; Neckles et al. 1985). Emergent species are perennials that can tolerate permanent inundation and also a year or more free of standing water (van der Valk and Davis 1980). Submersed aquatic species are normally found in standing water, but they can survive short drawdowns. These four water-depth tolerance types are designated as type-M for mudflat, type-W for wet meadow, E for emergent, and S for submersed. In the extended model, a wet-meadow species, VS-W-I, is distinguishable from an emergent species, VS-E-I on the basis of their water-depth tolerances, and the wet-meadow species can be found growing at higher elevations in the wetland than emergent species.

In prairie wetlands nearly all of the recruitment of emergent species that occurs during drawdowns is from seed banks (van der Valk and Davis 1976a, 1978a). The life-history model (van der Valk 1981) assumes that mean seedling densities of species during a drawdown are the same as their mean seed densities in the seed bank under drawdown conditions (van der Valk et al. 1989). Seed-bank data from two contiguous cells (Table 7.8) were used to predict the composition of the vegetation during the drawdown years in these cells. A comparison of the predicted mean densities of species and their actual densities in 1983 indicates that mean emergent density was overestimated by ≈300%, whereas the mean density of annuals was underestimated by >90%. The elevations at which species were expected to be found in the cells, however, were predicted accurately (Table 7.6). Attempts to apply the life-history model to other wetlands have also indicated that the model has significant shortcomings (Smith and Kadlec 1983, 1985).

Several studies have been conducted to investigate the reason for the discrepancy between the life-history model predictions and actual seedling densities in the field. Possible sources of error fall into two classes: (1) errors associated with the determination of the composition of the seed bank; and (2) errors caused by environmental factors (temperature, salinity, soil moisture) not considered in the model. Seedling assays were used to estimate the density of viable seed in the MERP cells (Pederson 1983). This technique involves collecting a surface sediment sample and placing it in trays in a greenhouse under both flooded and drawdown conditions and counting the number of seedlings (van der Valk and Davis 1978a; van der Valk and Pederson 1989; van der Valk et al. 1992). This approach always gives biased results because the samples are exposed to only a limited set of environmental conditions and not all the viable seeds may germinate. As part of a seed-bank study of a North Dakota prairie pothole, Poiani and Johnson (1988) compared the seedling assay technique for estimating seed densities to the seed assay techniques in which seeds are actually removed mechanically from the soil by sifting. Their results suggest that, although the two methods do not give identical results, they are comparable. Another similar comparison of seed-bank assay techniques by van der Valk and Rosburg (1997) showed similar results. In short, the differences between predicted and actual seedling densities (Table 7.8) do not seem to be due to the method used to estimate seed densities.

Galinato and van der Valk (1986) investigated the effects of temperature and salinity on seed germination of emergents and annuals recruited during drawdowns in the Delta

Table 7.8. Predicted and actual mean seedling densities (per 10 m^2) during the drawdown in 1983 in two MERP experimental cells

Species	Predicted	Actual
Emergents		
Scirpus lacustris	2,500	730
Typha glauca	4,200	780
Scolochloa festucacea	170	64
Phragmites australis	42	5
Scirpus maritimus	25	8
Carax atherodes	75	770
Mean	1,200	390
Annuals		
Atriplex patula	33	7,000
Aster laurentius	25	47
Chenopodium rubrum	83	360
Ranunculus sceleratus	250	27
Rumex maritimus	330	28
Mean	140	1,500
Wet meadow perennials		
Lycopus asper	75	1
Mentha arvensis	83	2
Stachys palustris/Teucrium canadense	33	72
Sonchus arvensis	17	15
Cirsium arvense	49	13
Urtica dioica	17	59
Mean	46	27

Source: From van der Valk and Pederson (1989).

Marsh. They found that the germination of many emergents and annual species is affected by both temperature and salinity. Other studies of seed germination of wetland plants showed that soil moisture also can significantly affect seed germination during drawdowns (Naim 1987; van der Valk and Pederson 1989).

A study of recruitment of species from a seed bank in the Delta Marsh under a variety of soil moisture and salinity conditions (Table 7.9) indicates that both soil moisture and salinity can have significant effects on seed germination. Annual seed germination is inhibited by flooded conditions or very wet soils, whereas emergent germination is inhibited by dry soils. Seabloom et al. (1998) conducted a comparable study with similar results. They exposed a composite seed bank from Iowa prairie potholes to several water depth, soil moisture, and temperature treatments and found that both soil moisture and temperature significantly affected species recruitment. Meredino et al. (1990) and Meredino and Smith (1991) experimentally studied the effect of the date of a drawdown on recruitment from the seed bank in the Delta Marsh. Their results indicate that recruitment of seedlings from the seed bank is significantly affected by time of year. Seed-bank samples drawn down in May had 600% more seedling than those drawn down in August.

In short, the results of studies of seed germination and seedling recruitment from seed banks indicate that environmental conditions have a major effect on recruitment patterns from seed banks. Consequently, the discrepancy between field seedling densities and predicted densities from seed-bank assays (Table 7.8) is due to soil moisture, temperature, and salinity effects on seed germination that are not taken into account in the life-history model.

Logistic Regression Models

A set of logistic regression models for the dominant emergent species in the Delta Marsh was developed by de Swart et al. (1994) to predict their distribution in the three water-level treatments during the reflooding years. These models are patterned after those developed by Barendregt et al. (1986, 1993). Unlike the life-history model, the logistic regression models are not based directly on morphological, reproductive, or life-span characteristics of individual species. They are instead based on correlations between the occurrence of species and environmental factors (salinity, soil texture, nutrient levels, water depth,

Table 7.9. Mean number of seedlings (per tray) of species recruited from a composite seed bank under five soil-moisture regimes and watered with either fresh or brackish water

Species	Water chemistry	Soil Moisture Regime					Mean
		FL	SA	W1	W2	W3	
Mudflat annuals							
Aster	F	24	82	115	91	16	66
laurentius	B	0.4	50	97	66	17	46[a]
Atriplex	F	0	7	8	5	5	6
patula	B	0	5	14	4	6	5
Chenopodium	F	0	122	95	68	52	78
rubrum	B	0	105	181	55	54	67
Wet-meadow species							
Carex	F	0	2	7	4	3	3
atherodes	B	0	1	1	4	2	2[a]
Hordeum	F	0	1	3	4	2	2
jubatum	B	0	1	2	5	1	2
Scripus	F	1	6	2	1	0.3	2
maritimus	B	1	6	3	1	1	2
Emergents							
Scirpus	F	20	20	14	8	0.2	12
lacustris	B	22	18	14	5	0.1	11
Typha	F	153	189	102	20	0	91
glauca	B	72	78	71	6	0.1	44[a]

Note: The soil moisture regimes were flooded (FL), saturated soil (SA), watered every day (W1), watered every other day (W2), and watered every third day (W3). Fresh water (F) had a conductivity of <400 µmhos, whereas brackish water (B) had a conductivity of >2,500 µmhos. For all species there was a significant ($P < 0.05$) soil moisture treatment effect.

[a] Significant ($P < 0.05$) water-chemistry effect.

etc.) suggested to determine their distribution in the field. By collecting a large number of vegetation samples and associated environmental measurements, a logistic regression model can be developed that gives the probability of finding a species under various environmental conditions.

For the MERP cells, most of the factors that control the distribution of emergents were assumed to be uniform (e.g, climate, soil texture [Wilson and Keddy 1985], nutrient levels, wave action [Keddy 1983], salinity, etc.) or to be of no consequence (herbivory [Jefferies et al. 1979], competition [Buttery and Lambert 1965; Grace and Wetzel 1981; Gaudet and Keddy 1995], pathogens, etc.). Water depth seemed to be the only environmental factor determining the distribution of the emergent species. In fact, no observable factors such as wave action or pathogens affected the distribution of emergents during the course of this study. There was some muskrat herbivory, but its effects are thought to have been inconsequential (Clark and Kroeker 1993; see Chapter 11).

Univariate logistic regression techniques were used to estimate the probabilities of occurrence of the dominant emergent species in the MERP cells at various water depths under steady-state conditions. This was done using the 1980 vegetation and topographic data from all 10 cells. The 1980 vegetation in the Delta Marsh was assumed to be in a steady-state condition because mean water levels in Lake Manitoba and the Delta Marsh had been controlled for nearly 20 years (see Chapter 2). Like an earlier model by Bukata et al. (1988), these univariate logistic regression models use only basin morphometry (i.e., water depth) to determine where species will be found after a long-term change in mean water level. To test this hypothesis, the actual frequencies of occurrence of emergent species in selected elevation ranges in the three water-level treatments from 1985 to 1989 were compared to their expected probabilities of occurrence in these elevation ranges to see if they converged toward their predicted probabilities of occurrence over time (i.e., by 1989).

In Figure 7.5, the 1980 logistic regression models of *P. australis*, *Typha* spp., *S. lacustris*, and *S. festucacea* are plotted as well as the data used to calculate them. The actual distributions of emergent species in the medium and high treatments were not predicted by the logistic regression models, and there was only limited evidence that predicted and actual distributions were converging by 1989. Even the predicted probabilities of finding *T. glauca* or *P. australis* in various elevation ranges in the normal treatments in 1989 were not comparable to their actual frequencies of occurrence. There are several possible reasons for the discrepancy between predicted and actual distributions. First, there may be very long lag times (>5 years) in the response of emergent species to changes in water levels. Second, the distribution of emergent species in a wetland may not be primarily a function of one environmental variable, water depth. Third, some species may be unable to adjust to changes in water level of the magnitude used in this study.

As noted in the section on increasing water levels, there is typically a 3-year lag time in the response of emergents to a new water level (Millar 1973; van der Valk and Davis 1980; van der Valk and Squires 1992; van der Valk et al. 1994) in northern prairie wetlands. Thus, predicted probabilities and actual frequencies of species are not expected to be similar until 1988 and 1989. There was some convergence toward the predicted distributions

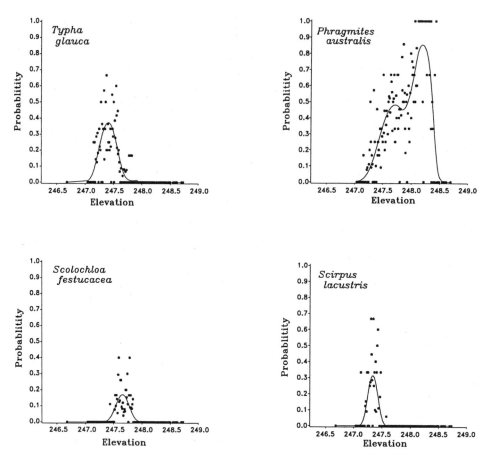

Figure 7.5   Logistic regression models of the four dominant emergents constructed using 1980 baseline data to predict the probability of occurrence of the four dominant emergents along elevation gradients in the MERP cells (de Swart et al. 1994).

for *P. australis* and *T. glauca* by 1988 and 1989, but it was far from complete. Sixty percent of the time the predicted probabilities and actual frequencies were significantly different in 1988 and 1989 for *P. australis*. For *T. glauca*, it was 85%. As noted, neither *P. australis* nor *T. glauca* migrated upslope as predicted in the medium and high treatments from 1985 to 1989. The logistic regression models assume that upslope migration occurs and that the emergent species eventually will be found over the same depth range in reflooding water-level treatments that they were during the baseline years.

Even though it was possible to model the distribution of all four emergent species using only water-depth data, the maximum probability of occurrence of a species in some of the models was low (Figure 7.5). For example, for *S. festucacea*, the maximum probability of occurrence was only 0.157. This suggests that water depth alone may not be sufficient to predict the distribution of this species and implies that some other factor or factors affect its distribution. This could be an environmental factor (soil salinity, soil fertility, soil texture, etc.), biological factor (plant pathogens, competition, allelopathy, etc.), or distur-bance (grazing, fire, ice scour, etc.). Spence (1982), in a review of factors responsible for zonation patterns in lakes, cites numerous situations in which wetland zonation is con-trolled by several factors simultaneously. Even within a wetland, as the analysis of the development of zonation patterns described previously in this chapter revealed, factors controlling the distribution of emergent species may vary from species to species and pos-sibly for a species from place to place. There is also unequivocal evidence that environ-mental conditions during the drawdown years affected initial recruitment of species in the MERP cells. It seems likely that these initial recruitment patterns had not yet been com-pletely obliterated by 1989, and this makes the use of univariate models, which assume steady-state or equilibrium conditions, inappropriate.

In summary, the hypothesis that the frequencies of occurrence of species in selected ele-vation ranges would increasingly become similar to those predicted by logistic regression models by 1989 was not born out by the MERP data (de Swart et al. 1994). This actually occurred <40% of the time for any species. These models were not adequate to predict the response of emergents to an increase in water level of either 30 or 60 cm. Emergent species did not migrate upslope in the medium and high treatments as predicted. Emergent species either died out (*S. festucacea* and *S. lacustris*) after 3 years because they could not survive permanent flooding, stayed where they were (*P. australis*) because they were unable to move upslope through clonal growth, or became more widespread (*T. glauca*) only because of the expansion of small local populations already established in 1985 in areas dominated formerly by other species. Because the underlying assumption that species migrate upslope when water level is raised 30 cm or more is not valid, realized niche models, like the logistic regression models that assume migration does occur, can-not reliably predict the response of wetland emergent species to an increase in water level of 30 cm or more, at least in the short term.

Spatial Models

The first spatially explicit model of vegetation dynamics in prairie wetlands was devel-oped by Poiani and Johnson (1993) by using a geographical information system (GIS) for a pothole in North Dakota. In their model, the wetland was covered by a grid of cells and in each cell a vegetation submodel, which determines the responses of six classes of prairie wetland species to water-level changes, is linked to a hydrology submodel that esti-mates water depths from basin morphometry, precipitation, and evapotranspiration data for that cell. Although it does not deal with individual species, but with vegetation class-es (e.g., deep-marsh emergents), the vegetation submodel in their computer simulation model has many features in common with the van der Valk (1981) life-history model. The Poiani and Johnson (1991, 1993) model successfully simulated changes in vegetation cover over a wet–dry cycle. A comparable computer simulation model also was developed by Ellison and Bedford (1995) that was used to model the impact of hydrologic changes on the vegetation of a sedge meadow–shallow marsh complex in Wisconsin.

A spatial model of the vegetation of the MERP cells (Figure 7.6) was developed by

Seabloom (1997). As in the Poiani and Johnson model, each cell in the MERP complex was covered with a grid. This is a cellular automaton model in which adult and seed distributions and a set of rule-based functions determine the fate of each emergent species and the annual guild in each cell in the grid. To avoid circularity, rules for each species were developed from non-MERP studies whenever possible.

In the Seabloom model, the state of a cell at any given time is a function of its past history as well as the history of adjacent cells in the grid. Species can colonize a cell from the seed bank or through clonal growth from an adjacent cell. Only one emergent species, however, can persist in a cell. A species can come to dominate a cell for one of three reasons: (1) it was the dominant species at the initiation of the model run; (2) it colonized an empty cell; or (3) it germinated from the seed bank and outcompeted seedlings of all other species. Species can be eliminated from a cell because they reach the end of their life span or because of water-depth effects. Water depth in a cell can affect seed germination and the growth of adult plants. The latter is implemented using a Gaussian water-depth response curve. The means and standard deviations for the emergent species' water depth response curves were derived from the depth distributions of the emergent species in the 1980 baseline year in the MERP cells. The 1980 distribution of species in the cells also was used to initiate the model. The initial seed bank for each species was simulated by randomly distributing seeds along the water-depth gradient. Water depths in the model were adjusted for the water level in the MERP cell being modeled.

The results of the spatial model are vegetation maps that show the predicted distribution of each emergent species and the annual guild. The predicted and actual distributions of species in each cell during each year of the MERP study were then compared using both

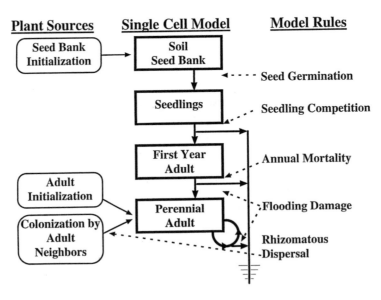

Figure 7.6 Model structure for a single cell in the spatial model. From Seabloom (1997).

univariate and multivariate statistical tests. The predicted and actual vegetation maps were most similar during the deep-flooding years, were less similar during the drawdown years and were least similar during the first few years of reflooding. How well the predicted maps resembled the actual maps varied not only by year but also by species and water-level treatment. Seabloom (1997) also compared the de Swart et al. (1994) logistic regression models to that of the spatial model. This was done by generating vegetation maps using the logistic regression models for each cell for each year during the reflooding year and comparing the predicted maps to the actual maps as was done with the spatial model. Their results indicated that during the reflooding years, the spatial model's predictions consistently had less error than the logistic regression models.

The spatial models of prairie wetland vegetation dynamics make it possible to quantify the potential impacts of water-level changes on the vegetation of these wetlands. This alone makes them much more useful than the qualitative life-history model that simply predicts the overall composition of the vegetation of the wetland at any time. The results of the spatial model indicate that landscape geometry was a major factor controlling the distribution of emergent species. This was because dispersal of species was constrained by both the composition of the seed bank and by the clonal spread of species. These are two factors that logistic regression models cannot directly take into account. Clonal growth of species, especially that of *Typha*, was an important mechanism for colonization of adjacent cells in the spatial model. Site history also is an important factor. Logistic regression models tend to overestimate the abundance of species because they do not take into account the local extirpation of emergent species and their inability to become reestablished until the next drawdown.

The Poiani and Johnson (1993) and the Seabloom (1997) spatial models represent only the pioneering stages of spatial vegetation dynamics modeling of prairie wetlands. They are both still relatively simplistic models. The Poiani and Johnson model linked a vegetation submodel to a hydrologic submodel, but this link is missing in the Seabloom model, which does not contain a hydrologic submodel. Although the Seabloom model has a more sophisticated vegetation submodel than the Poiani and Johnson model which only simulated the fate of vegetation types when water levels changed, the Seabloom model ignores known effects of soil moisture, temperature, and salinity on seed germination. It currently also does not allow the coexistence of adults of two or more emergent species in a cell, even though this is common in some stages of the wet–dry cycle. In short, much more work is needed before a reliable, general model of the vegetation dynamics of prairie wetlands will be available.

CONCLUSIONS

The results of the MERP experimental study of wet–dry cycle confirmed much of what was known about these cycles from earlier field studies. Water-level changes drive these cycles with high water primarily responsible for the elimination of emergents and drawdowns for their reestablishment. The MERP results also confirmed that there are lag times of 2 to 3 years before the effects of higher water levels or reflooding after a drawdown can be detected in the field. As in other prairie wetlands, it was recruitment from the seed bank during the drawdown that enabled the MERP cells to quickly recover from the extirpation of the emergents during the deep-flooding years. Unlike other prairie wetlands with water-level fluctuations whose seeds banks are relatively uniform over their elevation gradients, the MERP seed bank had a distinct maximum seed density around 247.5 m for nearly all species. This was due to the regulation of water levels in the Delta Marsh since the early 1960s that resulted in the windrowing of seeds year after year at the same water line. Although earlier field studies had indicated that environmental conditions effect recruitment patterns during drawdowns, the MERP studies, both field and laboratory, demonstrated how significant environmental conditions (soil moisture, temperature, and salinity) are for recruitment of emergents and annuals during drawdowns.

Increasing water levels by 30 and 60 cm for 5 years, did not hasten the onset of the degenerating stage of the wet–dry cycle. In both treatments, the degenerating stage began in 1988. All three water-level treatments lost significant emergent cover and production after 3 years of reflooding due to the extirpation of short-lived emergent species not able to tolerate permanent inundation. In the medium and high treatments, the onset of the degenerating stage was marked by the loss of long-lived emergent species growing in water depths beyond their tolerance. This was due in large part because emergent species were unable to migrate upslope in the medium and high treatments.

Arguably, the most important contribution made by MERP to our understanding of vegetation dynamics in wetlands concerns the genesis of vegetation zones. Although much has been written on this topic, this seems to be the only field study of the early stages in the formation of new zones. Zonation patterns in prairie wetlands are the collective result of seed distribution, seedling recruitment, seedling mortality, and adult mortality patterns. Zonation is not the result of just one overriding factor such as competition, seed dispersal, or the ecological tolerances of the species. In fact, different combinations of factors seem to be responsible for the final distribution of each species and differences in environmental conditions from place to place or year to year will be reflected in the distribution of species along the new coenocline.

Much progress has been made in the modeling of the vegetation dynamics of prairie wetlands since the initiation of MERP. Besides the simple, qualitative model that uses life-history characteristics of each species to predict the composition of the vegetation during a wet–dry cycle, several new quantitative models have been developed. One of these, the logistic regression model, seems less well suited for prairie potholes than the two spatially explicit models developed by Poiani and Johnson (1993) and Seabloom (1997). The Poiani and Johnson (1993) and especially the Seabloom (1997) models of the vegetation dynamics are based on the same concept that was the foundation of the van der Valk (1981) model, environmental filters. Environmental filters (e.g., water level) at any time determine which species can become established, which can persist, and which are extirpated. Various experimental studies of recruitment from the seed banks done as part of MERP (van der Valk and Pederson 1989; Meredino and Smith 1991; Meredino et al.

1990) as well as studies on initial recruitment patterns in other wetlands (Dean and Hurd 1980; Weiher and Keddy 1995; Seabloom et al. 1998) suggest that this is a productive framework for examining factors that control the composition of vegetation (van der Valk 1981, 1982; Johnstone (1986); Shipley and Keddy (1991); Keddy 1992). Neither the Poiani and Johnson (1993) nor the Seabloom (1997) models are the final word in vegetation dynamics models for prairie wetlands. A more sophisticated model is still needed that combines a hydrological model able to predict water-level changes as in the Poiani and Johnson (1993) model with a vegetation model comparable to the one that is in the Seabloom (1997) model. The vegetation model, however, also needs to incorporate the effects of temperature, soil moisture, and salinity on vegetation recruitment, seedling growth, and adult mortality.

RESEARCH NEEDS

Although a great deal about the vegetation dynamics of prairie wetlands has been learned in the last 20 years, there are still many areas that need further study. In fact, virtually every aspect of the recruitment and growth of emergent species needs additional work. Very little is known about seed production, dispersal, and longevity. All three have significant implications for understanding seed inputs and seed densities in the seed bank. Comparative studies of the water-depth tolerances of wetland species, especially emergent species, are needed along the lines of Squires and van der Valk (1992). Future studies should last at least 4 years, plus another year to acclimate the plants prior to the onset of the water-depth treatments. A better understanding of what controls the primary production of the various guilds of species in prairie wetlands also is needed.

Perhaps the most significant gap in our information about wetland species, particularly emergents, is about their clonal growth rates. Almost nothing is known about the rate at which different species can grow at different water depths. Such studies need to be done over the entire range of water depths that each species can be found. Better comparative information on clonal growth rates is essential for the next generation of vegetation dynamics models. Studies of the annual production of established stands of emergent species at different water depths also would be very informative.

Besides studies that focus on the short-term changes in vegetation during a wet–dry cycle, studies also are needed of long-term changes over multiple cycles. Historical information from the Delta Marsh suggests that the dominant emergent species in the marsh have changed several times in the last 60 or 70 years. The most recent is ongoing and involves the replacement of *P. australis* by *T. glauca* (de Geus 1987), suggested to be due to the elimination of water-level fluctuations, but this hypothesis was not supported by the MERP results. Why do such long-term shifts in dominant emergents occur? There are several possibilities that need to be investigated. Increased inputs of nutrients from farm runoff have been implicated in the replacement of dominant emergents by *Typha* species in other wetlands such as the Florida Everglades (Davis 1994; Urban et al. 1993). An older explanation for the spread of cattails is the development of new aggressive races of *Typha* due to hybridization (Smith 1967) that can outcompete other emergent species.

ACKNOWLEDGMENTS

MERP was conceived by Bruce Batt, then of the Delta Waterfowl and Wetlands Research Station, and Pat Caldwell, then of Ducks Unlimited Canada. Without their efforts to develop funding and a scientific team, MERP would never have been possible. During MERP, several people were employed in collecting and processing macrophyte samples, often under less than ideal conditions. I would like to thank every one of these MERPies for their dedication and contribution to this project. I especially would like to thank my former graduate students: Nina Bicknese, Marita Galinato, Mohammad Hosseini, Roger Pederson, Louisa Squires, and Chip Welling. It was primarily through their talents and efforts that MERP was able to contribute so much to our understanding of the vegetation of prairie wetlands Finally, I would like to thank Henry Murkin who kept MERP on track, both during and after the field phase. Without him, Bruce and Pat's vision would never have become a reality. This is Paper No. 100 of the Marsh Ecology Research Program, a joint project of Ducks Unlimited Canada and the Delta Waterfowl and Wetlands Research Station.

LITERATURE CITED

Anderson, M.G., and R.C. Jones. 1976. Submerged Aquatic Vascular Plants of East Delta Marsh. Winnipeg: Manitoba Department of Renewable Resources and Transportation Services.

Anonymous 1974. Flood Control. Winnipeg: Water Resources Division, Department of Mines, Resources, and Environmental Management, Government of Manitoba.

Barendregt, A., J.T. de Smidt, and M.J. Wassen. 1986. The impact of groundwater flow on wetland communities. In *Colloques Phytosociologiques. XIII. Vegetation et Geomorphologogie.* (Ed.) J.-M. Gehu, pp.603-612. Berlin, Germany: J. Cramer.

Barendregt, A., M.J. Wassen, and J.T. de Smidt. 1993. Hydroecological modeling in a polder landscape: a tool for wetland management. In *Landscape Ecology of a Stressed Environment.* (Eds.) C.C. Vos and P. Opdam, pp.79-99. London, UK: Chapman & Hall.

Bertness, M.D., and A.M. Ellison. 1987. Determinants of pattern in a New England salt marsh plant community. Ecological Monographs 57:129-147.

Boutin, C., and P.A. Keddy. 1993. A functional classification of wetland plants. Journal of Vegetation Science 4:591-600.

Bukata, R.P., J.E., Bruton, J.J. Jerome, and W.S. Harris. 1988. An evaluation of the impact of persistent water level changes on the areal extent of Georgian Bay/North Channnel marshlands. Environmental Management 12:359-368.

Buttery, B.R., and J.M. Lambert. 1965. Competition between *Glyceria maxima* and *Phragmites communis* in the region of the Surlingham Broad. I. The competition mechanism. Journal of Ecology 53:163-181.

Clark, W.R., and D.W. Kroeker. 1993. Population dynamics of muskrats in experimental marshes at Delta, Manitoba. Canadian Journal of Zoology 71:1620-1628.

Dabbs, D.L. 1971. A study of *Scirpus acutus* and *Scirpus validus* in the Saskatchewan River Delta. Canadian Journal of Botany 49:143-153

Davis, S.M. 1994. Phosphorus inputs and vegetation sensitivity in the Everglades. In *Everglades:*

The Ecosystem and its Restoration. (Eds.) S.M. Davis and J.C. Ogden, pp.357-378. Delray Beach: St. Lucie Press.

Dean, T.A., and L.E. Hurd. 1980. Development in an estuarine community: the influence of early colonists on late arrivals. Oecologia (Berl.) 46:295-301.

de Geus, P. M. 1987. Vegetation changes in the Delta Marsh, Manitoba between 1948–80. M.S. thesis, University of Manitoba, Winnipeg, Manitoba.

de Swart, E.O. A.M., A.G. van der Valk, K.J. Koehler, and A. Barendregt. 1994. Experimental evaluation of realized niche models for predicting responses of plant species to a change in environmental conditions. Journal of Vegetation Science 5:541-552.

Ellison, A.M., and B.L. Bedford. 1995. Response of a wetland vascular plant community to disturbance: a simulation study. Ecological Applications 5:109-123.

Galinato, M.I. 1985. Seed germination studies of dominant wetland species of the Delta Marsh. M.S. thesis, Iowa State University, Ames, Iowa.

Galinato, M.I., and A.G. van der Valk. 1986. Seed germination traits of annuals and emergents recruited during drawdowns in the Delta Marsh, Manitoba, Canada. Aquatic Botany 26:89-102.

Gaudet, C.L., and P.A. Keddy. 1995. Competitive performance and species distribution in shoreline plant communities: a comparative approach. Ecology 76:280-291.

Grace, J.B. 1989. Effects of water depth on *Typha latifolia* and *Typha domingensis*. American Journal of Botany 76:762-768.

Grace, J.B., and R.G. Wetzel. 1981. Habitat partitioning and competitive displacement in cattails (*Typha*): experimental field studies. American Naturalist 118:463-474.

Harris, S.W., and W.H. Marshall. 1963. Ecology of water-level manipulations on a northern marsh. Ecology 44:331-343.

Jefferies, R..L., A. Jensen, and K.F. Abraham. 1979. Vegetational development and the effect of geese on the vegetation at LaPerousse Bay, Manitoba. Canadian Journal of Botany 56:1439-1450.

Johnson, W.C., T.L. Sharik, R.A. Mayes, and E.P. Smith. 1987. Nature and cause of zonation discreteness around glacial prairie marshes. Canadian Journal of Botany 65:1622-1632.

Johnstone, I.M. 1986. Plant invasion windows: a time based classification of invasion potential. Biological Review 61:369-394.

Kadlec, J.A. 1962. Effects of a drawdown on a waterfowl impoundment. Ecology 43:267-281.

Kantrud, H.A., J.B. Millar, and A.G. van der Valk. 1989a. Vegetation of wetlands of the prairie pothole region. In *Northern Prairie Wetlands.* (Ed.) A.G. van der Valk, pp. 132-158. Ames: Iowa State University Press.

Kantrud, H.A., G.L. Krapu, and G.A. Swanson. 1989b. Prairie basin wetlands of the Dakotas: a community profile. U.S. Fish and Wildlife Service, Biological Report 85(7.28). 116pp.

Kautsky, L. 1988. Life strategies of aquatic soft bottom macrophytes. Oikos 53:126-135.

Keddy, P.A. 1983. Shoreline vegetation in Axe Lake, Ontario: effects of exposure on zonation. Ecology 64:331-344.

Keddy, P.A. 1985. Plant zonation on lakeshores in Nova Scotia, Canada: a test of the resource specialization hypothesis. Journal of Ecology 72:797-808.

Keddy, P.A. 1992. Assembly rules and response rules: two goals for predictive community ecology. Journal of Vegetation Science 3:157-164.

Leck, M.A. 1989. Wetland seed banks. In *Ecology of Soil Seed Banks.* (Eds.) M.A. Leck, V.T Parker, and R.L. Simpson, pp. 283-305. San Diego: Academic Press.

Lieffers, V.J., and J.M. Shay. 1981. The effects of water level on growth and reproduction of *Scirpus maritimus* var. *paludosus* on the Canadian prairies. Canadian Journal of Botany 59:118-121.

Löve, A., and D. Löve. 1954. Vegetation of a prairie marsh. Bulletin of the Torrey Botanical Club 81:16-34.

Macauley, A.J. 1973. Taxonomic and ecological relationships of *Scirpus acutus* Muhl. and *S. validus* Vahl. (Cyperacea) in southern Manitoba. Ph.D. dissertation, University of Manitoba, Winnipeg.

McKee, K.L., I.A. Mendelssohn, and D.M. Burdick. 1989. Effect of long-term flooding on root metabolic response in five freshwater marsh plant species. Canadian Journal of Botany 67:3446-3452.

Merendino, M.T., and L.M. Smith. 1991. Influence of drawdown date and reflood depth on wetland vegetation establishment. Wildlife Society Bulletin 19:143-150.

Merendino, M.T., L.M. Smith, H.R. Murkin, and R.L. Pederson. 1990. The response of prairie wetland vegetation to seasonality of drawdown. Wildlife Society Bulletin 18:245-251.

Millar, J.B. 1973. Vegetation changes in shallow marsh wetlands under improving moisture regime. Canadian Journal of Botany 51:1443-1457.

Murkin, E.J., and H.R. Murkin. 1989. Marsh Ecology Research Program: Long-term monitoring procedures manual. Technical Bulletin 2. Delta Waterfowl and Wetlands Research Station, Portage la Prairie, Manitoba.

Murkin, H.R. 1989. The basis for food chains in prairie wetlands. In *Northern Prairie Wetlands.* (Ed.) A.G. van der Valk, pp. 316-328. Ames: Iowa State University Press.

Naim, P.A., 1987. Wetland seed banks: implications in vegetation management. Ph.D. dissertation, Iowa State University, Ames.

Neckles, H.A. 1984. Plant and macroinvertebrate response to water regime in a whitetop marsh. M.S. thesis, University of Minnesota, St. Paul, Minnesota.

Neckles, H.A., J.W. Nelson, and R.L. Pederson. 1985. Management of whitetop (*Scolochloa festucacea*) marshes for livestock forage and wildlife. Technical Bulletin No. 1. Delta Waterfowl and Wetlands Research Station, Portage la Prairie, Manitoba.

Neill, C. 1990a. Effects of nutrients and water levels on species composition in prairie whitetop *Scolochloa festucacea* marshes. Canadian Journal of Botany 68:1015-1020.

Neill, C. 1990b. Effects of nutrients and water levels on emergent macrophyte biomass in a prairie marsh. Canadian Journal of Botany 68:1007-1014.

Neill, C. 1992. Life history and population dynamics of whitetop (*Scolochloa festucacea*) shoots under different levels of flooding and nitrogen supply. Aquatic Botany 42:241-252.

Neill, C. 1993a. Growth and resource allocation of whitetop (*Scolochloa festucacea*) along a water depth gradient. Aquatic Botany 46:235-246.

Neill, C. 1993b. Seasonal flooding, soil salinity and primary production in northern prairie marshes. Oecologia (Berl.) 95:499-505.

Neill, C., and J.C. Cornwell. 1992. Stable carbon, nitrogen, and sulfur isotopes in a prairie marsh food web. Wetlands 12:217-224.

Pederson, R.L. 1981. Seedbank characteristics of the Delta Marsh, Manitoba: Applications for wetland management. In *Selected Proceedings of the Midwest Conference on Wetland Values and Management.* (Ed.) B. Richardson, pp.61-69. Navarre: The Freshwater Society.

Pederson, R.L. 1983. Abundance, distribution, and diversity of buried viable seed populations in the Delta Marsh, Manitoba. Ph.D. dissertation, Iowa State University, Ames, Iowa.

Pederson, R.L., and A.G. van der Valk. 1985. Vegetation change and seed banks in marshes: ecological and management implications. Transactions of the North American Wildlife and Natural Resource Conference 49:271-280.

Pip, E., and J. Stepaniuk. 1988. The effect of flooding on wild rice, *Zizania aquatica* L. Aquatic Botany 32:283-290.

Poiani, K.A., and W.C. Johnson. 1988. Evaluation of the emergence method in estimating seed bank composition of prairie wetlands. Aquatic Botany 32:91-97.

Poiani, K.A., and W.C. Johnson. 1989. Effect of hydroperiod on seed-bank composition in semi-permanent prairie wetlands. Canadian Journal of Botany 67:856-864.

Poiani, K.A., and W.C. Johnson. 1991. Global warming and prairie wetlands. BioScience 41:611-618.

Poiani, K.A., and W.C. Johnson. 1993. A spatial simulation model of hydrology and vegetation dynamics in semi-permanent prairie wetlands. Ecological Applications 3:279-293.

Scoggan, H.J. 1978–1979. *The flora of Canada.* Parts 1–4. Ottawa: National Museum of Natural Science.

Seabloom, E.W. 1997. Vegetation dynamics in prairie wetlands. Ph.D. dissertation, Iowa State University, Ames, Iowa.

Seabloom, E.W., A.G. van der Valk, and K.A. Moloney. 1998. The role of water depth and soil temperature in determining initial composition of wetland coenoclines. Plant Ecology 138:203-216.

Shay, J.M., and C.T. Shay. 1986. Prairie marshes in western Canada, with specific reference to the ecology of five emergent macrophytes. Canadian Journal of Botany 64:443-454.

Shipley, B., and P.A. Keddy. 1987. The individualistic and community-unit concepts as falsifiable hypotheses. Vegetatio 69:47-55.

Shipley, B., and P.A. Keddy. 1991. Mechanisms producing plant zonation along a water depth gradient: a comparison with the exposure gradient. Canadian Journal of Botany 69:1420-1424.

Shipley, B., P.A. Keddy, D.R.J. Moore, and K. Lemsky. 1989. Regeneration and establishment strategies of emergent macrophytes. Journal of Ecology 77:1093-1110.

Sjöberg, K., and K. Danell. 1983. Effects of permanent flooding on *Carex-Equisetum* wetlands in northern Sweden. Aquatic Botany 15:275-286.

Smith, L.M., and J.A. Kadlec. 1983. Seed banks and their role during drawdown of a North American marsh. Journal of Applied Ecology 20:673-684.

Smith, L.M., and J.A. Kadlec. 1985. The effects of disturbance on marsh seed banks. Canadian Journal of Botany 63:2133-2137.

Smith, S.G. 1967. Experimental and natural hybrids in North American *Typha* (Typhaceae). American Midland Naturalist 78:257-287.

Snow, A., and S.W. Vince. 1984. Plant zonation in an Alaskan salt marsh. II. An experimental study of the role of edaphic conditions. Journal of Ecology 71:669-684.

Spence, D.H.N. 1982. The zonation of plants in freshwater lakes. Advances in Ecological Research 12:37-125.

Squires, L. 1991. Water-depth tolerances of emergent plants in the Delta Marsh, Manitoba, Canada. M.S. thesis, Iowa State University, Ames, Iowa.

Squires, L., and A.G. van der Valk. 1992. Water-depth tolerances of the dominant emergent macrophytes of the Delta Marsh, Manitoba. Canadian Journal of Botany 70:1860-1867.

Stevenson, S.C., and P.F. Lee. 1987. Ecological relationships of wild rice, *Zizania aquatica*. 6. The effects of increases in water depth on vegetative and reproductive production. Canadian Journal of Botany 65:2128-2132.

Stewart, R.E., and H.A. Kantrud. 1971. Classification of natural ponds and lakes in the glaciated prairie region. Washington, U.S. Fish and Wildlife Service. Resource Publ. 92.

Stockey, A., and R. Hunt. 1994. Predicting secondary succession in wetland mesocosms on the basis of autecological information on seeds and seedlings. Journal of Applied Ecology 31:543-559.

Thomas, A.G., and J.M. Stewart. 1969. The effect of different water depths on the growth of wild rice. Canadian Journal of Botany 47:1525-1531.

Urban, N.H, S.M. Davis, and N.G. Aumen. 1993. Fluctuations in sawgrass and cattail densities in Everglades Water Conservation Area 2A under varying nutrient, hydrologic and fire regimes. Aquatic Botany 46:203-223.

van der Valk, A.G. 1981. Succession in wetlands: a Gleasonian approach. Ecology 62:688-696.

van der Valk, A.G. 1982. Succession in temperate North American wetlands. In *Wetlands: Ecology and Management*. (Eds.) B. Gopal, R.E. Turner, R.G. Wetzel, and D.F. Whigham, pp.169-179. Jaipur: International Scientific Publications, India.

van der Valk, A.G. 1985. Vegetation dynamics of prairie glacial marshes. In *Population Structure of Vegetation*. (Ed.) J. White, pp.293-312. The Hague: Junk.

van der Valk, A.G. 1986. The impact of litter and annual plants on recruitment from the seed bank of a lacustrine wetland. Aquatic Botany 24:13-26.

van der Valk, A.G. 1991. Response of wetland vegetation to a change in water level. In *Wetland Management and Restoration–Proceedings of a Workshop*, Sweden 12–15 1990. September (Ed.) C.M. Finlayson and T. Larsson, pp.7-16. Solna: Swedish Environmental Protection Agency, Sweden.

van der Valk, A.G. 1994. Effects of prolonged flooding on the distribution and biomass of emergent species along a freshwater wetland coenocline. Vegetatio 110:185-196.

van der Valk, A.G., and C.B. Davis. 1976a. Seed banks of prairie glacial marshes. Canadian Journal of Botany 54:1832-1838.

van der Valk, A.G., and C.B. Davis. 1976b. Changes in the composition, structure, and production of plant communities along a perturbed wetland coenocline. Vegetatio 32:87-96.

van der Valk, A.G., C.B. Davis. 1978a. The role of the seed banks in the vegetation dynamics of prairie glacial marshes. Ecology 59:322-335.

van der Valk, A.G., and C.B. Davis. 1978b. Primary production of prairie glacial marshes. In *Freshwater Marshes*. (Eds.) R.E. Good, D.F. Whigham, and R.L. Simpson, pp.21-37. New York: Academic Press.

van der Valk, A.G., and C.B. Davis. 1980. The impact of a natural drawdown on the growth of four emergent species in a prairie glacial marsh. Aquatic Botany. 9:301-322.

van der Valk, A.G., and R.L. Pederson. 1989. Seed banks and the management and restoration of natural vegetation. In *Ecology of Soil Seed Banks*. (Eds.) M.A. Leck, V.T. Parker, and R.L. Simpson, pp.329-346. New York: Academic Press.

van der Valk, A.G., and T.R. Rosburg. 1997. Seed bank composition along a phosphorus gradient in the northern Florida Everglades. Wetlands 17:228-236.

van der Valk, A.G., and L. Squires. 1992. Indicators of flooding derived from aerial photography in northern prairie wetlands. In *Ecological Indicators, Volume 1*. (Eds.) D.H. McKenzie, D.E. Hyatt, and V.J. McDonald, pp.593-602. New York: Elsevier Applied Science.

van der Valk, A.G., and C.H. Welling. 1988. The development of zonation in freshwater wetlands: an experimental approach. In *Diversity and Pattern in Plant Communities*. (Eds.) H.J. During, M.J.A. Werger, and J.H. Willems. pp.145-158. The Hague, Netherlands: SPB Academic Publishing.

van der Valk, A.G., C.H. Welling, and R.L. Pederson. 1989. Vegetation change in a freshwater wetland: a test of a priori predictions. In *Freshwater Wetlands and Wildlife*. (Eds.) R.R. Sharitz and J.W. Gibbons, pp.207-217. Oak Ridge: US DOE Office of Scientific and Technical Information.

van der Valk. A.G., R.L. Pederson, and C.B. Davis. 1992. Restoration and creation of freshwater wetlands using seed banks. Wetlands Ecology and Management 1:191-197.

van der Valk, A.G., L. Squires, and C.H. Welling. 1994. An evaluation of three approaches for assessing the impacts of an increase in water level on wetland vegetation. Ecological Applications 4:525-534.

Voigts, D.K. 1976. Aquatic invertebrate abundance in relation to changing marsh vegetation. American Midland Naturalist 95:313-322.

Walker, J.M. 1959. Vegetation studies on the Delta Marsh, Manitoba. M.S. thesis, University of Manitoba, Winnipeg, Manitoba.

Walker, J.M. 1965. Vegetation changes with falling water levels in the Delta Marsh, Manitoba. Ph.D. dissertation, University of Manitoba, Winnipeg, Manitoba.

Wallsten, M., and P.-O. Forsgren. 1989. The effects of increased water level on aquatic macrophytes. Journal of Aquatic Plant Management 27:32-37.

Waters, I., and J.M. Shay. 1990. A field study of the morphometric response of *Typha glauca* shoots to a water depth gradient. Canadian Journal of Botany 68:2339-2343.

Weiher, E., and P.A. Keddy. 1995. The assembly of experimental wetland plant communities. Oikos 73:323-335.

Weller, M.M. 1981. *Freshwater Marshes*. Minneapolis: University of Minnesota Press.

Weller, M.W., and L.H. Fredrickson. 1974. Avian ecology of a managed glacial marsh. Living Bird 12:269-291.

Weller, M.W., and C.S. Spatcher. 1965. Role of habitat in the distribution and abundance of marsh birds. Iowa Agriculture and Home Economics Experiment Station, Special Report 43, Ames, Iowa.

Welling, C.H. 1987. Reestablishment of perennial emergents macrophytes during a drawdown in a lacustrine marsh. M.S. thesis, Iowa State University, Ames, Iowa.

Welling, C.H., R.L. Pederson, and A.G van der Valk. 1988a. Recruitment from the seed bank and the development of zonation of emergent vegetation during a drawdown in a prairie wetland. Journal of Ecology 76:483-496.

Welling, C.H., R.L. Pederson, and A.G. van der Valk. 1988b. Temporal patterns in recruitment from the seed bank during drawdowns in a prairie wetland. Journal of Applied Ecology 25:999-1007.

Wilson, S.D., and P.A. Keddy. 1985. Plant zonation on a shoreline gradient: physiological response curves of component species. Journal of Ecology 73:851-860.

Yamasaki, S. 1990. Population dynamics in overlapping zones *of Phragmites australis* and *Miscanthus sacchariflorus*. Aquatic Botany 36:367-377.

Yamasaki, S., and I. Tange. 1981. Growth responses of *Zizania latifolia*, *Phragmites australis*, and *Miscanthus sacchariflorus* to varying inundation. Aquatic Botany 10:229-239.

8

Algae in Prairie Wetlands

Gordon G. C. Robinson
Sharon E. Gurney
L. Gordon Goldsborough

ABSTRACT

Algae are an abundant and structurally diverse component of prairie wetland ecosystems. Planktonic algae occupy the water column and benthic algae colonize sediments and submersed surfaces of aquatic macrophytes. Prior to the Marsh Ecology Research Program (MERP) experiment, the quantitative contribution of algae to primary production in wetlands was largely unknown, as was their response to water-level fluctuations. In this chapter, we describe the four algal assemblages occurring in wetlands (phytoplankton, epipelon, epiphyton, and metaphyton), review their roles in the wetland food web, and consider the effects of hydrodynamics, wind, nutrients, light, temperature, and anthropogenic factors on their abundance. Finally, we discuss the results of the MERP experiment in the context of a new wetland model, examine the implications for wetland management, and identify gaps in our knowledge of wetland algae.

INTRODUCTION

Compared with a rich literature on algae of other aquatic habitats, there is little published information on algal assemblages of freshwater wetlands (Crumpton 1989; Vymazal 1994; Goldsborough and Robinson 1996). This discrepancy is particularly so for prairie wetlands, which are important as nutrient sinks, refugia for plant and animal biodiversity, and nurseries for waterfowl and lacustrine fish. This review considers algal ecology in wetlands generally, and focuses specifically on the nature and roles of algal assemblages in prairie "lacustrine lagoon marshes" (NWWG 1998). These wetlands are characterized by sheltered, open and lakelike, shallow water bodies vegetated by emergent macrophytes such as *Typha, Scirpus, Phragmites,* and *Scholochloa*; submersed *Potamogeton* species; *Myriophyllum, Ceratophyllum,* and *Utricularia*; and floating mats of *Lemna* species. We draw on algal research conducted at Delta Marsh, a large prairie wetland adjacent to Lake Manitoba, since the early 1970s. We also integrate results from our investigations of algal biomass and productivity (Robinson et al. 1997a,b) conducted in Delta Marsh as part of the Marsh Ecology Research Program (MERP) (Murkin et al. 1984; see Chapter 1). To put the results in a broader context, we incorporate, where appropriate, other published information from studies in the littoral zone of shallow lakes. Finally, we modify a hydrologically driven conceptual model of van der Valk and Davis (1978) to account for the varying predominance of algal assemblages in space and time within a lacustrine marsh.

ALGAL ASSEMBLAGES IN WETLANDS

Planktonic algae are usually the largest contributors to total primary production in deep aquatic systems, such as lakes and oceans. The productivity (and roles) of benthic algae increases progressively with declining water depth and increasing extent of the littoral zone, with freshwater wetlands presenting an extreme of littoral development. Most algae in such wetlands normally occur in close association with submersed substrata. These are often collectively referred to as "periphyton" although this term refers more generally to a diverse assemblage of autotrophs and heterotrophs in which algae are only one constituent. For purposes of this review, we consider four algal assemblages common to prairie wetlands that, collectively, constitute a conspicuous, but generally neglected, feature of those wetlands. These assemblages are phytoplankton, epipelon, epiphyton, and metaphyton.

Phytoplankton

Phytoplankton are algae entrained in the water column. Consideration of the phytoplankton species composition in wetlands is skewed because most research has been done in eutrophic wetlands during the summer when they are dominated by cyanobacteria (Shamess et al. 1985). For example, phytoplankton composition in prairie potholes resembles that of eutrophic lakes, being dominated by such genera as *Aphanizomenon*, *Microcystis*, *Anabaena*, or *Coelosphaerium*. Less nutrient-rich wetlands typically contain more diverse assemblages of filamentous cyanobacteria, diatoms, colonial and filamentous green algae, cryptomonads, and dinoflagellates (Murkin et al. 1991). Given the shallowness of most wetlands and the abundance of benthic algae, phytoplankton often contain varying proportions of tychoplankton (detached periphyton). Consequently, species composition of wetland phytoplankton is determined in part by species present in the epiphyton and epipelon available for detachment. In lakeshore wetlands, the composition of lake phytoplankton is probably a further determinant.

Epipelon

Epipelon (epipelic algae) inhabits soft surficial sediments. It is comprised primarily of mobile diatoms (e.g., the genera *Pinnularia*, *Cymatopleura*, *Caloneis*, *Nitzschia*, and *Navicula*), cyanobacteria (e.g., *Oscillatoria* and *Arthrospira*), desmids (e.g., *Closterium*) and numerous flagellates (euglenoids and chlamydomonads). Members of this assemblage characteristically display diurnal migrations to the illuminated sediment surface by day, retreating into the sediment by night (Round 1981). Stable sediments may become encrusted with nonmigratory gelatinous mats of cyanobacteria and diatoms (plocon), which comprise a functionally and structurally distinct assemblage from epipelon. No systematic study of the ecology of plocon mats has been conducted in prairie wetlands. However, preliminary measurements in shallow channels at Delta Marsh indicate their abundance may be >100 times that of epipelon (Goldsborough, unpublished data).

Epiphyton

Epiphyton (epiphytic algae) colonizes the aquatic surfaces of submersed and emergent macrophytes. Its development typically involves a relatively orderly succession of prostrate diatoms (e.g., *Cocconeis*, *Achnanthes,* and *Epithemia*), roseate diatoms (e.g., *Synedra*), pedunculate diatoms (e.g., *Gomphonema* and *Cymbella*), heterotrichous filamentous Chlorophyceae (e.g., *Chaetophora* and *Coleochaete*) and attached cyanobacteria (e.g., *Gloeotrichia*). As the assemblage develops, it may be differentiated into loosely and tightly associated components (Haines et al. 1987) and often contains entanglements of solitary and chain-forming diatoms (e.g., *Fragilaria*, *Tabellaria*, and *Diatoma*).

Metaphyton

Metaphyton, also known as "flab" (Hillebrand 1983), is an assemblage of entangled filamentous and parenchymatous chlorophytes (Chaetophorales, Zygnematales, Cladophorales, Oedogoniales, and Ulvales) with associated members of the Xanthophyceae and filamentous cyanobacteria (e.g., *Scytonema* and *Tolypothrix*). It has been described as being associated with, but not confined by, submersed surfaces (Turner et al. 1995). Metaphyton most often has its origins in the loosely attached epiphyton, and is similarly entangled with diatoms and its own epiphytic flora. Metaphytic mats can occupy any position in the water column, although most commonly they occur at the water surface, and even partially raised above the surface, being buoyed by trapped gas bubbles.

ROLE OF ALGAE IN PRAIRIE WETLANDS

Primary Production

Algae are significant contributors to total primary production in prairie wetlands and, as such, they may support much of secondary production in wetland ecosystems. However, evaluation of their functional significance is severely limited because there is little information on algal productivity in situ, and largely circumstantial evidence of grazing. Without specific information on these topics, knowledge of wetland food webs is limited to the point where management decisions are based on assumptions rather than verifiable information (Murkin 1989). For example, it is suggested that prairie wetland food webs are "fueled" primarily by macrophytic detritus. Yet, the importance of algal production to consumers should be considered because algae are available more or less consistently throughout the year and are relatively easily ingested compared with macrophytic tissue which, because of low nutritional value (high carbon:nitrogen ratio), and the presence of tough cell walls, lignin, and other secondary products, is relatively unacceptable to many grazers (Brönmark 1989).

Wetland macrophyte biomass is typically higher than algal biomass (Table 8.1). However, annual and perennial macrophytes complete, at most, only one reproductive cycle in a year, whereas turnover times of algal assemblages are measured in days. For example, estimates of algal turnover times derived from the MERP experiments ranged from 2 to 91 days (Table 8.2). Thus, a comparison of annual algal primary productivity with that of macrophytes provides a dramatically different picture. If one assumes that the maximum macrophyte biomass presented in Table 8.1 also represents annual net productivity (g/m²/yr), then equivalent algal values of 362–813 g C/m²/yr (Table 8.3) are notably higher. Therefore, we conclude, based on results from Delta Marsh, that algal productivity in prairie wetlands can be as significant as macrophyte productivity.

The conclusion that algal production is an important source of organic material in wetlands is supported by work done in estuarine salt marshes (Sullivan and Moncreiff 1988; Pinckney and Zingmark 1993; Vernberg 1993), tundra ponds (Stanley and Daley 1976), and dystrophic wetlands (Schalles and Shure 1989). However, the applicability of this conclusion to temperate freshwater wetlands generally cannot be assessed because there has been no directly comparable study of algal biomass and productivity in wetlands other than Delta Marsh. The diversity of units used to measure algal biomass or productivity confounds comparisons. More seriously, productivity or biomass data are typically not expressed on an areal basis, and productivity is commonly integrated over periods of hours or days, but seldom in years. Available comparable data (Table 8.4) confirm that the ranges upon which the previously stated conclusion is based overlap, with few exceptions, ranges reported elsewhere. Algal productivity used in reaching this conclusion was esti-

Table 8.1. Maximum biomass (g C/m²) of algae and macrophytes in MERP cells during the reflooding phase (1985–1989)

Primary producer	Biomass (g C/m²)	% total
Phytoplankton	1	<1
Epipelon	1	<1
Epiphyton	10	2
Metaphyton	75	18
Emergent macrophytes (aboveground)	108	26
Emergent macrophytes (belowground)	203	49
Submersed macrophytes	18	4
Total biomass	416	

Note: Macrophyte dry weight values (van der Valk, herein) were converted to carbon units on the basis of 45% carbon content (Murkin 1989). Algal biomass was converted from chlorophyll a measurements by using a carbon:chlorophyll ratio of 130 (Hosseini 1986).

Table 8.2. Turnover time (days) of algal biomass in low-, medium-, and high-flooded wetland cells (1985–1989)

Year	Phytoplankton			Epipelon			Epiphyton			Metaphyton		
	Low	Medium	High	Low	Medium	High	Low	Medium	High	Low	Medium	High
1985	3.8	3.7	6.4	9.8	18.2	32.0	7.5	14.4	20.4	33.1	41.6	48.1
1986	2.2	3.4	6.5	13.0	10.6	78.0	9.9	11.9	16.0	23.8	33.5	36.3
1987	3.2	3.8	5.4	10.0	17.1	91.0	8.0	11.0	14.6	23.9	30.0	40.4
1988	3.2	3.6	6.6	13.6	9.3	27.5	8.0	11.5	16.3	22.0	27.9	36.1
1989	3.1	3.8	5.8	14.7	17.4	23.5	9.1	8.1	18.8	22.1	30.9	40.0
Overall mean		4.3			25.7			12.4			32.6	

Note: Data were calculated by dividing algal biomass (mg Chl-a/m^2; converted to carbon units as in Table 8.1) by the rate of daily productivity (mg C/m^2/d) over the same time interval.

Table 8.3.   Annual productivity (g C/m²/yr) of planktonic and benthic algae in low-, medium- and high-flooded wetland cells, as estimated using turnover times (Table 8.2) calculated on the basis of a 200-day annual ice-free period

	Phytoplankton	Epipelon	Epiphyton	Metaphyton	Total
Low cells	34	6	201	573	813
Medium cells	42	9	163	474	688
High cells	45	3	71	243	362
Mean	40	6	145	430	621

Table 8.4.   Maximum algal biomass and annual primary productivity in freshwater wetlands. Data are expressed on an areal basis (per square meter of wetland surface)

Assemblage	Biomass (mg Chl a/m²)	Productivity (g C/m²/yr)	Source
Phytoplankton	104		Reeder (1994)
		366	Reeder and Mitsch (1989)
		2–11	Robarts et al. (1995)
	11	44	This study
Epipelon and plocon	2		Campeau et al. (1994)
	8		Gabor et al. (1994)
	5		Murkin et al. (1991)
	4		Murkin et al. (1994)
	435		Shamess (1980)
	757		Goldsborough (unpublished data)
	5	7	This study
Epiphyton		85	Cronk and Mitsch (1994)
	650		Goldsborough (1993)
		49	Hooper and Robinson (1976)
	74	278	This study
Metaphyton	229[a]		Hosseini and van der Valk (1989b)
	838[a]		Wu and Mitsch (1998)
	17[a]		Richardson and Schwegler (1986)
	597	474	This study

[a]Converted from dry-weight measurements by using a Chl/biomass ratio of 0.0025 (Goldsborough, unpublished data).

mated from curvilinear relationships between biomass-normalized carbon fixation rates and incident irradiance derived for each assemblage (Figure 8.1). Photosynthetic parameters derived from these relationships compare well with those reported elsewhere (Table 8.5).

A comparison of contributions by individual algal assemblages to total algal biomass and productivity (Figures 8.2 and 8.3) in the three flooding treatments (1985–1989) of the MERP experiment (see Chapter 1) demonstrates clearly that metaphyton is the dominant assemblage, followed by epiphyton, phytoplankton, and epipelon (Table 8.1). This is in

Table 8.5. Parameters of P versus I relationships for wetland algae; α = photosynthetic efficiency $((\mu g\ C/\mu g\ Chl\text{-}a/h)/(\mu mole/m^2/s))$; I_k = irradiance at onset of P_{max} ($\mu mole/m^2/s$); P_{max} = light saturated specific photosynthetic rate ($\mu g\ C/\mu g\ Chl\text{-}a/h$)

Assemblage	α	I_k	P_{max}	Source
Phytoplankton				
	0.007–0.008	253–446	2.0–3.0	Gurney and Robinson (1988)
	0.092	303	3–79	Kotak (1990)
		235–392	6–14	Robinson (1988)
	<0.001–0.056	282± 40	0.2–23.4	This study
Epipelon	0.002–0.003[a]	579–834	1.0–3.0[a]	Pinckney and Zingmark (1993)
		237–563	0.8–4	Robinson (1988)
	0.001–0.010	407±75	0.3–4.9	This study
Epiphyton	0.016	430	3–21	Kotak (1990)
		198–659	0.5–4	Robinson (1988)
	<0.001–0.022	292–17	0.1–7.9	This study
Metaphyton	0.002–0.005	36–306	0.2–0.7	Gurney and Robinson (1988)
	0.007[a]	344[c]	2.0	Richmond (1992)
	<0.005–0.001[b]	186–>460[c]	0.8[a,d]	Eiseltová and Pokorný (1994)
		322[c]		Simpson and Eaton (1986)
		161[c]		Adams and Stone (1973)
			0.8[a,d]	Auer et al. (1983)
			3.2[a,d]	Wood (1975)
	<0.001–0.010	399 ± 95	0.1–10.9	This study

[a] Cassimilation from O_2 evolution by using a photosynthetic quotient of 1.
[b] Derived from P_{max} and I_k.
[c] Assuming 4.6 $\mu mole/m^2/s$ = 1.0 W/m^2.
[d] Assuming Chl a content is 0.25% of dry weight.

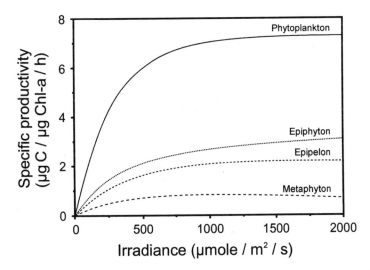

Figure 8.1   Curvilinear relationship, fitted by nonlinear regression, between biomass-normalized photo-synthetic rate ($\mu g\ C/\mu g\ Chl\text{-}a/h$) and irradiance ($\mu mole/m^2/s$) for phytoplankton, epiphyton, epipelon and metaphyton in the MERP cells. Curves are based on cumulative data collected between 1985 and 1987.

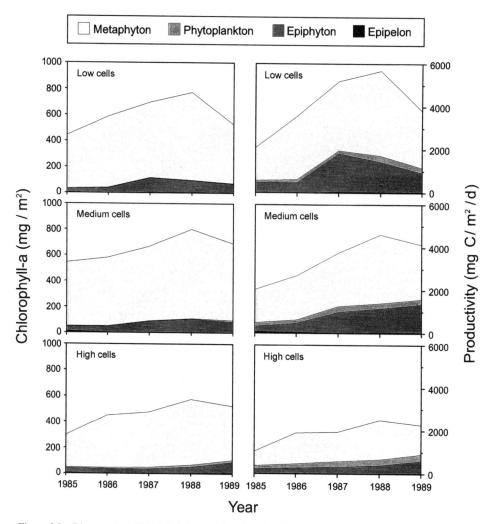

Figure 8.2   Biomass (mg Chl-a/m²; left panels) and productivity (mg C/m²/d; right panels) of benthic and planktonic algae in low water (cells 3, 4, 7 and 8; top panels), medium water (cells 1, 5, 9; middle panels), and high water (cells 2, 6, 10; bottom panels) over the 5-year reflooding phase of the MERP experiment (1985–1989).

spite of the fact that, in terms of maximum rate of production and photosynthetic efficiency (carbon fixed per unit of chlorophyll), phytoplankton was the most photosynthetically efficient assemblage, followed by epiphyton, epipelon, and metaphyton. The irradiance at which photosynthesis was saturated was similar for all assemblages (Table 8.5; Figure 8.1).

Nutrient Cycling

Algal assemblages of wetlands must play a significant role in nutrient dynamics, if for no other reason than they possess a competitive advantage over other wetland plants in nutri-

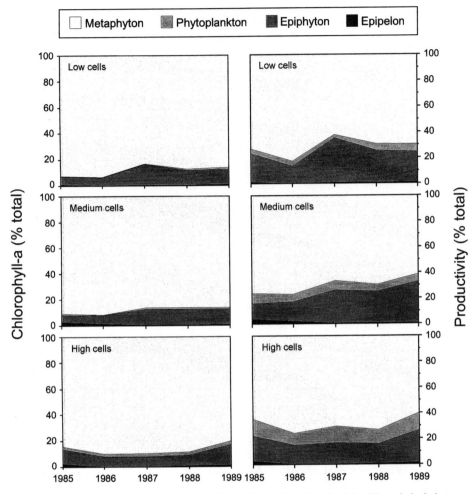

Figure 8.3 Biomass (% total algal chlorophyll/m^2; left panels) and productivity (% total algal photosyn-thesis/m^2/d; right panels) of benthic and planktonic algae in low water (cells 3, 4, 7, and 8; top panels), medium water (cells 1, 5, 9; middle panels), and high water (cells 2, 6, 10; bottom panels) over the 5-year reflooding phase of the MERP experiment (1985–1989).

ent assimilation from the water column and surficial sediments, and thus act as a short-term nutrient sink (Howard-Williams 1981; Grimshaw et al. 1993). In addition, their rel-atively short turnover times cause them also to be effective nutrient sources and to poten-tially exert considerable control over cycling and availability.

Interstitial waters of wetland sediments have higher nutrient concentrations than over-lying waters (Figure 8.4; Kadlec 1986), and are thus major potential sources of nutrients for both macrophytes and algae. However, epipelon assimilates nutrients from sediments so it may reduce substantially the quantity of nutrients diffusing to the water column (Hansson 1988, 1989). Carlton and Wetzel (1988) showed that oxygenation of surficial sediments by epipelic algal photosynthesis inhibits phosphorus release, whereas darkened

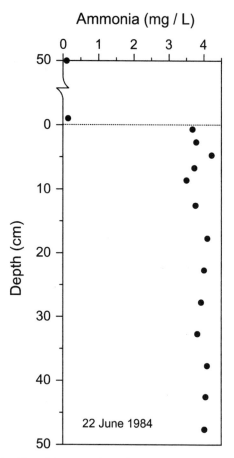

Figure 8.4 Ammonia-N (mg/l) profile across the sediment–water interface of Crescent Pond in Delta Marsh (Goldsborough, unpublished data), determined using an equilibrated "peeper" sampler (Hesslein 1976).

surficial sediments become deoxygenated and phosphorus release to overlying waters is enhanced. Similarly, efflux of ammonia from sediments is stimulated when epipelon is inhibited by mercuric chloride treatment (Jansson 1980); the importance of algae in regulating efflux rate is shown by the fact that the amount of nitrogen released into the water column when epipelon is inhibited equals the amount incorporated by algae in the absence of the inhibitor. Inhibition of epipelon by using herbicides that block photosynthesis also increases the efflux of ammonia, phosphorus, and silicon from freshwater marsh sediments (Figure 8.5; Goldsborough and Robinson 1985; Gurney and Robinson 1989). This phenomenon may affect competition for sedimentary nutrients in situations where sediments are major sources (Hansson 1990).

Herbivory

The high productivity of wetland algae implies that they may play a major role in providing energy for wetland herbivores. There have been few studies on effects of invertebrate

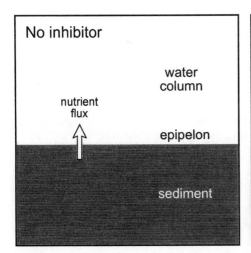
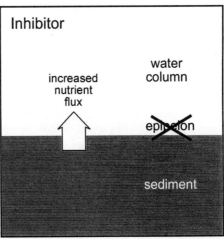

Figure 8.5   Hypothesized effect of an epipelon layer on wetland sediments on the rate of efflux of nutrients to the water column, and the impact of a photosynthetic inhibitor on that flux.

and vertebrate grazers on algae in prairie wetlands; however, there is ample evidence of grazing potential from a broader range of habitats. For example, copepod and cladoceran feeding on phytoplankton is well documented (Porter 1977). Microcrustaceans (cladocerans, copepods, and ostracods) also affect the abundance and taxonomic composition of epiphyton (Hann 1991). Other common epiphyton grazers include snails (Brönmark 1989), chironomid larvae (Mason and Bryant 1975; Cattaneo 1983), and oligochaetes (Cattaneo 1983; Kairesalo and Koskimies 1987), all of which can reduce its biomass substantially (Cattaneo 1983; Hann 1991). Aquatic invertebrates are commonly associated with metaphyton mats in large numbers (Power 1990; Ross and Murkin 1993), although it is unclear whether this reflects direct herbivory of the mats or simply their use as habitat and refugia from predators. Several grazers have been observed to feed on *Cladophora*, a common metaphytic green alga (Dodds and Gudder 1992). However, Neill and Cornwell (1992) concluded, based on the dissimilarity of the carbon- and nitrogen-stable isotopic "signatures" of metaphyton from one of the MERP cells compared with those of potential grazers, that dietary use of metaphyton was limited (Figure 8.6); minnows were the only grazer sampled that may have consumed metaphyton. It seems probable that small, attached epiphytes (e.g., diatoms) on metaphytic filaments are the most accessible grazing material but these epiphytes make up a small component of the total algal biomass in metaphyton mats. Nonetheless, elevated rates of caddisfly emergence have been associated with increased metaphyton abundance in flooded wetlands (Ross and Murkin 1993), and increased abundance of cladocerans, copepods, and chironomids has been associated with increased metaphyton production following nutrient enrichment (Gabor et al. 1994). Increased invertebrate growth rate and reproduction associated with increased algal biomass led Campeau et al. (1994) to conclude that the algal contribution to secondary production in freshwater wetlands is significant. Analyses of stable isotopes of carbon, sulfur, and nitrogen in salt marshes indicate that algae and *Spartina* (cord grass) are the two major sources of organic matter for the marsh fauna (Peterson and Howarth 1987; Sullivan and Moncreiff 1990). Indeed, benthic and planktonic algae in salt marshes may be the major food sources for both invertebrates and fish, with *Spartina* and *Juncus* playing comparatively minor roles (Sullivan and Moncreiff 1990) although substantial amounts of epipelon production may be lost through burial (Stanley 1976), and hence be unavailable for consumption. Other vertebrates also may feed on benthic algae; frog tadpoles graze actively on filamentous green algae (Dickman 1968) and dabbling ducks have been observed feeding on metaphyton mats (Murkin 1989). Recent results of stable isotope analyses from North Dakota potholes emphasize the importance of algae as a base of the wetland food web (Figure 8.6).

Other Roles

There are other potential roles played by algal assemblages in wetlands. Given the predominance of metaphyton in the MERP cells (Figures 8.2 and 8.3), this assemblage must have profound influences on various ecosystem features. Most obviously, proliferation of a dense floating mat reduces irradiance in the water column substantially (Figure 8.7; Eiseltová and Pokorný 1994), with corresponding effects on subsurface temperature, oxygen, and nutrients (Hillebrand 1983). These effects have been speculated to be responsible for increased caddisfly emergence in metaphyton-dominated MERP cells (Ross and Murkin 1993). Although there are limited direct data on impacts of metaphyton mats in wetlands, they are probably similar to those in other aquatic systems in which they flourish. For example, Turner et al. (1995) suggested that profuse metaphyton in an experimentally acidified lake might have the following effects: (1) create nutrient demand sufficient to impact other algae; (2) release significant amounts of dissolved organics that could adsorb contaminants, chelate metals, and serve as substrates in microbial processes; (3) change redox conditions sufficiently to influence nutrient cycling; (4) increase light attenuation with depth; (5) reduce turbulence via effects on boundary layers around surfaces; and (6) act as a refugium for invertebrates, although organisms could experience diurnal anoxic conditions.

ENVIRONMENTAL FACTORS AFFECTING WETLAND ALGAE

Hydrodynamics

Hydrodynamic factors in wetlands are extremely important in regulating algal distribution and abundance. Water depth essentially defines wetlands (Kadlec 1979) and strongly influences their biological processes, but a consideration of this factor can be complicated. Prairie wetlands have historically been subject to irregular periods of flooding and drought to which they are adapted (Good et al. 1978). Flooding reduces macrophyte abundance and releases nutrients from dead vegetation, litter, and sediments that may subsequently stimulate algal production. Drought and exposure of sediments are needed to allow recruitment of vegetation from seed banks (van der Valk 1986) to complete a cycle.

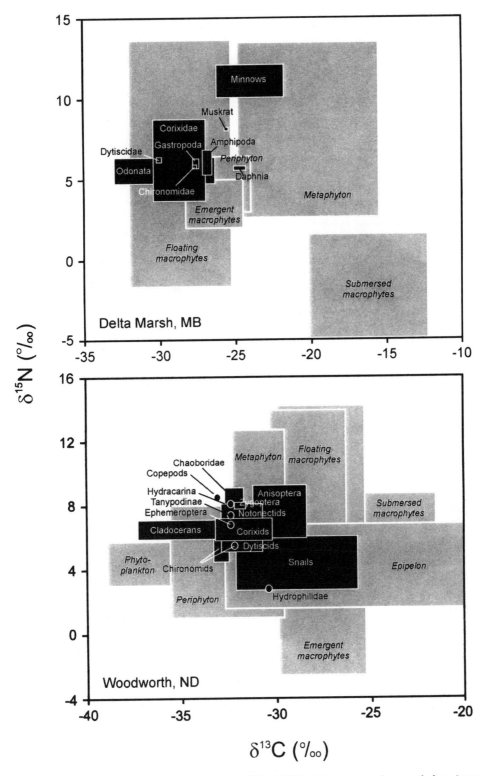

Figure 8.6 Comparison of the isotopic ratios for δ¹³C and δ¹⁵N (°/₀₀) in macrophytes and algae (gray boxes) in prairie wetlands with those of potential invertebrate and vertebrate grazers (black boxes or points). Dimensions of a given box are defined by the range of values for each isotope. Plants and grazers with overlapping isotope ranges are assumed to represent a trophic link in the wetland food web. Delta Marsh data, modified from Neill and Cornwell (1992), were collected from the MERP cells in 1988 (minnows were collected from the adjacent unenclosed marsh). Woodworth data were collected between 1994 and 1995 from a series of wetlands in North Dakota by D.A. Wrubleski (unpublished data).

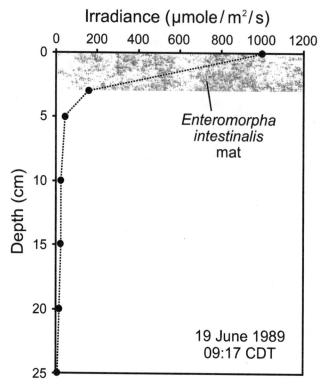

Figure 8.7   Vertical extinction profile of photosynthetically active radiation (PAR; μmole/m²/s) through a 3-cm-thick floating mat of *Enteromorpha intestinalis* in a MERP cell (Robinson et al., unpublished data).

This regime forms the basis of the hydrologically driven pulse stability model of van der Valk and Davis (1978). However, this model does not account for some other realities of lakeshore wetlands that influence algal assemblages. For example, effects of flooding may depend on the size and exposure of the wetland, which, in turn, affect the extent of wind-driven water turbulence. The depth to which wetlands are flooded also has an effect, as may the duration of flooding, drought, and drawdown events. This is an important consideration because of the increased stabilization of water levels by humans to optimize agriculture, recreation, and shoreline activities, and the generation of hydro-electricity. Lacustrine lagoon marshes can be subject to brief but dramatic fluctuations in water level due to wind-induced setups of their adjoining lakes.

During the deep-flooding phase of the MERP experiment (1981–1982; see Chapter 4), when the wetland cells were flooded to simulate a lake marsh state (van der Valk and Davis 1978), phytoplankton biomass and productivity per unit volume of water was highest in the unflooded control (Hosseini and van der Valk 1989a). In contrast, epiphyton biomass and productivity and metaphyton biomass were significantly higher in flooded cells, with this difference becoming more pronounced with length of flooding for metaphyton

(Hosseini and van der Valk 1989a, b). Dissolved nutrients levels were not elevated after flooding, perhaps due to their rapid incorporation into algal biomass, or their movement into sediment interstitial water (Kadlec 1986). This suggests that epiphyton and metaphyton were the assemblages most responsive to increased nutrient availability. This interpretation is supported by higher nitrogen content per unit of algal biomass after flooding (Hosseini and van der Valk 1989b). Following the deep-flooding period when eight MERP cells were drawn down (1983–1984; see Chapter 5), there was no significant difference in epiphyton production between the control and previously flooded cells, suggesting that the effect of increased nutrient availability was short-lived.

During the period when the MERP cells were reflooded to one of three water levels (1985–1989; see Chapter 5) to simulate a regenerating marsh (van der Valk and Davis 1978), the only algal assemblage whose biomass per unit of substratum area (or water volume for phytoplankton) varied with water depth was metaphyton. For metaphyton, highest algal biomass occurred in shallow water among emergent macrophytes (Gurney and Robinson 1988), and it decreased as water depth increased and macrophyte cover decreased. When algal biomass and productivity were expressed on an areal basis (i.e., per unit of wetland area), however, both parameters increased with decreasing water depth for metaphyton, epiphyton, and epipelon and decreased with decreasing water depth for phytoplankton (Table 8.6). For all assemblages, biomass and productivity were related to depth-mediated availability of habitat. Macrophytic substratum decreased and open-water and exposed-sediment areas increased with increasing water depth and time (Figure 8.8). The net result of this depth effect was that phytoplankton production in MERP cells declined in water <22 cm in depth; epiphyton production increased as depths declined from 34 cm to 7 cm; epipelon production increased in water depths up to 29 cm but declined to lowest values at ≈45 cm; and metaphyton production increased as depth declined to 13 cm (Figure 8.9). To fully examine impacts of water depth on algae in the MERP cells, it would be useful to compare algal biomass and productivity during the deep-flooding phase (1981–1982; Hosseini and van der Valk 1989a,b) with that during the reflooding phase (1985–1989). Unfortunately, data from the earlier period were not expressed in terms of wetland surface area so they are not comparable to those from the later period. Comparisons based simply on substratum surface area would be spurious, as the availability of substratum varied with water depth and time.

The duration of wetland drawdown affects the quantity and species composition of macrophytes present at the time of reflooding. Consequently, this duration could affect the quantity of algal biomass in the reflooded wetland due to differences in substratum availability and suitability, and degree of nutrient competition and shading. Effects of 1 versus 2 years of drawdown in the four low-flooded MERP cells on algal growth were equivocal (Table 8.7). Cells drawn down for 2 years supported less phytoplankton and epipelon in the year after reflooding (1985) than those drawn down for 1 year but differences in epiphyton biomass (higher) and metaphyton biomass (lower) were not significant. Trends were the same when all data (1985–1989) were pooled but only the higher epiphyton biomass of cells drawn down for 2 years was statistically significant.

Table 8.6. Mean daily productivity (mg C/m² wetland area per day) of planktonic and benthic algae in low-, medium-, and high-flooded wetland cells from 1985 to 1989

	Phytoplankton			Epipelon			Epiphyton			Metaphyton		
	Low	Medium	High	Low	Medium	High	Low	Medium	High	Low	Medium	High
1985	76	160	134	24	77	26	465	247	205	1,530	1,551	676
1986	137	153	174	1	54	1	478	410	282	2,952	2,064	1,445
1987	125	264	250	44	19	1	1,791	973	325	3,224	2,497	1,371
1988	280	237	271	44	46	17	1,399	1,120	381	3,950	3,202	1,833
1989	229	229	298	46	41	31	881	1,317	572	2,659	2,545	1,366
Mean	169	209	225	32	47	15	1,003	813	353	2,863	2,372	1,338

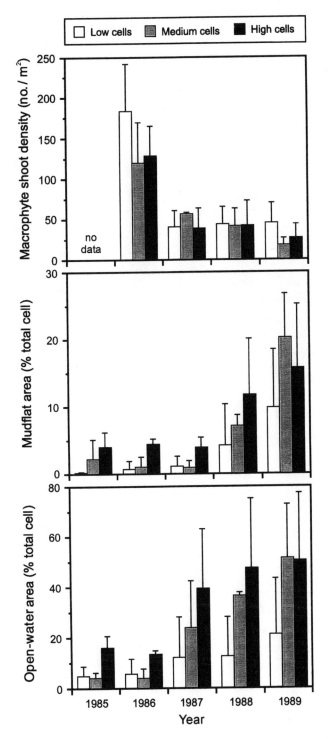

Figure 8.8 Changes in macrophyte density (top panel), unvegetated mudflat area (middle panel), and open-water area (bottom panel) in low- (white bars), medium- (gray bars), and high-water (black bars) treatments over the 5-year reflooding phase of the MERP experiment (1985–1989). Unpublished macrophyte data were kindly provided by A.G. van der Valk.

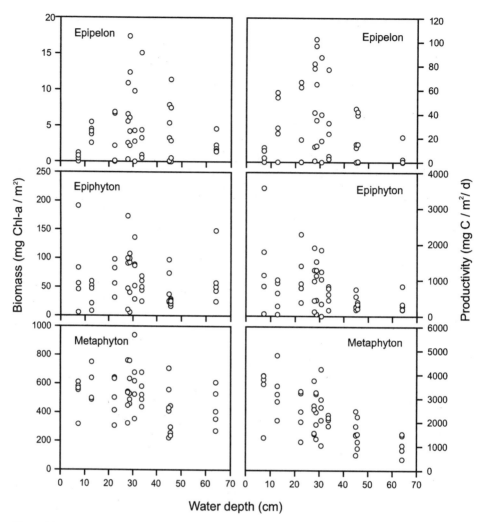

Figure 8.9   Relationships between epipelon, epiphyton, and metaphyton biomass (left panels; mg Chl-a/m²) and productivity (right panels; mg C/m²/d) (yearly means) and individual mean cell depth (cm) in the MERP cells (1985–1989). Phytoplankton was excluded due to autocorrelation of its biomass with water depth.

Wind

Winds can be sufficiently intense for convective circulation to resuspend sediments in shallow water bodies (Carper and Bachmann 1984). Kotak and Robinson (unpublished data) found that 53–90% of all wind events at an open, shallow (<1 m) site in Delta Marsh had sufficient velocity to cause sediment resuspension; these critical-velocity events were most frequent in May and declined through June, July, and August. A direct, curvilinear relationship between suspended particulate concentration in the water column and turbidity or light extinction was demonstrated using littoral enclosures in which turbulence was

Table 8.7. Comparison of mean algal biomass (mg Chl-a/m^2) in low-marsh cells following drawdown for 1 year (1984; cells 3 and 7) or 2 years (1983 and 1984; cells 4 and 8)

	1985 data only			1985–1989 data combined		
	1 year	2 years	p-value	1 year	2 years	p-value
Phytoplankton	5	2	0.015	5	4	ns
Epipelon	4	1	0.002	4	4	ns
Epiphyton	56	83	ns	56	91	0.028
Metaphyton	573	538	ns	573	539	ns

Note: ns, not significantly different at $p = 0.05$.

increased experimentally (Kotak and Robinson 1991). The most significant effects of this phenomenon are likely to include physical detachment of attached algal cells (Hoagland 1983) and decreased subsurface irradiance and photosynthesis (Hellström 1991). Goldsborough et al. (1986) found that photosynthesis of periphyton colonizing vertically positioned acrylic rods in Delta Marsh decreased by 75% over a depth of 40 cm, closely paralleling the light extinction profile. The irradiance at which photosynthesis by algal assemblages in MERP cells was maximal ranged from 282 to 407 μmole/m^2/s (Table 8.5), indicating that any light reduction below these values would limit algal production. On the positive side, turbulent resuspension of sediments can increase nutrient levels (Hamilton and Mitchell 1988), and turbulence may erode nutrient-depleted boundary layers around substrata on which algae grow. Under conditions of nutrient limitation both processes would benefit wetland algae.

The extent of wind-induced turbulence depends on the size of a water body and the degree of protection afforded by emergent vegetation (Dieter 1990; Brix 1994). In large, shallow, and exposed wetland water bodies, wind effects may be such that macrophytes are inhibited along with epiphyton and metaphyton, creating a situation dominated by phytoplankton and tychoplankton of epipelic origin. Phytoplankton flourishes in an environment where sedimentary loss of algal cells is reduced and nutrient levels are high. This would probably be exacerbated under prolonged stable water levels. However, many wetland water bodies are small and sheltered, and thus much less exposed to high winds. Under such situations, and the presence of macrophytic vegetation, wind-induced turbulence would be severely reduced by boundary-layer effects of vegetation, so relatively still-water conditions should prevail. Losee and Wetzel (1993) demonstrated that turbulent flow rates of ≈30 cm/s are dissipated by submersed macrophyte beds. Under these conditions abundant phytoplankton would not be expected due to sedimentary losses; epiphyton and metaphyton are likely to predominate, although localized nutrient limitation may exist in the microenvironment of dense epiphyton and metaphyton masses due to uneroded boundary layers. The above situation typifies the MERP cells as they existed from 1985 to 1989.

Nutrients

As indicated, nutrient availability to algae may be closely linked to hydrological conditions (e.g., flooding of highly vegetated areas, wind-induced turbulence), so nutrient supply probably varies considerably between wetlands. In oligotrophic wetlands, phytoplankton can exhibit extreme nitrogen and phosphorus deficiency (Murkin et al. 1991). In such cases, experimental nutrient additions do increase phytoplankton, epiphyton, and metaphyton biomass (Campeau et al. 1994) although nutrient deficiency (assessed by production of alkaline phosphatase, and carbon, nitrogen, and phosphorus composition ratios) persists, at least for phytoplankton and epiphyton (Murkin et al. 1994). Despite more eutrophic conditions, nutrient deficiency also occurs in Delta Marsh; Hooper-Reid and Robinson (1978) reported nitrogen, phosphorus, and silicon deficiency in periphyton growing on artificial substrata in a small marsh pond, although it developed late in the growing season when algal biomass was high. This situation may reflect a microenvironmental biomass-mediated condition, as indeed may all measures of deficiency in epiphyton and metaphyton. In addition to this caveat, the validity of nutrient-deficiency indicators is questionable in the absence of investigation of other possible limiting conditions (e.g., light availability). For example, parameters describing the relationship of irradiance and photosynthesis by epipelon, phytoplankton, epiphyton, and metaphyton in MERP cells (Table 8.5) show considerable unaccountable variability. It is known that this relationship is sensitive to variation in nutrient supply (Fee et al. 1987). Given these factors, we suggest that the nutrient relations of algae in prairie wetlands are insufficiently understood. Superficially, given the proliferation of algal biomass, the general fertility of prairie landscapes, extensive anthropogenic sources of nutrients, and the internal cycling potential, it is difficult to conceive of widespread nutrient limitation. Many prairie potholes support cyanobacterial blooms characteristic of phosphorus-rich hypertrophic lakes (Barica 1990), in which temporary nitrogen deficiency is compensated for by cyanobacterial nitrogen fixation, thus mitigating against general nutrient deficiency. It is unlikely that epipelon would be nutrient limited, having access to sediment interstitial nutrient levels and being in close proximity to a highly reducing environment. Metaphytic mats have considerable potential for internal cycling of nutrients. This may occur to a lesser extent in epiphyton so nutrient limitation of epiphyton and metaphyton is possibly only related to the thickness of the algal mat through which nutrients must diffuse and not to general nutrient availability. Nutrient limitation may be most likely to develop in phytoplankton, due to competition with other wetland plants, and only then under still-water conditions. We suggest that the matter of scale is worthy of consideration, in that nutrient availability in the microenvironments of benthic algae is not adequately assessed by bulk chemical analysis. Several advances have been made using ion-selective and quantum microelectrodes within benthic mats in other aquatic habitats (Revsbech and Jorgensen 1986); their application in freshwater wetlands is warranted. A further consideration on the relationship of wetland algal growth and nutrient supply is that any such relationship probably involves interactions with other factors. For example, McCormick and Stevenson (1989) found that nutrient additions had no overall effect on the biomass of readily grazed components of epiphyton (the "overstory") because any increase in algae was overwhelmed by increases in snail grazing pressure. However, nutrient additions stimulated the understory even as grazing pressure increased.

Light

The light environment in prairie wetlands is extremely variable. Not only are wetlands exposed to a seasonal cycle of incident radiation but also light reaching the water surface varies with the degree of shading by overlying macrophyte canopies (Brix 1994). Stands of *Phragmites* reduce light by up to 95% during periods of maximum growth (Higashi et al. 1981; Roos and Meulemans 1987). A range of 0–90% reduction of available light was found in the range of vegetation densities in MERP cells (Figure 8.10). The greatest shading phenomenon probably occurs in floating *Lemna* beds where transmittance may be ≈0.1% of surface irradiance (Goldsborough 1993). The degree of macrophyte shading is variable, with part of this variability being related to water depth. Less shading occurs in deeper water (Figure 8.10) because the canopy opens as the emergent plant assemblage changes and biomass decreases with increasing depth (Figure 8.8). Exposure and size of a wetland water body also influence the degree of macrophyte shading. At one extreme, large, highly exposed, turbulent water bodies may be practically devoid of macrophytic vegetation; in contrast, sheltered, small water bodies may be densely vegetated and possibly covered by *Lemna* mats.

Extinction of light within a wetland water column is variable, being dependent primarily on macrophyte density, metaphyton abundance, and turbulence caused by winds (Kotak and Robinson 1991; Klarer and Millie 1992) or sediment disruption by benthivorous fish such as carp (*Cyprinus carpio*) (Meijer et al. 1990; Cline et al. 1994). These sources of variability in extinction are not necessarily independent because macrophytes can dampen turbulence (Losee and Wetzel 1993; Brix 1994) and macrophyte density is, in turn, under the influence of water depth and, in the extreme case, turbulence itself. Furthermore, intense activity by carp can reduce macrophyte density (King and Hunt 1967). In sheltered MERP cells over a depth range of 7 to 64 cm, light extinction was highest in shallow water (Figure 8.10). Combined with the inverse relationship of surface shading and depth noted previously, the unexpected result was that light availability in the water column did not vary with depth, remaining at ≈35 to 45% of incident photosynthetically active radiation (PAR) at all sites. The applicability of this phenomenon in shallow, turbid wetlands is generally unknown.

Given the spatial and temporal variability in the light climate of wetlands, it is difficult to draw any general conclusions on its influence on algal assemblages. Although there was no consistent evidence of a depth effect on the biomass of epiphyton, epipelon, or phytoplankton per unit of substratum area or volume in the MERP cells, vertical distribution of epiphytes is influenced by depth in some environments (Wetzel 1983). Indeed, some studies of wetland epiphyton have found that algal development is restricted among dense macrophytic vegetation, possibly due to combined effects of intense shading and physical abrasion (Hooper and Robinson 1976; Gabor et al. 1994; Murkin et al. 1994). Further evi-

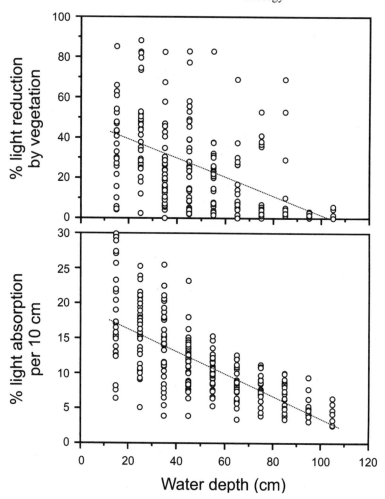

Figure 8.10   Relationship between the reduction of incident PAR (expressed as a percentage of incident PAR) by macrophytic vegetation shading and water depth in MERP cells (top panel), and relationship between the absorption of incident radiation within the water column and water depth in the MERP cells (bottom panel).

dence comes from manipulative studies in which wetland epiphyton, epipelon, and phy-toplankton biomass increased following macrophyte removal (Wrubleski 1991). Photosynthesis-saturating irradiances for algae in the MERP cells ranged from 282 to 407 µmole/m²/s (Table 8.5). This, coupled with the fact that instantaneous surface irradiance in the geographical range of prairie wetlands often falls below 500 µmole/m²/s in the autumn, on cloudy days, and early and late in each day, it is probable that algae are light limited for portions of every day. Light limitation of epiphyton and metaphyton may occur even more frequently when their biomasses are high, due to self-shading. Irradiance below a thick metaphyton mat in the MERP cells commonly fell below 20% of surface values (Figure 8.7). As epiphyton biomass in Delta Marsh increases, photosynthesis per unit biomass (i.e., specific productivity) decreases (Goldsborough and Robinson 1996). This may be related to light attenuation within the assemblage itself. A similar phenomenon was reported within metaphytic masses in the littoral zone of an acidified lake (Turner et al. 1995). Conversely, epiphyton situated near the water surface might experience photoinhibitory irradiances during periods of high water clarity, as indicated by results of species-specific photosynthesis measurements (Robinson and Pip 1983).

Temperature

It is generally assumed that the characteristic shallowness and susceptibility to turbulence of wetland water bodies precludes the development of prolonged vertical temperature gradients. However, *Lemna* and metaphyton mats developing in sheltered, calm waters cause rapid light extinction (Figure 8.7), leading to sharp thermal profiles within short distances of the water surface (Goldsborough 1993; Hillebrand 1983). On a smaller scale, and one that is appropriate for epiphyton response, Dale and Gillespie (1976) recorded substantial diurnal temperature changes associated with substrata. An investigation of temperature profiles within the dark-colored surfaces of illuminated marsh sediments by using closely spaced microthermocouples demonstrated highly transitory warming of sediment surfaces (Robinson, unpublished data).

Algal metabolic processes are temperature dependent, often with Q_{10} values in excess of 2 (Stewart 1974). Prairie wetlands are typically subject to spatial (e.g., through metaphyton mats) and seasonal variations in temperature of at least 25°C, which implies that temperature fluctuations do affect algal growth over time. Both parameters of photosynthesis–irradiance relationships, the photosynthetic efficiency (α) and maximum rate of specific productivity (P_{max}), and indeed I_k, which is derived from α and P_{max}, are temperature dependent (Prezelin and Ley 1980; Tilzer et al. 1986). Yet, temperature does not explain any significant amount of the observed variability of these parameters for wetland algae (Table 8.8).

Table 8.8. Coefficient of determination (r^2) for the multiple regression of parameters α (photosynthetic efficiency), β (photoinhibition), or P_{max} (light-saturated photosynthetic rate) on ambient water temperature, algal biomass (chlorophyll *a*), and 10-day irradiance history for each algal assemblage in the flooded wetland cells

Algal assemblage	r^2		
	α	β	P_{max}
Phytoplankton	0.024	0.065	0.134
Epipelon	0.132	0.040	0.251
Epiphyton	0.231	0.067	0.211
Metaphyton	0.248	0.193	0.196

Anthropogenic Factors

There are few quantitative data on effects of human activities on wetland algae. It is inevitable that prairie wetlands will be contaminated, to varying degrees, by herbicides, insecticides, eroded sediments, and fertilizers given their proximity to areas subject to chemical-intensive agriculture (Neely and Baker 1989; Goldsborough and Crumpton 1998). The extent to which these additions impact algae directly, through herbicidal toxicity, and indirectly, through regulation of grazer abundance, is chemical- and concentration-dependent. Work in Delta Marsh with a broad suite of agricultural herbicides has demonstrated their ability to reduce the biomass of epiphyton, epipelon, and macrophytes, alter the trajectory of algal community succession (Goldsborough and Robinson 1986), and stimulate sediment efflux of nutrients (Goldsborough and Robinson 1985; Gurney and Robinson 1989), leading to proliferation of phytoplankton (Goldsborough, unpublished data). Nutrients in chemical fertilizer and biological waste stimulate metaphyton (Campeau et al. 1994; Gabor et al. 1994; Murkin et al. 1994; McDougal et al. 1997) which, if ungrazed, reduce the energy supply to higher wetland consumers. Stabilization of water levels in lakes to which some prairie wetlands connect has promoted sedimentation and encroachment by emergent and terrestrial vegetation into open water (Shay and Shay 1986), altering the availability and suitability of substrata for algae. Finally, the impact of plant and animal species introduced into prairie wetlands by humans (e.g., *Lythrum salicaria*, purple loosestrife; common carp) may have numerous, yet undocumented, ecological effects.

Variability

Throughout much of this consideration of physical and chemical factors that influence algal assemblages there is a recurring theme of variability, which is worth considering in a larger context. There is increasing evidence that variability is inversely related to the size of aquatic ecosystems (Fee et al. 1987), with large, deep lakes having the least variability in controlling parameters and the greatest stability. We submit that wetlands represent the opposite extreme, and consequently perhaps the greatest challenge in understanding the sources of variability that exist. An excellent example of this situation is seen in the extensive, and for the most part unaccountable, variability in the relationship between algal photosynthetic rate and irradiance (Table 8.8). As long as the sources of variability in these relationships are unresolved, we cannot characterize fully the nature and magnitude of roles played by algae in wetlands.

CONCEPTUAL MODEL OF WETLAND ALGAE

Considerable debate has centered on the importance of "bottom-up," as opposed to "top-down" control of lake ecosystems (Carpenter et al. 1985). To the extent that biomanipulative experiments involving removal of zooplanktivorous fish have led to changes in macrophytes and algal communities in shallow lakes (Hanson and Butler 1994), it is clear that herbivory plays a major role in determining primary producer biomass in these ecosystems. However, to our knowledge, there have been no experiments conducted to test whether manipulations of primary producer abundance in wetlands lead ultimately to changes in the biomass of subsequent trophic levels. It may be overly simplistic to expect either mechanism to predominate; wetland algae may be alternately regulated by bottom-up (nutrient enrichment) and top-down (grazing) controls, depending on conditions at a given time and place. A robust model of their abundance should account for both possibilities.

Murkin (1989) described the algae in northern prairie wetlands, in the context of their role in the aquatic food web, according to a model developed for ephemeral Iowa wetlands by van der Valk and Davis (1978). According to the van der Valk and Davis (VD) model, prairie wetlands experience regular or intermittent drawdowns due to drought, leaving them at a dry marsh state. When reflooding occurs, macrophytes established from the sediment seed bank produce a regenerating marsh in which there is little open water and abundant plant cover. Gradually, high-water stress reduces macrophyte abundance in the deepest portions of the wetland, creating a degenerating marsh. Macrophytes can be subsequently destroyed by high water, disease, insects, muskrat feeding, and natural senescence, leading to an open lake marsh with emergent macrophytes around its periphery. Drought then cycles the wetland back to the dry marsh. As applied to Delta Marsh, unperturbed channels and flooded areas most closely match the degenerating marsh and regenerating marsh states, respectively. Wet-sedge meadows exist at the periphery of discrete water bodies, whereas large, open bays with few emergent macrophytes are the closest, albeit imperfect, match to the lake marsh state.

By considering algal habitat requirements relative to availability of those habitats in each stage of the VD model, we can predict the relative abundances of epipelon, epiphyton, metaphyton, and phytoplankton during wetland development (Figure 8.11). The dry marsh, lacking sufficient submersed substrata to support epiphyton or metaphyton, and having insufficient water for phytoplankton, is dominated by epipelon and plocon crusts on moist, exposed sediments. There have been no quantitative studies of algae in dry wetlands; however, their contribution to total primary production is probably low, given that the overlying macrophyte canopy can reduce light sufficiently that algae at the sediment surface are severely light limited. Following wetland flooding, profuse macrophyte development provides abundant colonizable substrata for algae that, by using nutrients liberated from reflooded soil and decomposing vegetation, flourish and detach from macrophytes, yielding metaphyton. Hence, metaphyton blooms may be symptomatic of the early stages in the reflooding of a dry marsh (the regenerating marsh). Compared with a dry marsh, production in the regenerating state is high. With passing time, nutrients in the water column are assimilated by profuse algal mass so further growth is limited. Combined with reductions in macrophyte biomass as the wetland progresses to the degenerating marsh state, this reduces the potential for metaphyton growth. Therefore, epiphyton assumes dominance at the degenerating marsh stage, although total algal production is probably somewhat lower than at the regenerating state. Finally, as macrophytes are eliminated through senescence, disease, herbivory by muskrats, shading by epiphytes and

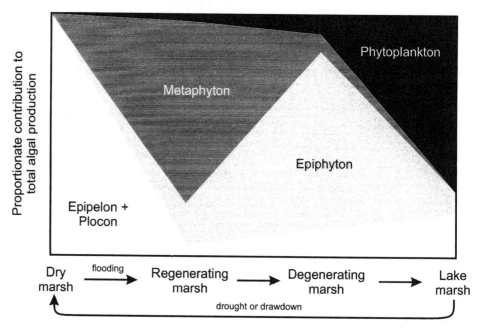

Figure 8.11   Predicted contributions to total algal production by epipelon + plocon, epiphyton, metaphyton, and phytoplankton at four stages in wetland development, according to the model of van der Valk and Davis (1978).

metaphyton, and flood stress, loss of substrata causes epiphyton to decline and phytoplankton to assume dominance. Although its biomass may be high during periodic blooms, overall algal biomass is probably lower than in any stage but the dry marsh.

The VD model is a useful basis for a conceptual model of wetland algal abundance because it defines four states, each dominated by epipelon, epiphyton, metaphyton, or phytoplankton (Figure 8.11). However, there are several shortcomings of the model that we suggest compromise its generality. Perhaps most critically, the model assumes that hydrological change drives wetland ontogeny. For shallow potholes such as those studied by van der Valk and Davis (1978), this may be a fair assumption. It may be less appropriate for wetlands that undergo limited fluctuations in water level, such as those in climatically more stable neotropical or subarctic regions, or in wetlands such as Delta Marsh, where water level is artificially stabilized. In such situations, transitory developmental stages can persist for decades. Many of the shallow pothole lakes in western Manitoba, studied by Barica (1975) and others in the early 1970s, have supported massive phytoplankton blooms for >25 years, possibly because, having reached the lake marsh state, they have not undergone drawdown to return to a dry marsh. Studies of anthropogenically impacted wetlands in the United States (Hanson and Butler 1994) and England (Moss 1983) also have demonstrated that permanent displacement of benthic algae by phytoplankton is common. However, phytoplankton dominance is not the only stable endpoint in wetlands that lack periodic major changes in water level. A shallow, isolated pond within Delta Marsh has supported dense metaphyton mats over much of its surface intermittently since at least the 1970s (Brown 1972), whereas the main marsh channels have supported dense epiphyton on submersed macrophytes, with no perceptible metaphyton. Another difficulty with the VD model is its implied assumption that developmental stages occur in a fixed sequence that precludes reversal or redirection of development due to anthropogenic and other environmental influences. Consequently, the model cannot accommodate the impacts of cultural nutrient loading and biomanipulations. The role of algae in wetlands, and the possibility that algal assemblages modify the outcome of wetland development, also is not considered in the VD model. For example, shade from dense epiphyton and phytoplankton can reduce submersed macrophyte growth and abundance (Sand-Jensen 1977; Phillips et al. 1978; Losee and Wetzel 1983). We also have evidence that occlusion of the water surface by metaphyton mats caused the loss of *Potamogeton* spp. in MERP cells (Robinson et al., unpublished data).

Our modification of the VD model (Goldsborough and Robinson 1996) explicitly recognizes the role and function of algae in wetlands, while considering that wetland development can comprise four alternative stable states (cf. alternative equilibria of Scheffer et al. [1993]), dominated by epipelon, epiphyton, metaphyton, or phytoplankton (Figure 8.12). Unlike the VD model, this scheme enforces no directionality on shifts from one state to another, and it assumes that development of a specific wetland depends on the combined, interacting effects of nutrient loading, grazing pressure, and water-column stability. These factors determine the duration that a specific state will persist. The endpoints of wetland development, the lake state and dry state, are essentially the same as those in the VD model. The other two states, open state and sheltered state, are less clearly correlated because they each encompass some features of the regenerating and degenerating states. The environmental cues dictating the persistence and transition of wetlands from one state to another are similar to those listed for the VD model (Figure 8.11) and are discussed by Goldsborough and Robinson (1996).

In the context of the reflooding phase of the MERP experiment (1985–1989), this conceptual model describes a transition between three of the four states. Prior to reflooding in early 1985, the MERP cells were at the dry state. Upon addition of water and immediate loss of wet-meadow vegetation and more gradual loss of emergent macrophytes, there was a short-lived period in which epiphyton flourished on remaining macrophytes (the open state). Although there is no evidence that ambient nutrient levels increased due to flooding (Kadlec 1986), it is possible that rapid assimilation by quickly growing algae masked any increase in nutrients in the water column. It is particularly interesting that the MERP cells did not stabilize at this epiphyton-dominant state (that characterizes many of the unenclosed, well-vegetated channels in Delta Marsh) but progressed to the sheltered state in which metaphyton flourished. The stimulus for metaphyton proliferation is not clear. It may relate to wetland reflooding because abundant metaphyton has developed in newly flooded wetlands elsewhere (Wu and Mitsch 1998). There is also abundant anec-

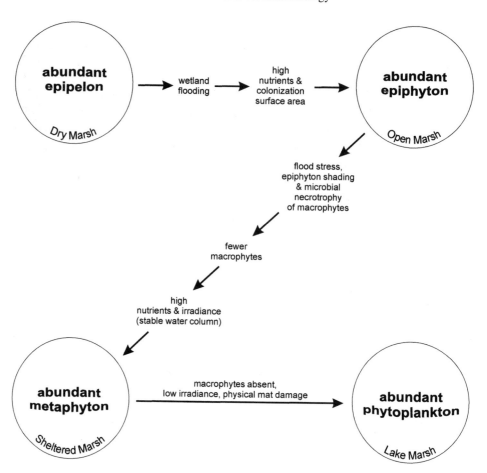

Figure 8.12 Conceptual model, modified from Goldsborough and Robinson (1995), of the progression of MERP cells from a drawndown dry state, through a transient macrophyte–epiphyte open state, to a more persistent metaphyton-dominated sheltered state. There is some evidence that deep-flooded cells were progressing to phytoplankton-dominated lake state although the transition was not completed during the reflooding phase (1985–1989) of the MERP experiment.

dotal and some quantitative evidence that species comprising metaphyton mats respond positively to high irradiance and nutrient status (Whitton 1970; McDougal et al. 1997), and they may have few grazers to keep their biomass in check (Dodds 1991; Neill and Cornwell 1992). To some extent, the rapid light extinction and intense nutrient competition resulting from the establishment of metaphyton mats in and on the water surface, combined with their relative freedom from grazing and the absence of large consumers (e.g., fish) from MERP cells, may cause them to become self-stabilizing, preventing their displacement by substratum-associated epiphytes or phytoplankton. This might explain why profuse metaphyton mats persisted throughout the 5-year reflooding phase, although evidence of decline was observed in the fifth year. However, metaphyton is at least partially dependent on continued macrophyte cover (and its epiphytes) for algal inoculum and stabilization of the water column, as metaphytic mats were least well developed in turbulent, open-water areas of the MERP cells (Gurney and Robinson 1988). Although none of the MERP cells developed phytoplankton blooms of the magnitude occurring in other nutrient-rich shallow waters in Manitoba (Barica 1975) and elsewhere in Delta Marsh (Hosseini 1986; Goldsborough, unpublished data), there was some evidence that the deepest cells supported slightly more phytoplankton (Table 8.6) and that a transition to the lake state was in progress during the final 2 years of the experiment (1988–1989). Its progress might have been accelerated had the depth of flooding been greater, leading to more rapid elimination of macrophytes and consequent loss of metaphyton.

MANAGEMENT IMPLICATIONS

The proposed model provides a convenient framework to evaluate impacts of both intentional and accidental human intervention in prairie wetland ecosystems. Included are effects of water-level manipulation (and stabilization) on all components of the ecosystem; imposition of bottom-up ecosystem control through input of nutrients from sewage, agricultural fertilizers and soil; and top-down biomanipulation of the ecosystem through introduction or removal of piscivorous fish and other top predators. From the standpoint of providing energy to wetland herbivores, epiphyton is probably the most desirable algal assemblage because it consists of small, easily assimilable cells (more so than large filamentous green algae comprising metaphyton), is relatively abundant compared with epipelon and plocon, and is less prone to erratic fluctuations in biomass characteristic of phytoplankton-dominated shallow systems. Also implicit in a wetland dominated by epiphyton are abundant submersed and emergent macrophytes. Hence, maintenance of the open state is a desirable endpoint of wetland management and can be achieved through controlled flooding of a dry marsh, although prevention of metaphyton dominance, such as occurred in MERP cells, would probably require that populations of large consumers be augmented to maintain grazing pressure and physical disturbance. For example, we propose that the absence of metaphyton from unmanipulated areas of Delta Marsh is due in part to the pronounced physical disturbance of sediments and macrophytes by carp that invaded the marsh in the 1960s. They may have replaced the disturbance lost when water levels on the adjacent lake were stabilized, coincidentally, at about the same time. The depth of flooding, however, should be controlled carefully to avoid eliminating macrophytes that stabilize the water column and provide substratum for epiphytes, in the absence of which phytoplankton flourish. Control of epiphyton abundance, by grazers, physical disturbance, and wind-induced variation in water transparency is also necessary to ensure they do not cause macrophytes to decline. Nutrient inputs should be minimized, where possible because they promote metaphyton development (Gabor et al. 1994; Murkin et al. 1994; McDougal et al. 1997) and can lead to excessive biological oxygen demand either diurnally or during decomposition.

RESEARCH NEEDS

A full consideration of the contribution by the MERP project to wetland ecology must recognize its limitations. Differences in water level between low, medium, and high cells during the reflooding phase were less than internal variation in topography within a given cell, with the result that it was difficult to isolate depth-related effects between cells from depth-related effects within cells. Heterogeneity between cells comprising one "treatment" lead necessarily to spatial heterogeneity in algal biomass that confounded attempts to generalize from the results. The restriction on top-predators imposed by the use of pumps to fill and maintain water levels in MERP cells (no fish such as carp to provide physical disturbance or minnows to consume metaphyton) meant that the ecosystems created within cells did not truly simulate those existing in the unenclosed marsh. Although it is unclear how significant top-down mechanisms of ecosystem control are in prairie wetlands, exclusion of top fish-consumers may have significantly affected observed algal responses.

A consideration of the role of algae in freshwater wetlands lags well behind most other areas of wetland research. Consequently, many areas of productive study remain to be addressed. The following list outlines some particularly pertinent and important issues for future wetland algal research:

1. The wetlands that predominate much of the southern prairie provinces are sufficiently saline to be considered salt marshes. Compared with a rich literature on the algae of coastal salt marshes, comparatively little is known about the algal assemblages of inland saline wetlands.

2. Most research on freshwater wetland algae has focused on epiphyton. Consequently, conclusions about the significance of algae in dry wetlands are largely speculative. Work is needed on the epipelon and plocon of freshwater wetlands, with particular focus on their contribution to total primary production and the nitrogen budget, via their nitrogen fixation, of wet meadows and other semidry systems.

3. Most existing data on algal abundance in wetlands derive from short-term studies and are expressed relative to substratum surface area. Quantification of annual trends in algal primary production, calculated on a wetland areal basis, is needed to provide reference data for assessment of changes in the wetland food web due to external nutrient loading, introduced plant and animal species, climatic change, increases in ultraviolet radiation, etc.

4. Additional photosynthesis–irradiance data, and clearer understanding of the environmental controls on photosynthetic parameters, are needed to permit detailed modeling of primary production in wetlands.

5. The niche requirements of metaphyton and their interaction with macrophyte stands should be clarified. This should include consideration of the degree to which metaphyton biomass is used by wetland consumers. More generally, the wetland food web must be more clearly defined, with particular focus on the use of wetland algae, as opposed to macrophyte detritus, by invertebrate and vertebrate grazers. The role of detritivorous fish such as carp in providing the necessary physical disturbance for maintenance of epiphy-

ton–macrophyte dominance in open wetlands should be investigated.

6. Given the increasing likelihood of contamination of prairie wetlands by a wide range of anthropogenic substances, research on their impact on wetland plants generally, and algae specifically, is needed to address the potential implications for ecological sustainability of the wetland ecosystem.

ACKNOWLEDGMENTS

This project was supported financially by the Marsh Ecology Research Program (MERP), a collaborative project of the Delta Waterfowl and Wetlands Research Station and Ducks Unlimited Canada. GGCR acknowledges the additional support of the University of Manitoba, and GGCR and LGG acknowledge the support of Research Grants from the Natural Sciences and Engineering Research Council of Canada.

Teams of field assistants ("Merpies") helped conduct the field sampling program and Colleen Barber, Leslie Goodman, and Nicole Armstrong assisted with data collation and analyses. Logistic support for sample collection and analysis was provided by Henry Murkin, Bruce Batt, the Delta Waterfowl and Wetlands Research Station, Ducks Unlimited Canada through the Institute for Wetland and Waterfowl Research, and the University Field Station (Delta Marsh). Dale Wrubleski permitted the use of his unpublished stable isotope data.

This is Paper 101 of the Marsh Ecology Research Program, Paper 261 of the University Field Station (Delta Marsh), and Paper 8 of the Prairie Wetland Ecology Team.

LITERATURE CITED

Adams, M.S., and W. Stone. 1973. Field studies of photosynthesis of *Cladophora glomerata* (Chlorophyta) in Green Bay, Lake Michigan. Ecology 54:853-862.

Auer, M.T., J.M. Graham, L.E. Graham, and J.A. Kranzfelder. 1983. Factors regulating spatial and temporal distribution of *Cladophora* and *Ulothrix* in the Laurentian Great Lakes. In *Periphyton of Freshwater Ecosystems*. (Ed.) R.G. Wetzel, pp.135-145. The Hague, Netherlands: Dr W. Junk Publishers.

Barica, J. 1975. Collapses of algal blooms in prairie pothole lakes: their mechanism and ecological impact. Verhandlungen Internationale Vereinigung für Theoretische und Angewandte Limnologie 19:606-615.

Barica, J. 1990. Seasonal variability of N:P ratios in eutrophic lakes. Hydrobiologia 191:97-103.

Brix, H. 1994. Functions of macrophytes in constructed wetlands. Water Science and Technology 29:71-78.

Brönmark, C. 1989. Interactions between epiphytes, macrophytes and freshwater snails: a review. Journal of Mollusc Studies 55:299-311.

Brown, D.J. 1972. Primary production and seasonal succession of the phytoplankton component of Crescent Pond, Delta Marsh, Manitoba. M.S. thesis, University of Manitoba, Winnipeg, Manitoba.

Campeau, S., H.R. Murkin, and R.D. Titman. 1994. Relative importance of algae and emergent plant litter to freshwater marsh invertebrates. Canadian Journal of Fisheries and Aquatic Science 51:681-692.

Carlton, R.G., and R.G. Wetzel. 1988. Phosphorus flux from lake sediments: effect of epipelic algal oxygen production. Limnology and Oceanography 33:562-570.

Carpenter, S.R., J.F. Kitchell, and J.R. Hodgson. 1985. Cascading trophic interactions and lake productivity. Bioscience 35:634-639.

Carper, G.L., and R.W. Bachmann. 1984. Wind resuspension of sediments in a prairie lake. Canadian Journal of Fisheries and Aquatic Sciences 41:1763-1767.

Cattaneo, A. 1983. Grazing on epiphytes. Limnology and Oceanography 28:124-132.

Cline, J.M., T.L. East, and S.T. Threlkeld. 1994. Fish interactions with the sediment-water interface. Hydrobiologia 275/276:301-311.

Cronk, J.K., and W.J. Mitsch 1994. Periphyton productivity on artificial and natural surfaces in constructed freshwater wetlands under different hydrologic regimes. Aquatic Botany 48:325-341.

Crumpton, W.G. 1989. Algae in northern prairie wetlands. In *Northern Prairie Wetlands*. (Ed.) A.G. van der Valk, pp.188-203. Ames: Iowa State University Press.

Dale, H.M., and T. Gillespie. 1976. The influence of floating vascular plants on the diurnal fluctuations of temperature near the water surface in early spring. Hydrobiologia 49:245-256.

Dickman, M. 1968. The effect of grazing by tadpoles on the structure of periphyton community. Ecology 49:1188-1190.

Dieter, C.D. 1990. The importance of emergent vegetation in reducing sediment resuspension in wetlands. Journal of Freshwater Ecology 5:467-473.

Dodds, W.K. 1991. Community interactions between the filamentous alga *Cladopora glomerata* (L.) Kützing, its epiphytes, and epiphytic grazers. Oecologia (Berl.) 85:572-580.

Dodds, W.K., and D.A. Gudder. 1992. The ecology of *Cladophora*. Journal of Phycology 28:415-427.

Eiseltová , M., and J. Pokorný. 1994. Filamentous algae in fish ponds of the Trebon Biosphere Reserve-ecophysiological study. Vegetatio 113:115-170.

Fee, E.J., R.E. Hecky, and H.A. Welch. 1987. Phytoplankton photosynthesis parameters in central Canadian lakes. Journal of Plankton Research 9:305-316.

Gabor, T.S., H.R. Murkin, M.P. Stainton, J.A. Boughen, and R.D. Titman. 1994. Nutrient additions to wetlands in the interlake region of Manitoba, Canada: effects of a single pulse addition in spring. Hydrobiologia 280:497-510.

Goldsborough, L.G. 1993. Diatom ecology in the phyllosphere of the common duckweed (*Lemna minor* L.). Hydrobiologia 269/270:463-471.

Goldsborough, L.G., and W.G. Crumpton. 1998. Distribution and environmental fate of pesticides in prairie wetlands. Great Plains Research 8:73-95.

Goldsborough, L.G., and G.G.C. Robinson. 1985. Effect of an aquatic herbicide on sediment nutrient flux in a freshwater marsh. Hydrobiologia 122:121-128.

Goldsborough, L.G., and G.G.C. Robinson. 1986. Changes in periphytic algal community structure as a consequence of short herbicide exposure. Hydrobiologia 139:177-192.

Goldsborough, L.G., and G.G.C. Robinson. 1996. Pattern in Wetlands. In *Algal Ecology in Freshwater Benthic Ecosystems*. (Eds.) R.J. Stevenson, M.L. Bothwell and R.L. Lowe. New York: Academic Press.

Goldsborough, L.G., G.G.C. Robinson, and S.E. Gurney. 1986. An enclosure/ substratum system for in situ ecological studies of periphyton. Archiv für Hydrobiologie 106:373-393.

Good, R.E., D.F. Whigham, and R.L. Simpson. 1978. *Freshwater Wetlands: Ecological Processes and Management Potential*. New York: Academic Press.

Grimshaw, H.J., M. Rosen, D.R. Swift, K. Rodberg, and J.M. Noel. 1993. Marsh phosphorus concentrations, phosphorus content and species composition of Everglades periphyton communities. Archiv für Hydrobiologie 128:257-276.

Gurney, S.E., and G.G.C. Robinson. 1988. The influence of water level manipulation on metaphyton production in a temperate freshwater marsh. Verhandlungen Internationale Vereinigung für Theoretische und Angewandte Limnologie 23:1032-1040.

Gurney, S.E., and G.G.C. Robinson. 1989. The influence of two triazine herbicides on the productivity, biomass and. community composition of freshwater marsh periphyton. Aquatic Botany 36:1-22.

Haines, D.W., K.H. Rogers, and F.E.J. Rogers. 1987. Loose and firmly attached epiphyton, their relative contributions to algal and bacterial carbon productivity in a *Phragmites* marsh. Aquatic Botany 29:169-176.

Hamilton, D.P., and S.F. Mitchell. 1988. Effects of wind on nitrogen, phosphorus, and chlorophyll in a shallow New Zealand lake. Verhandlungen Internationale Vereinigung für Theoretische und Angewandte Limnologie 23:624-628.

Hann, B.J. 1991. Invertebrate grazer-periphyton interactions in a eutrophic marsh pond. Freshwater Biology 26:87-96.

Hanson, M.A., and M.G. Butler. 1994. Responses to food web manipulation in a shallow waterfowl lake. Hydrobiologia 279/280:457-466.

Hansson, L.-A. 1988. Effects of competitive interactions on the biomass development of planktonic and periphytic algae in lakes. Limnology and Oceanography 33:121-128.

Hansson, L.-A. 1989. The influence of a periphytic biolayer on phosphorus exchange between substrate and water. Archiv für Hydrobiologie 115:21-26.

Hansson, L. 1990. Quantifying the impact of periphyton algae on nutrient availability for phytoplankton. Freshwater Biology 24:265-273.

Hellström, T. 1991. The effect of resuspension on algal production in a shallow lake. Hydrobiologia 213:183-190.

Hesslein, R.H. 1976. An in situ sampler for close interval pore water studies. Limnology and Oceanography 21:912-914.

Higashi, M., T. Miura, K. Tanimizu, and Y. Iwasa. 1981. Effect of the feeding activity of snails on the biomass and productivity of an algal community attached to a reed stem. Verhandlungen Internationale Vereinigung für Theoretische und Angewandte Limnologie 21:590-595.

Hillebrand, H. 1983. Development and dynamics of floating clusters of filamentous algae. In *Periphyton of Freshwater Ecosystems*. (Ed.) R.G. Wetzel, pp.31-39. The Hague, Netherlands: Dr W. Junk Publishers.

Hoagland, K.D. 1983. Short-term standing crop and diversity of periphytic diatoms in a eutrophic reservoir. Journal of Phycology 19:30-38.

Hooper, N.M., and G.G.C. Robinson. 1976. Primary production of epiphytic algae in a marsh pond. Canadian Journal of Botany 54:2810-2815.

Hooper-Reid, N.M., and G.G.C. Robinson. 1978. Seasonal dynamics of epiphytic algal growth in a marsh pond: composition, metabolism, and nutrient availability. Canadian Journal of Botany 56:2441-2448.

Hosseini, S.M. 1986. The effects of water level fluctuations on algal communities of freshwater marshes. Ph.D. dissertation, Iowa State University, Ames, Iowa.

Hosseini, S.M., and A.G. van der Valk. 1989a. Primary productivity and biomass of periphyton and phytoplankton in flooded freshwater marshes. In *Freshwater Wetlands and Wildlife*. (Eds.) R.R. Sharitz and J.W. Gibbons, pp.303-315. USDOE Symposium Series No. 61, Oak Ridge: USDOE Office of Scientific and Technical Information.

Hosseini, S.M., and A.G. van der Valk. 1989b. The impact of prolonged, above-normal flooding on metaphyton in a freshwater marsh. In *Freshwater Wetlands and Wildlife*. (Eds.) R.R. Sharitz and J.W. Gibbons, pp.317-324. USDOE Symposium Series No. 61, Oak Ridge: USDOE Office of Scientific and Technical Information.

Howard-Williams, C. 1981. Studies on the ability of a *Potamogeton pectinatus* community to remove dissolved nitrogen and phosphorus compounds from lake water. Journal of Applied Ecology 18:619-637.

Jansson, M. 1980. Role of benthic algae in transport of nitrogen from sediment to lake water in a shallow clearwater lake. Archiv für Hydrobiologie 89:101-109.

Kadlec, J.A. 1979. Nitrogen and phosphorus dynamics in inland freshwater wetlands. In *Waterfowl and Wetlands: An Integrated Review*. (Ed.) T.A. Bookhout, pp.17-41. Madison: The Wildlife Society.

Kadlec, J.A. 1986. Effects of flooding on dissolved and suspended nutrients in small diked marshes. Canadian Journal of Fisheries and Aquatic Science 43:1999-2008.

Kairesalo, T., and I. Koskimies. 1987. Grazing by oligochaetes and snails on epiphytes. Freshwater Biology 17:317-324.

King, D.R., and G.S. Hunt. 1967. Effect of carp on vegetation in a Lake Erie marsh. Journal of Wildlife Management 31: 181-188.

Klarer, D.M., and D.F. Millie. 1992. Aquatic macrophytes and algae at Old Woman Creek estuary and other Great Lakes coastal wetlands. Journal of Great Lakes Research 18:622-633.

Kotak, B.G. 1990. The effects of water turbulence on the limnology of a shallow, prairie wetland. M.S. thesis, University of Manitoba, Winnipeg, Manitoba.

Kotak, B.G., and G.G.C. Robinson. 1991. Artificially-induced water turbulence and the physical and biological features within small enclosures. Archiv für Hydrobiologie 122:335-349.

Losee, R.F., and R.G. Wetzel. 1983. Selective light attenuation by the periphyton complex. In *Periphyton of Freshwater Ecosystems*. (Ed.) R.G. Wetzel, pp.89-96. The Hague, Netherlands: Dr W. Junk Publishers.

Losee, R.F., and R.G. Wetzel. 1993. Littoral flow rates within and around submersed macrophyte communities. Freshwater Biology 29:7-17.

Mason, C.F., and R.J. Bryant. 1975. Periphyton production and grazing by chironomids in Alderfen Broad, Norfolk. Freshwater Biology 5:271-277.

McCormick, P.V., and R.J. Stevenson. 1989. Effects of snail grazing on benthic algal community structure in different nutrient environments. Journal of the North American Benthological Society 8:162-172.

McDougal, R.L., L.G. Goldsborough, and B.J. Hann. 1997. Responses of a prairie wetland to inorganic nitrogen and phosphorus: production by planktonic and benthic algae. Archiv für Hydrobiologie 140:145-167.

Meijer, M.A., M.W. deHaan, A.W. Breukelaar, and H. Buiteveld. 1990. Is reduction of the benthivorous fish an important cause of high transparency following biomanipulation in shallow lakes? Hydrobiologia 200/201:303-315.

Moss, B. 1983. The Norfolk Broadland: experiments in the restoration of a complex wetland. Biological Review 58:521-561.

Murkin, H.R. 1989. The basis for food chains in prairie wetlands. In *Northern Prairie Wetlands*. (Ed.) A.G. van der Valk, pp.316-338. Ames: Iowa State University Press.

Murkin, H.R., B.D.ÿJ. Batt, P.J. Caldwell, C.B. Davis, J.A. Kadlec, and A.G. van der Valk. 1984. Perspectives on the Delta Waterfowl Research Station - Ducks Unlimited Canada Marsh Ecology Research Program. Transactions of the North American Wildlife and Nature Research Conference 49:253-261.

Murkin, H.R., J.B. Pollard, M.P. Stainton, J.A. Boughen, and R.D. Titman. 1994. Nutrient additions to wetlands in the interlake region of Manitoba, Canada: effects of periodic additions throughout the growing season. Hydrobiologia 280:483-495.

Murkin, H.R., M.P. Stainton, J.A. Boughen, J.B. Pollard, and R.D. Titman. 1991. Nutrient status of wetlands in the interlake region of Manitoba, Canada. Wetlands 11:105-122.

National Wetlands Working Group (NWWG) 1998. The Canadian Wetland Classification System, 2nd edition. Wetlands Research Centre, Waterloo: University of Waterloo.

Neely, R.K., and J.L. Baker. 1989. Nitrogen and phosphorus dynamics and the fate of agricultural runoff. In *Northern Prairie Wetlands*. (Ed.) A.G. van der Valk, pp.92-131. Ames: Iowa State University Press.

Neill, C., and J.C. Cornwell. 1992. Stable carbon, nitrogen, and sulfur isotopes in a prairie marsh food web. Wetlands 12:217-224.

Peterson, B.J., and R.W. Howarth. 1987. Sulfur, carbon, and nitrogen isotopes used to trace organic matter flow in the salt-marsh estuaries of Sapelo Island, Georgia. Limnology and Oceanography 32:1195-1213.

Phillips, G.L., D. Eminson, and B. Moss. 1978. A mechanism to account for macrophyte decline in progressively eutrophicated freshwaters. Aquatic Botany 4:103-126.

Pinckney, J., and R.G. Zingmark. 1993. Modeling the annual production of intertidal benthic microalgae in estuarine ecosystems. Journal of Phycology 29:396-407.

Porter, K.G. 1977. The plant-animal interface of freshwater ecosystems. American Scientist 65:159-170.

Power, M.E. 1990. Effects of fish in river food webs. Science 250:811-814.

Prezelin, B.B., and A.C. Ley. 1980. Photosynthesis and chlorophyll-a fluorescence rhythms of marine phytoplankton. Marine Biology 55:295-307.

Reeder, B.C. 1994. Estimating the role of autotrophs in nonpoint source phosphorus retention in a Laurentian Great Lakes coastal wetland. Ecological Engineering 3:161-169.

Reeder, B.C., and W.J. Mitsch 1989. Seasonal patterns of planktonic and macrophyte productivity of a freshwater coastal wetland. In *Wetlands of Ohio's Coastal Lake Erie: A Hierarchy of Systems*. (Ed.) W.J. Mitsch, pp.49-68. Columbia: Ohio Sea Grant.

Revsbech, N.P., and B.B. Jorgensen. 1986. Microelectrodes: their use in microbial ecology. Advances in Microbial Ecology 9:293-352.

Richardson, C.J., and B.R. Schwegler 1986. Algal bioassay and gross productivity experiments using sewage effluent in a Michigan wetland. Water Resources Bulletin 22:111-120.

Richmond, K.-A. 1992. A comparison of photosynthesis of metaphyton in eutrophic littoral waters with that of an acidified lake. B.S. thesis, University of Manitoba, Winnipeg, Manitoba.

Robarts, R.D., D.B. Donald, and M.T. Arts. 1995. Phytoplankton primary production of three temporary northern prairie wetlands. Canadian Journal of Fisheries and Aquatic Sciences 52: 897-902.

Robinson, G.G.C. 1988. Productivity-irradiance relationships of the algal communities in the Delta Marsh: a preliminary report. University of Manitoba Field Station (Delta Marsh) Annual Report 23:100-110.

Robinson, G.G.C., and E. Pip. 1983. The application of nuclear track autoradiographic technique to the study of periphyton photosynthesis. In *Periphyton of Freshwater Ecosystems*. (Ed.) R. G. Wetzel, pp.267-273. The Hague, Netherlands: Dr. W. Junk Publishers.

Robinson, G.G.C., S.E. Gurney, and L.G. Goldsborough. 1997a. Response of benthic and planktonic algal biomass to experimental water level manipulation in a prairie lakeshore wetland. Wetlands 17:167-181.

Robinson, G.G.C., S.E. Gurney, and L.G. Goldsborough. 1997b. The primary productivity of benthic and planktonic algae in a prairie wetland under controlled water-level regimes. Wetlands 17:182-194.

Roos, P.J., and J.T. Meulemans. 1987. Under water light regime in a reedstand—short-term, daily, and seasonal. Archiv für Hydrobiologie 111:161-169.

Ross, L.C.M., and H.R. Murkin. 1993. The effect of above-normal flooding of a northern prairie marsh on *Agraylea multipunctata* Curtis (Trichoptera: Hydroptilidae). Journal of Freshwater Ecology 8:27-35.

Round, F. 1981. The Ecology of Algae. Cambridge: Cambridge University Press, UK.

Sand-Jensen, K. 1977. Effect of epiphytes on eelgrass photosynthesis. Aquatic Botany 3:55-63.

Schalles, J.F., and D.J. Shure. 1989. Hydrology, community structure, and productivity patterns of a dystrophic Carolina Bay wetland. Ecological Monographs 59:365-385.

Scheffer, M., S.H. Hosper, M.-L. Meijer, B. Moss, and E. Jeppesen. 1993. Alternative equilibria in shallow lakes. Trends in Ecology and Evolution 8:275-279.

Shamess, J.J. 1980. A description of the epiphytic, epipelic and planktonic algal communities in two shallow eutrophic lakes in southwestern Manitoba. M.S. thesis, University of Manitoba, Winnipeg, Manitoba.

Shamess, J.J., G.G.C. Robinson, and L.G. Goldsborough. 1985. The structure and comparison of periphytic and planktonic algal communities in two eutrophic prairie lakes. Archiv für Hydrobiologie 103:99-116.

Shay, J.M., and C.T. Shay 1986. Prairie marshes in western Canada, with specific reference to the ecology of five emergent macrophytes. Canadian Journal of Botany 64:443-454.

Simpson, P.S., and J.W. Eaton. 1986. Comparative studies of the submerged macrophyte *Elodea canadensis* and the filamentous algae *Cladophora glomerata* and *Spirogyra* sp. Aquatic Botany 24:1-12.

Stanley, D.W. 1976. Productivity of epipelic algae in tundra ponds and a lake near Barrow, Alaska. Ecology 57:1015-1024.

Stanley, D.W., and R.J. Daley. 1976. Environmental control of primary productivity in Alaskan tundra ponds. Ecology 57:1025-1033.

Stewart, W.D.P. (Ed.) 1974. *Algal Physiology and Biochemistry.* Berkeley: University of California Press.

Sullivan, M.J., and C.A. Moncreiff. 1988. Primary production of edaphic algal communities in a Mississippi salt marsh. Journal of Phycology 24:49-58.

Sullivan, M.J., and C.A. Moncreiff. 1990. Edaphic algae are an important component of salt marsh food-webs: evidence from multiple stable isotope analyses. Marine Ecology Progress Series 62:149-159.

Tilzer, M.M., M. Elbrachter, W.W. Gleskes, and B. Beese. 1986. Light-temperature interactions in the control of photosynthesis in antarctic phytoplankton. Polar Biology 5:105-111.

Turner, M.A., G.G.C. Robinson, B.E. Townsend, B.J. Hann, and J.A. Amaral. 1995. Ecological effects of blooms of filamentous green algae in the littoral zone of an acid lake. Canadian Journal of Fisheries and Aquatic Science 52:2264-2275.

van der Valk, A.G. 1986. The impact of litter and annual plants on recruitment from the seed bank of a lacustrine wetland. Aquatic Botany 24:13-26.

van der Valk, A.G., and C.B. Davis. 1978. The role of seed banks in the vegetation dynamics of prairie glacial marshes. Ecology 59:322-335.

Vernberg, F.J. 1993. Salt-marsh processes: a review. Environmental Toxicology and Chemistry 12:2167-2195.

Vymazal, J. 1994. *Algae and Element Cycling in Wetlands.* Boca Raton: Lewis Publishers.

Wetzel, R.G. 1983. *Limnology, 2nd edition.* New York: Saunders College Publishing.

Whitton, B. 1970. Biology of *Cladophora* in freshwaters. Water Research 4:457-476.

Wood, K.G. 1975. Photosynthesis of Cladophora in relation to light and CO_2 limitation; $CaCO_3$ precipitation. Ecology 56:479-484.

Wrubleski, D.A. 1991. Chironomidae (Diptera) community development following experimental manipulation of water levels and aquatic vegetation. Ph.D. dissertation, University of Alberta, Edmonton, Alberta.

Wu, X., and W.J. Mitsch. 1998. Spatial and temporal patterns of algae in newly constructed freshwater wetlands. Wetlands 18:9-20.

9

Invertebrates in Prairie Wetlands

Henry R. Murkin
Lisette C.M. Ross

Abstract

Long- and short-term water level changes in wetlands can cause significant shifts in the dominant species of wetland vegetation. Changes in vegetation due to flooding and drying can affect the complexity of aquatic habitats in the form of stem materials, root masses, and decaying plant material. Our understanding of how these changes in habitat impact marsh invertebrates is limited. Prior to the Marsh Ecology Research Program (MERP), few studies examined the interannual variation in invertebrate populations as wetlands move through the wet–dry cycle. This chapter incorporates the results of the MERP invertebrate studies with current information available on prairie wetland invertebrates. We review the factors affecting invertebrate distribution and abundance in prairie wetlands, their trophic structure in these systems, and the effect of the wet–dry cycle on invertebrate community characteristics and on regulation of survival within and among years.

Introduction

As knowledge of the structure and function of prairie wetlands expands, increasing effort has focused on aquatic invertebrates in these systems (Rosenberg and Danks 1987; Murkin and Wrubleski 1988). Much of this interest results from the need for a better understanding of the role of invertebrates in the trophic dynamics of prairie wetlands (Murkin 1989). For example, the importance of invertebrates for feeding waterfowl has been documented in many food-habit studies (Krapu and Reinecke 1992; see Chapter 10). A variety of other species representing all classes of vertebrates also feeds on invertebrates in prairie wetlands (Clark 1978; Murkin and Batt 1987), and this often results in invertebrates acting as vectors of disease and parasites for many species of vertebrate consumers (Wobeser 1981). The role of invertebrates in litter processing and nutrient cycling in wetlands also has drawn attention in recent years (Bicknese 1987; Campeau et al. 1994). An important objective of the Marsh Ecology Research Program (MERP) involved monitoring invertebrate responses to the simulated wet–dry cycle in the MERP experimental cells (see Chapter 1 and Appendix 1; Ross and Murkin 1989). This chapter incorporates information from the MERP invertebrate studies with other current information in a review of the functional role of invertebrates in prairie wetlands. We also summarize factors affecting invertebrate distribution and abundance in prairie wetlands.

AQUATIC INVERTEBRATES AND THE TROPHIC STRUCTURE OF PRAIRIE WETLANDS

Invertebrates as Primary Consumers

High macrophyte productivity in freshwater wetlands, coupled with generally low rates of direct herbivory on prairie wetland macrophytes, ensures that a large proportion of the organic matter within these systems occurs as standing or fallen plant litter. Invertebrates constitute the principal link between these primary production and detrital resources and higher-order consumers. The generalization that freshwater wetlands are detritus based stems from the premise that wetland trophic structure proceeds from plant detritus, to microorganisms, to a variety of invertebrate consumers, and then on to vertebrate consumers (Murkin 1989). These concepts were developed from observations in salt marshes (Odum and Heald 1975) and streams (Vannote et al. 1980), where invertebrates play a major role in litter processing, and have been applied to freshwater wetlands without critical evaluation of differences in system processes (Nelson and Kadlec 1984). Although a variety of prairie wetland invertebrates are classified as detritivores, the magnitude of their role in overall secondary production in these systems is unknown.

Plant litter in prairie wetlands and aquatic systems is colonized by a wide range of bacteria, fungi, and other microorganisms soon after it enters the water column as coarse particulate organic matter (CPOM) (Berrie 1976). These microorganisms are essentially the first-level consumers of plant litter, however, the litter and microbes combined are generally considered the base of detrital food chains. Secondary production occurs at higher trophic levels (Darnell 1976; Benke 1984). Colonization of the litter by microbes can increase its nutritive quality for secondary consumers (Ward and Cummins 1979; Motyka et al. 1985, Brönmark 1989). Microbes take up nutrients from the surrounding water column, further increasing the overall nutrient content of the litter (Berrie 1976; Polunin 1984). Studies in MERP showed that many litter types accumulated nitrogen (N) and phosphorus (P) over time (Figure 9.1). Species such as common reed (*Phragmites australis*) increased its N and P content by >200% for varying periods following submergence (Murkin et al. 1989). It has been argued that nutrient composition of the original litter may be of minor importance to detritivores compared with nutrients associated with the colonizing microbes (Findlay and Tenore 1982; Lawson et al. 1984; Graca et al. 1993). Some litter may not be suitable for invertebrate consumption until microbial colonization has taken place (Lawson et al. 1984). Campeau et al. (1994) found that chironomids in a Manitoba wetland colonized submersed bulrush (*Scirpus* spp.) litter only after enough time had passed to allow microbial colonization to occur.

It has been suggested that invertebrate detritivores do not actually digest the litter they ingest, rather they simply assimilate nutrients associated with the microbes and then egest

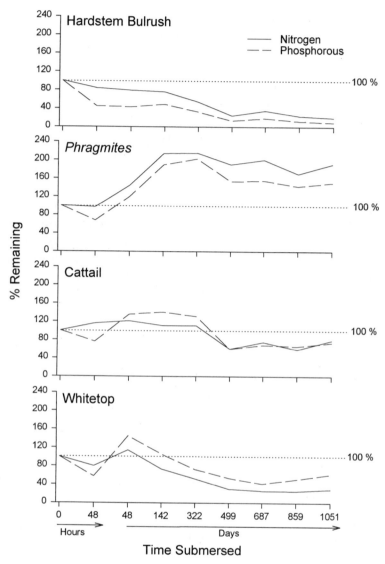

Figure 9.1   Percentage of nitrogen and phosphorus remaining in submersed litter of the dominant emergent plant species in the Delta Marsh (after Murkin et al. 1989).

the litter in smaller particle sizes, which are then recolonized by microbes (Montague et al. 1981). However, Graca et al. (1993) showed that some detritivores may assimilate litter material and thus avoid the microbial component of the detritus. Both processes (assimilation of microbial material and the litter itself) probably contribute to secondary production in detrital systems.

The concept of functional groups of invertebrates in streams has been described by

Cummins (1973) and Cummins and Klug (1979). Nelson (1982) first applied this concept to litter processing in freshwater wetlands. Shredders and grazers are the first invertebrate functional groups to respond to CPOM. In prairie wetlands, shredders include the amphipods *Hyalella azteca* and *Gammarus lacustris* (Nelson 1982; Wen 1992). *Gammarus* species are also important shredders in other aquatic systems (Gee 1988; Graca et al. 1993). *Hyalella* feed primarily on CPOM and play an important role in the direct conversion of coarse organic matter and associated epiphytes into fine organic matter (Wen 1992). They also ingest algae and bacteria associated with sediments (Hargrave 1970; de March 1981). Scrapers on detrital material in prairie wetlands include a variety of snail species (Pip 1978). As discussed previously, shredder and grazer functional groups reduce litter particle size through their feeding activities. As litter particle size is reduced, fine particulate organic matter (FPOM) becomes available to another set of consumers. Filter feeders remove FPOM from within the water column, whereas collectors actively gather fine particles from the surfaces of submersed substrates. Within prairie wetlands, the family Chironomidae has representatives in both of these functional groups. Continued processing results in smaller and smaller particle size until the material resistant to decomposition is finally incorporated into the sediments (Nelson et al. 1990).

There has been some disagreement regarding the impact of invertebrates on the rates of litter breakdown in wetlands. Invertebrates have been shown to increase decomposition rates of leaf litter in lotic systems (Cummins 1974; Barnes et al. 1986; Bird and Kaushik 1992; Stewart 1992), salt marshes (Newell and Barlocher 1993), and estuaries (Robertson and Mann 1980). However, Jackson et al. (1986) found that <20% of the litter in salt marshes was processed by secondary invertebrate consumers. Macrophyte decomposition studies in wetlands have produced conflicting results. Small, yet significant, impacts by invertebrates on macrophyte decomposition were reported by Danell and Sjoberg (1979), Polunin (1984), and Rogers and Breen (1983), but these results were not supported by the MERP study (Bicknese 1987) and others (Mason and Bryant 1975). Regardless, it appears that invertebrates play a minor role in overall litter decomposition rates in wetlands.

A neglected part of the detrital pool in aquatic systems and particularly wetlands is the dissolved organic matter (DOM) released during decomposition or digestion of organic matter by consumer organisms (Mann 1988). Organic complexes and dissolved molecules may be abundant in wetlands (Bowen 1984). The DOM may be the primary food source for microflagellates and Protozoa (Linley and Newell 1984; Taylor et al. 1985), but the role of DOM in the trophic structure of wetlands requires further investigation.

Invertebrate herbivory is generally considered to be relatively unimportant in the trophic dynamics of wetlands (Parsons and de la Cruz 1980; Simpson et al. 1983). Simpson et al. (1979) concluded that invertebrate herbivory on macrophytes in freshwater tidal marshes was minimal. Although living macrophytes are not ingested directly by many invertebrate consumers (Otto 1983; Brönmark 1989; Newman 1991), there are several studies documenting invertebrates feeding directly on living submersed and emergent vegetation in wetlands (Berg 1950a, b; McDonald 1955; Pip and Stewart 1976; Newman and Maher 1995). Skuhravy (1978) reported that insects damaged about one-third of the *Phragmites* stems in a wetland in Czechoslovakia, reducing annual primary productivity by 10 to 20%.

Penko and Pratt (1986) suggest that the moth *Bellura obliqua* can remove 8–15% of the biomass in cattail stands in Minnesota wetlands over one growing season. They found ≈10% of the cattail plants were affected by the feeding of this moth. Beule (1979) documented the destruction of a large area of cattail by the moth *Leucania scirpicola* in a wetland in Wisconsin, although this appeared to be an isolated event. Foote et al. (1988) also reported significant herbivory by a variety of insect larvae on cattail and bulrush in an inland brackish wetland. However, there are few other examples of significant invertebrate herbivory on wetland macrophytes. Chambers et al. (1990) showed that low densities of the crayfish *Orconectes virilis* could reduce the growth of several species of submersed vegetation common in prairie wetlands, but the occurrence of crayfish in prairie ponds is limited to ponds associated with streams or larger lakes. Although the role of live macrophytes as a food resource for invertebrates in most aquatic systems is not well understood (Lodge 1991; Newman 1991), most evidence indicates that invertebrate herbivory on wetland macrophytes is a minor trophic pathway in these ecosystems (Simpson et al. 1979).

In contrast, invertebrate herbivory on algae may play an important role in wetland trophic dynamics (Fuller et al. 1986). Because most algal groups do not have high standing crops at any point in time (see Chapter 8), their role in sustaining secondary production is often overlooked. The MERP studies (Gurney and Robinson 1988; Hosseini and van der Valk 1989a, b) and others (Hooper and Robinson 1976; Shamess et al. 1985; Klarer and Millie 1992) showed that the rapid turnover rates of algal biomass result in high annual production in spite of the low standing crop at any one point in time. This annual production represents a significant pool of potential nutrients for a wide variety of consumers. In salt marshes where macrophyte detritus was generally considered to support the bulk of secondary production, algae are now recognized as an important factor in consumer production (Montague et al. 1981; Kitting 1984). In *Spartina* marshes, Levinton et al. (1984) showed that macrophyte litter contributed little to secondary production in these systems. Consumption of algae by invertebrate herbivores in prairie wetlands has received little attention (Murkin 1989), although many of the invertebrate groups common to prairie wetlands are known to feed on algae. Chironomidae are among the most abundant and widely distributed invertebrate group in prairie wetlands (Wrubleski 1987). In wetlands, chironomids use diverse feeding strategies, with algae being the predominant food items for many species (Kajak and Warda 1968; Fairchild et al. 1989; Table 9.1). Chironomids have the potential to reduce the abundance of epiphytic algae in shallow aquatic areas, often selecting specific taxa among the algal community (Cattaneo 1983; Hann 1991; Botts and Cowell 1992; Botts 1993). Some midge species are primarily filter feeders, with planktonic algae and detritus as primary food sources. Algae are the dominant component of invertebrate diets in the spring and summer when algal production is high (Lamberti and Moore 1984). Stomach contents of *Procladius nietus* larvae collected in the Delta Marsh contained primarily diatoms (Wrubleski, MERP unpublished data).

Some freshwater snails feed by scraping algae from submersed surfaces (Kairesalo and Koskimies 1987; Brönmark 1989). Many of the snails common to prairie wetlands (e.g., *Lymnaea*, *Physa*, and *Helisoma*) feed on epiphytic algal populations (Hunter 1980). Snail grazing can affect not only the abundance of the epiphytic community but also its species

Table 9.1. Literature references for habitat and food habits of dominant chironomid species from the MERP experimental cells (Wrubleski 1991)

Reference	Habitat	Food habits
Tanypus punctipennis		
Fellton (1940)	lake bottom	predacious; feeding on newly hatched and older chironomid larvae
Oláh (1976)	lake; open-water sediments	feed mainly on diatoms
Titmus and Badock (1981)	gravel pit–lake	feed mainly on unicellular algae
Corynoneura scutellata		
Kesler (1981)	free-living on submerged surfaces	grazers on periphyton
Cricotopus sylvestris		
Darby (1962)	tubes on the surface of the mud, bottom debris, or submersed vegetation	feed on diatoms, algal debris, and green algae
Menzie (1981)	on submersed plants when present or on sediments	not reported
Cricotopus ornatus		
Swanson and Hammer (1983)	tubes on the sediment, algal mats, and submersed vegetation	not reported
Chironomus tentans		
Sadler (1935)	tubes in the sediment	eat whatever is available, but algae comprise the bulk of the diet when present
Palmén and Aho (1966)	shallow water with large amounts of detritus	not reported
Hall et al. (1970)	tubes in the sediment	filters plankton during the day and searches the sediment surface at night for larger food items
Topping (1971)	soft ooze and detritus	not reported
Mason and Bryant (1975)	within dead *Typha* stems	brown detrital material of rotting *Typha* and no living algae
Dicrotendipes nervosus		
Fellton (1940)	restricted to green algal mats	not reported
Moore (1980)	organically rich lake sediments	detritus, very little algae
Paratanytarsus sp. 1		
Wrubleski (unpublished obs.)	tubes on flooded vegetation	grazing on algae
Glyptotendipes barbipes		
Kimerle and Anderson (1971)	tubes in the sediment	irrigation of the tubes supplies sestonic food particles, primarily algal cells
Driver (1977)	axiles of *Scolochloa*	not reported
Wrubleski and Rosenberg (1984)	miner of polystyrene foam	not reported

composition as well (Lodge 1986; Merritt and Wotton 1988; Brönmark 1989). This response is probably caused by a combination of selective grazing on certain species and other species being resistant to grazing pressure (Brönmark 1989). The snail–epiphytic algal interaction on submersed vegetation plays an important role in the survival and production of submersed vegetation. Grazing by snails reduces the epiphytic cover, thereby increasing available light and nutrients for the submersed vegetation (Rogers and Breen 1983; Lodge 1986). When snail predators increase in a pond, snail populations decline and epiphytic algae populations increase (Brönmark et al. 1992; Brönmark and Weisner 1996).

Cladocerans and copepods feed primarily on phytoplankton in the water column (Porter 1977). Cladocerans and copepods are abundant in the Delta Marsh (Smith 1968; Hann 1991), MERP experimental cells (Murkin et al. 1991a), and prairie wetlands in general (Kantrud et al. 1989). During population peaks, cladocerans consume >100% of the daily phytoplankton production in aquatic systems (Smirnov 1961). Hanson and Butler (1994) found that cladoceran filtration rates exceeded 100% of the water column. Algal species composition can be modified through the selective grazing of the most palatable species present (Porter 1977). Hann (1991) found that microcrustacean grazers (cladocerans, copepods, ostracods) in an isolated pond in the Delta Marsh effectively reduced periphytic algal biomass throughout the growing season. The removal of fish from a shallow waterfowl lake in Minnesota caused an increase in zooplankton populations and a subsequent decrease in phytoplankton populations (Hanson and Butler 1994).

There is increasing evidence that bacteria and fungi are potential food resources for invertebrates in aquatic systems. Cladocerans may switch between algae and bacteria, based on their availability at different times of the year (Kankaala 1988). Coveney et al. (1977) suggest that some species of zooplankton feed exclusively on bacteria. Stanhope et al. (1987) report that fungi may be an important factor in habitat selection for certain groups of aquatic invertebrates. Isopods raised in fungus-rich environments demonstrated superior growth and survival compared to isopods located in fungus-poor environments. Additional work on the role of these groups is needed to clarify their importance for wetland invertebrates.

Invertebrates as Predators

Prairie wetlands support a diverse community of invertebrate predators (Table 9.2). Common examples of invertebrate predators in these systems include a wide range of hemipterans (Scudder 1987), coleopterans (Larson 1987; Hanson and Swanson 1989), odonates (Hilton 1987), hirudinea (Wrona et al. 1979), and other taxa (Kantrud et al. 1989). Some of the larger invertebrate predators (Belostomatidae and Odonata) feed on small fish and tadpoles (Caldwell et al. 1980; Travis et al. 1985). Invertebrate predators play important roles in the trophic structure of lakes and bogs (Benke 1976; Elser et al. 1987; Arnott and Vanni 1993), but there is little information on their role in the trophic structure of prairie wetlands or other wetlands. Density and biomass of predacious invertebrates in prairie wetlands can be high and in some habitats exceed those of prey species at certain times of the year (Murkin and Kadlec 1986; Murkin et al. 1991a), indicating the major role of invertebrate predators in the trophic structure of these systems.

Table 9.2. Functional group classification of macroinvertebrate taxa collected in the Delta Marsh, Manitoba

Shredders	Scrapers
Amphipoda	Gastropoda
Talitridae	Physidae
Gammaridae	Lymnaeidae
Coleoptera	Planorbidae
Scirtidae	Coleoptera
	Scirtidae
Collectors–Filters	**Predators**
Nematoda	Rhynchobdellida
Oligochaeta	Glossiphoniidae
Cladocera	Trombidiformes
Ostracoda	Hydrachnidae
Copepoda	Odonata
Ephemeroptera	Lestidae
Caenidae	Coenagrionidae
Baetidae	Aeshnidae
Hemiptera	Libellulidae
Corixidae	Coleoptera
Trichoptera	Haliplidae
Hydroptilidae (*Agraylea multipunctata*)	Hydrophilidae
Leptoceridae	Gyrinidae
Phryganeidae	Dytiscidae
Polycentropodidae	Hemiptera
Diptera	Belostomatidae
Ceratopogonidae	Corixidae
Chironomidae	Notonectidae
Culicidae	
Ephydridae	
Tipulidae	
Coleoptera	
Scirtidae	

Invertebrates as Prey for Secondary Consumers

Numerous vertebrate species are permanent residents of wetlands or use wetlands for portions of their annual cycles. An important reason is the abundant invertebrate food resources provided by these habitats. Waterfowl feeding on invertebrates has been well-documented (Swanson and Meyer 1973; Murkin 1989; Krapu and Reinecke 1992) (see Chapter 10). Many species of waterfowl consume large quantities of invertebrates during the breeding season to meet the protein demands of gonadal development and egg laying. Calcium requirements for eggshell formation during laying also are met by feeding on invertebrates such as gastropods with their calcium-rich shells. Invertebrate food resources are an important factor in habitat selection by breeding waterfowl. Spring waterfowl densities on the MERP experimental cells and other wetland sites were positively correlated with invertebrate densities (Joyner 1980; Murkin et al. 1982; Murkin and Kadlec 1986). Lack of invertebrate foods on the breeding grounds can adversely affect clutch size and egg hatchability (Krapu 1981).

Waterfowl breeding chronology varies among species, resulting in differential exploitation of the invertebrate community over the annual season. Early-nesting species such as the mallard (*Anas platyrhynchos*) and pintail (*Anas acuta*) use early seasonal and temporary ponds and forage on associated invertebrate fauna. Later-nesting species are restricted to more permanent wetlands and lakes as the more temporary habitats dry. Krapu (1974) reported that early-nesting pintails fed primarily on *Anostraca* and *Oligochaeta* in temporary ponds, later-nesting and renesting hens fed on chironomid larvae in more permanent wetlands. Protein demands of juvenile waterfowl are also high due to rapid growth and feather development (Murkin and Batt 1987). Plant foods do not provide the necessary range of amino acids for growing ducklings and invertebrates fulfill this dietary requirement. Talent et al. (1982) reported that mallard hens selected brood-rearing areas with high invertebrate (chironomid) densities. Emerging adult insects are particularly important to young waterfowl feeding on the surface of the wetland (Swanson and Sargeant 1972; Swanson 1977). Duckling mortality has been recorded when invertebrate resources are too low (Street 1977). Cox et al. (1998) found that mallard duckling growth is greater as numbers of invertebrates increase and that duckling survival is positively related to growth.

Prairie wetlands provide a wide range of feeding sites for birds (Weller 1981). For example, pied-billed grebes (*Podilymbus podiceps*) feed on odonate larvae and other invertebrates in the water column (Munro 1941). Red-winged (*Agelaius phoeniceus*) and yellow-headed blackbirds (*Xanthocephalus xanthocephalus*), and sedge (*Cistothorus platensis*) and marsh wrens (*C. palustris*) are insectivores that search for prey on the exposed stems and leaves of emergent vegetation (Voigts 1976; Weller 1981). Common snipe (*Gallinago gallinago*) probe for invertebrates in the moist soil at the wetland edge (Tuck 1972). See Chapter 10 for additional examples of invertebrate use by wetland birds.

A variety of fish species use prairie wetlands, especially lacustrine marshes and wetlands associated with streams (Peterka 1989). The fish community in shallow, isolated prairie wetlands is limited primarily to fathead minnows (*Pimephales promelas*) and brook sticklebacks (*Culaea inconstans*). Both species are able to withstand the hypoxia and temperature fluctuations characteristic of shallow wetland systems (Suthers and Gee 1986; Peterka 1989; Hanson and Riggs 1995). High densities of fish in semipermanent and permanent prairie wetlands often result from human activity such as ditching or direct addition. A more diverse fish community develops in wetlands connected directly to lakes and river systems, including yellow perch (*Perca flavescens*) and northern pike (*Esox lucius*) among others (Suthers and Gee 1986; Hanson and Butler 1994). The use of these habitats by fish is partly related to the abundant invertebrate food resources available through parts of the year. Fish can influence aquatic invertebrate abundance and community structure in wetlands, and may reduce the suitability of wetland habitat for feeding waterfowl and other wetland birds (Hanson and Riggs 1995).

Invertebrates and Nutrient Cycling in Prairie Wetlands

Part II of this book suggests that invertebrates play a minor role in overall nutrient cycling within the MERP experimental cells. Studies often indicate that this may be true for freshwater wetlands in general (Valiela and Teal 1978; Kadlec 1979). The invertebrate community is only a small portion of the overall nutrient pool within these systems, especially compared with major constituents such as macrophytes (see Chapters 3–6). Kitchell et al. (1979) suggest that aquatic invertebrates affect nutrient cycling through translocation or transformation of nutrients without ever being a major pool themselves (translocation is the movement of nutrients within the system and transformation is the change of surface–volume ratios of various substrates in a manner that affects nutrient cycling processes). Schindler et al. (1993) studied the cycling of phosphorus (P) in piscivore-dominated and planktivore-dominated lakes. Zooplankton and *Chaoborous* were found to recycle pelagic P, whereas fish recycled P derived from predation on benthic and littoral organisms to pelagic algae.

As shown by the MERP studies in Part II of this book and in other studies (Kadlec 1979), the sediments in freshwater wetlands represent a large pool of nutrients. Invertebrates may play a role in the translocation of these nutrients into the overlying water column. For example, a variety of invertebrates, including chironomids can reach very high densities in the sediments of prairie wetlands (Murkin and Kadlec 1986), feeding within the sediments and excreting nutrients into the overlying water (Gallepp 1979; Matisoff et al. 1985; Fukuhara and Sakamoto 1987). Overall impacts of this translocation within the wetland may be minor, but local effects may be important.

High densities of emerging insects are common in prairie wetlands (Wrubleski and Rosenberg 1990). Such emergence represents a potential nutrient export from the water column and sediments; however, most adults remain in the wetland or return to lay eggs. Thus, the net translocation is probably minor compared with other nutrient fluxes within the wetland. Other researchers also have suggested that the emergence of adult aquatic insects represents a negligible loss of nutrients from aquatic systems (Vallentyne 1952; Paasivirta 1974).

The reduction in litter particle size as it is processed by the sequence of shredders, scrapers, and filter feeders certainly affects the surface area of the litter available for microbial colonization and subsequent nutrient release. This concept of transformation and associated nutrient cycling has been little studied in wetland systems.

FACTORS AFFECTING THE DISTRIBUTION AND ABUNDANCE OF INVERTEBRATES IN PRAIRIE WETLANDS

As demonstrated throughout this book, prairie wetlands are dynamic ecosystems exhibiting extreme spatial and temporal variability. Organisms living in these habitats must be adapted to the range of conditions occurring during prairie wet–dry cycles. Quantity and quality of water, duration of flooding, abundance and diversity of vegetation, and available food resources influence the distribution and abundance of invertebrates in prairie wetlands. High densities and diversity of invertebrates indicate that many species have adapted to the variety of conditions in these systems.

Water Regime

Water regime (depth and duration of flooding) is the principal factor controlling productivity of prairie wetlands (see Chapters 6–11). Annual variation in spring runoff, summer precipitation, and evapotranspiration produce cyclical fluctuations in the water levels within prairie basins (see Chapter 1). These wet–dry cycles result in progressive changes in the vegetation and, in turn, influence availability and complexity of habitats for invertebrates. The rate of cycling between the wet and dry stages is dependent on the overall depth of the basin. Shallow basins dry more quickly and frequently than deeper sites.

The ability to survive desiccation during dry periods is a major determinant of the distribution and abundance of invertebrate species in prairie wetlands. Invertebrates of intermittently flooded habitats possess structural, behavioral, and physiological adaptations for surviving dry conditions. Mobile species such as adult Corixidae can fly from drying ponds to find refuge in deeper, more permanent habitats (Fernando and Galbraith 1973) and then return when the basins are reflooded. Other taxa, such as many *Cladocera* species, produce drought-resistant eggs (ephippia) that hatch when water returns to the basin (Wiggins et al. 1980). Mosquitoes of the genus *Aedes* lay drought-resistant eggs in dry basins, the eggs hatch upon reflooding. Few adult forms of aquatic invertebrates can withstand the rigorous conditions of drying. Some snail species avoid desiccation by burrowing into the substrate (Pennak 1978), whereas others secrete a mucilaginous covering over the shell opening that hardens into an effective seal, allowing them to survive extended dry periods (Wiggins et al. 1980). Species unable to withstand drying, such as the amphipods *Hyalella* and *Gammarus*, cannot survive in ponds that dry periodically.

Stewart and Kantrud (1971) place prairie wetlands into several classes based on duration of flooding. These classes are useful when discussing the effect of water regime on invertebrate diversity in these systems. Each vegetation community (Figure 9.2) provides very different habitat to aquatic invertebrates in these systems.

Ephemeral ponds—These habitats are found in shallow depressions and are most obvious in open areas such as agricultural fields. Flooding duration is too short to allow development of wetland vegetation in the basin, but there may be some minor algal production especially in areas of productive soils. The short period of inundation also does not allow establishment of an extensive invertebrate community. Even groups that are adapted to extended dry periods normally do not have enough time to complete their life cycle during the short period that water is available. The aquatic invertebrates in these ponds are usually the adults of very mobile species (e.g., corixids) that depend on surrounding permanently flooded wetlands. Waterfowl commonly forage in ephemeral ponds, feeding on waste grain (LaGrange and Dinsmore 1989), earthworms (Krapu 1974), and possibly drought-resistant invertebrates belonging to the subclasses Branchiopoda and Malacostraca.

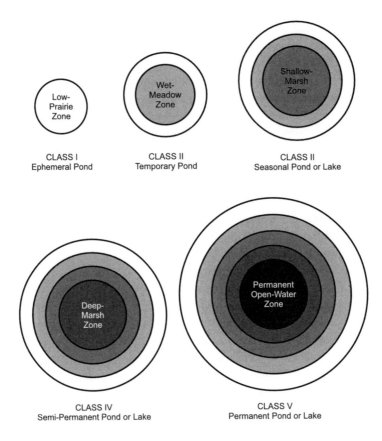

Figure 9.2   Vegetation zonation patterns in the five classes of northern prairie wetlands (after Stewart and Kantrud 1971).

Temporary ponds—Flooding duration and depth increase in temporary ponds. Invertebrate diversity increases because species with very short life cycles (e.g., copepods, ostracods) have time to complete their life cycles and reproduce. These species also attract some of the more mobile invertebrate predators (e.g., dytiscid adults) that can leave when the ponds dry up. Development of wetland vegetation in the deeper parts of these basins adds diversity to the submersed habitats available. Increased algal production and a diverse algal community result from longer inundation. The increased habitat and available food resources all contribute to an increased abundance and diversity of aquatic invertebrates.

Seasonal ponds—As flooding duration increases from temporary to seasonal ponds, more species of invertebrates have sufficient time to complete their life cycles. Drying is still an annual event, but many more species are able to survive and reproduce during the extended flooding. A shorter dry period improves survival of the species that remain in the pond during dry conditions (e.g., snails). A more diverse algal assemblage develops (Crumpton 1989) along with a more diverse vegetation community, providing a greater array of available niches for invertebrates. Besides the development of emergent vegetation stands, some submersed plant species also begin to appear. Whitetop *(Scholochloa festucacea)* wetlands are common seasonal wetlands in the prairies and support abundant invertebrates (Neckles et al. 1990), making these sites preferred feeding areas for waterfowl in the spring (Kaminski and Prince 1981). High densities of invertebrates in these wetlands are attributed to abundant detrital food resources that become available upon reflooding (Wiggins et al. 1980; Sklar 1985; Wylie and Jones 1986) and to reduced predator populations, especially fish, due to periodic drying (Williams 1985; Dodson 1987).

Semipermanent and permanent ponds—The longer duration of flooding in semipermanent and permanent ponds results in the development of a different invertebrate fauna than in ponds with annual drawdown periods. Invertebrate communities in newly flooded semipermanent ponds resemble those of more temporarily flooded wetlands (Neckles et al. 1990). The first invertebrates to appear are those that are drought resistant (Wiggins et al. 1980) or able to actively seek out newly available habitats (e.g., corixids, dytiscids) (Fernando and Galbraith 1973). With prolonged flooding, the invertebrate taxa adapted to periodic dry conditions are replaced by species adapted to continuously flooded conditions (Neckles et al. 1990). For example, in the Delta Marsh, ostracods and mosquitoes (Culicidae) were virtually eliminated from ponds with continuous flooding (Figure 9.3). Kantrud et al. (1989) also found changes in invertebrate taxa with varying hydrologic regimes. They report that the gastropod fauna of North Dakota wetlands changes from a primarily *Aplexa hydrorum–Gyraulus circumstriatus* association (species adapted to surviving dry periods) in seasonally flooded ponds to a *Helisoma trivolvis–Lymnaea stagnalis* association in ponds that are continuously flooded over several years. Species adapted to seasonal flooding are forced to the periphery of permanently flooded ponds where there is some annual drying of the substrate, whereas species adapted to permanent flood-

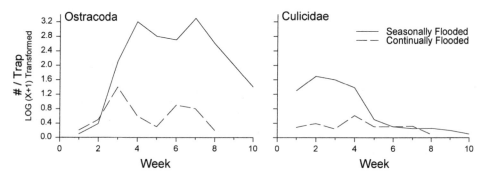

Figure 9.3   Order Ostracoda and family Culicidae (order Diptera) response in seasonally and continuously flooded ponds in the Delta Marsh, Manitoba (after Neckles et al. 1990).

ing are found in the deeper, central portions of the basin. Invertebrates that require water year-round for development and overwintering (some chironomid species and amphipods) probably enter permanent systems through both active and passive (attached to birds, mammals, etc.) dispersal (Daborn 1976; Swanson 1984). The recolonization of semipermanent ponds is affected by the proximity of permanent water bodies that serve as refugia for many invertebrate species (Bataille and Baldassarre 1993).

Habitat Type and Structure

Invertebrates use four basic habitats in prairie wetlands: the water surface, the water column, submersed surfaces within the water column, and the bottom substrate. The presence or absence of these habitats and their structure or complexity has important effects on invertebrate distribution and abundance within a wetland (Figure 9.4). Neustonic invertebrates live on the water surface and include primarily predacious taxa such as gerrids (water striders) (Spence and Andersen 1994). Nektonic invertebrates (free swimming) and zooplankton (free floating) use the water column. Benthic invertebrates are found on and within the bottom sediments and on the surfaces of submersed plants (Murkin et al. 1994). Availability and complexity of these habitats are related to the water depth and the vegetation development within the basin. Water depth not only establishes dimensions of the water column but also determines, as discussed throughout this book, characteristics of the vegetation community (see Figure 2.3 in Chapter 2). Vegetation, in turn, determines the complexity of submersed structures available for invertebrate colonization. Wrubleski (1987) found that some chironomid species were located on certain plants and at specific depths within the Delta Marsh (Table 9.3).

There is an increase in invertebrate habitat complexity when moving from shallower to deeper wetland classes (Figure 9.2). For example, shallow, ephemeral ponds contain a single zone of low-prairie vegetation, whereas deeper, semipermanent ponds include a deep wetland zone of aquatic emergents such as cattail and bulrush, a shallow wetland zone consisting primarily of grasses and sedges, a wet-meadow zone with plants such as *Juncus balticus* and *Hordeum jubatum*, and the low-prairie zone of mostly terrestrial plants (Stewart and Kantrud 1971). Permanent wetlands differ from semipermanent ponds by having a central, permanent open-water zone often dominated by a variety of submersed vegetation (Stewart and Kantrud 1971).

In prairie wetlands, habitat structure important to invertebrates is provided primarily by living and dead emergent vegetation, submersed vegetation, and metaphytic algae. This structure provides both living space, protection from wave action and water movement, security from predators (Bennett and Streams 1986), and submersed surfaces for the development of algal and microbial communities that serve as food (Brönmark 1989; Murkin 1989). The MERP study and others indicate that habitat types and plant species with complex structure enhance the diversity and abundance of invertebrates. Plants with the most complex physical structure provide the greatest surface area for invertebrate colonization (Krecker 1939; Dvörak and Best 1982). Wrubleski and Rosenberg (1990) found higher diversity and abundance of chironomids in areas of the Delta Marsh dominated by

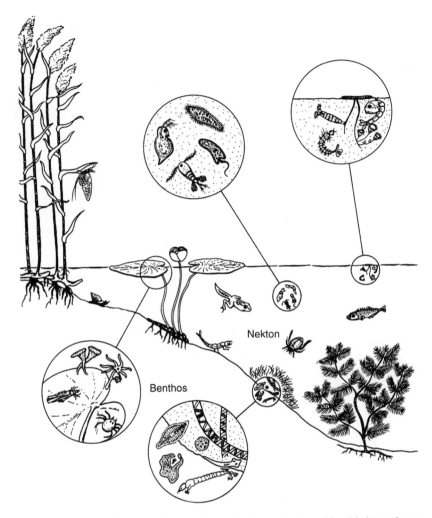

Figure 9.4   Cross section of a wetland water column showing nektonic and benthic invertebrates.

sago pondweed (*Potamogeton pectinatus*), with its abundant finely dissected leaves, than in cattail (*Typha* spp.) or bulrush (*Scirpus* spp.) stands, with their simple narrow linear leaves. Voigts (1976) also found higher densities of nektonic invertebrates associated with beds of submersed vegetation than in stands of emergent vegetation in Iowa wetlands. In the MERP cells, the microcaddisfly *Agraylea multipunctata* (order Trichoptera) was most abundant in the open-water areas dominated with sago pondweed. The pondweed provided the habitat needed for the development of late-instar larvae (Ross and Murkin 1993).

Both living and dead emergent vegetation are extremely important habitat components for invertebrates, especially during seasons when submersed vegetation is not available (Murkin et al. 1991a). Some invertebrate species require the submersed surfaces provided

Table 9.3. Dominant chironomid species from the Delta
Marsh, listed in decreasing order of relative abundance

Habitat	Species
Phragmites	*Pseudosmittia* spp.
	Limnophyes n. sp.
	L. hudsoni
	Paraphaenocladius nasthecus
Whitetop	*Limnophyes* n. sp.
	L. hudsoni
	Chironomus nr. *atroviridis*
	Corynoneura cf. *scutellata*
	Pseudosmittia spp.
Cattail	*Cricotopus sylvestris*
	Limnophyes n. sp.
	Paratanytarsus sp. 1
	Tanypus punctipennis
	L. hudsoni
Bulrush	*Paratanytarsus* sp. 1
	Limnophyes n. sp.
	L. hudsoni
	Glyptotendipes lobiferus
	Dicrotendipes nervosus
Pondweed	*Tanypus punctipennis*
	Cricotopus sylvestris
	Glyptotendipes barbipes
	Corynoneura cf. *scutellata*
	Chironomus tentans
Open water	*Chironomus* spp.
	Polypedilum simulans

Source: Wrubleski 1987.

by emergent vegetation. For example, the chironomid species *Glyptotendipes barbipes* and *Chironomus tentans* mine into the stems of emergent vegetation. In the MERP experimental cells, these two species were most prevalent in whitetop and *Phragmites*, plants that are easily mined because of their hollow stems. In another MERP study, Campeau et al. (1994) found that the habitat structure provided by submersed emergent plant litter was more important to invertebrate abundance than the food value of the litter itself. Availability of macrophyte litter varies depending on the decomposition rates of the species involved. Bulrush and whitetop litter decomposed very quickly in the MERP cells, whereas *Phragmites* and cattail decomposed more slowly, thereby providing habitat for much longer periods of time (Figure 9.5).

The presence or absence of vegetation also affects invertebrates living on the water surface. There are two types of water surface habitats: those associated with open-water areas with no vegetation and those within stands of emergent vegetation. For example, different

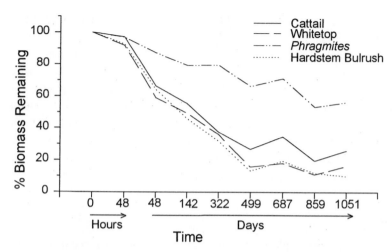

Figure 9.5   Mean percentage of dry weight remaining of litter submersed in the Delta Marsh, Manitoba (after Murkin et al. 1989).

species of water striders (Gerridae) are found in habitats with and without emergent vegetation; within a species there may be shifts between these two habitats at different times in the life cycle (Spence and Andersen 1994).

Many wetland studies report highest abundance peaks for invertebrates when submersed vegetation is interspersed with emergent vegetation (Voigts 1976; Driver 1977; Henson 1988; Kenow and Rusch 1989; Wrubleski and Rosenberg 1990). This combination of submersed surface types provides for the requirements of the broadest range of invertebrate groups. From the MERP studies, Murkin et al. (1991a) concluded that high abundance and diversity of invertebrates in prairie wetlands were due to the diversity and interspersion of vegetation types within the wetland basin. However, the differences in surface area provided by wetland vegetation extend beyond the differences between submersed and emergent vegetation. Even among the various species of submersed vegetation there are differences in the habitats provided and their subsequent use by invertebrates. For example, Voigts (1976) reported higher amphipod (primarily *Hyalella* spp.) densities in coontail (*Ceratophyllum demersum*) than in clasping-leaf pondweed (*Potamogeton richardsonii*). *Glyptotendipes lobiferus* is often associated with habitats containing sago pondweed (Berg 1950c). Ross and Murkin (unpublished data) found that *G. lobiferus* emergence declined when sago pondweed was replaced by bladderwort in the MERP experimental cells.

Water-level fluctuations alter plant species composition and associated habitat structure and have pronounced effects on invertebrates within wetlands. Elimination of emergent macrophytes in the MERP experimental cells following deep flooding resulted in decreases in epiphytic chironomid species such as *Corynoneura* cf. *scutellata* (Wrubleski 1991). Physidae, the most common snail family found in the MERP cells, also declined in areas where emergent vegetation was eliminated by high water (Murkin and Ross 1999) (Figure 9.6). Increasing water levels during the deep-flooding stage of MERP eliminated sago pondweed from the open-water areas of the experimental cells and decreased *G. lobiferus* densities (Figure 9.7).

Other factors that affect plant distribution and abundance also influence invertebrate diversity and densities. During MERP Wrubleski (1989) found that the removal of sago pondweed beds by feeding waterfowl changed the species composition of the chironomid community (Table 9.4). Elimination of the pondweed resulted in loss of habitat for the smaller epiphytic species. Feeding activities of carp would have similar effects on wetland habitats and invertebrate communities. Muskrat-feeding and lodge-building activities add large quantities of emergent plant litter to the water column (Nelson 1982), which also influences the amount of submersed habitat available for aquatic invertebrates.

Food Resources

The role that invertebrates play in the trophic structure of prairie wetlands and the foods consumed by various invertebrate groups were discussed previously in this chapter. The role of food resources in determining invertebrate distribution and abundance in these systems has received little attention. There are very few examples of the effects of changing food supplies on invertebrate fauna of prairie wetlands or wetlands in general (Murkin and

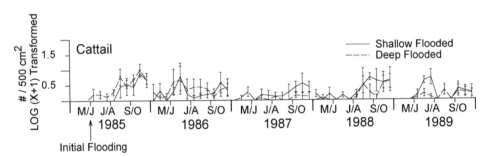

Figure 9.6   Mean number of Physidae (class Gastropoda) collected in shallow-flooded and deep-flooded cattails in the MERP experimental cells from 1985 to 1989.

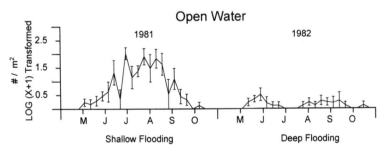

Figure 9.7   Number of *Glyptotendipes lobiferus* (family Chironomidae) emerging from shallow and deeply flooded open-water sampling sites in the MERP experimental cells.

Table 9.4. Mean (± SD) densities (number per trap) in emergence traps of the 10 most abundant Chironomid species collected from waterfowl exclosures and controls in Bone Pile Pond, Delta Marsh, Manitoba, Canada

Species	Exclosures (with sago pondweed)	Control (without sago pondweed)
Chironomus utahensis	0	20.0±14.5
Chironomus staegeri	0	35.0±41.1
Dicrotendipes nervosus	11.7±5.7	6.0±2.6
Parachironomus tenuicaudatus	10.3±7.5	6.7±2.5
Paratanytarsus sp.1	6.7±8.3	1.3±0.6
Paratanytarsus sp.3	25.3±31.1	5.3±3.1
Cricotopus sylvestris	274.0±196.7	77.7±39.6
Cricotopus ornatus	24.0±18.5	9.3±3.2
Cricotopus trifasciatus	10.0±6.1	1.0±1.7
Corynoneura cf. *scutellata*	51.7±51.9	1.0±1.0

Source: After Wrubleski 1989.

Wrubleski 1988). Abundance and species composition of macrophytes fluctuates within and between years based on water depth and duration of flooding. These plants represent large pools of nutrients for primary consumers, yet there are few, if any, examples of how changes in these pools affect herbivorous invertebrates.

Availability of submersed litter also varies considerably. Northern prairie wetlands experience two main periods of litter input: fall and spring. Some fragmentation and toppling of stems occur during the fall following normal senescence. However, this fall input is minor compared with the input during the following spring (Davis and van der Valk 1978a). Winter and spring storms cause extensive fragmentation of standing litter, so litter input to the water column is high at the time of spring thaw (Davis and van der Valk 1978b). In the MERP cells, maximum densities of herbivore–detritivore invertebrates did not correspond to these periods of maximum litter input. Therefore, factors other than the availability of litter as food or habitat were limiting invertebrate levels during these periods (Murkin 1983). Cool water temperatures probably limit invertebrate growth and reproduction during the spring and fall. The litter also requires some period of time for microbial colonization before it is suitable as a food resource for aquatic invertebrates. For example, during MERP Campeau et al. (1994) found that the response of detritivorous chironomids to newly submersed bulrush litter was delayed for several weeks. Thus, the effect of litter input on invertebrate abundance probably occurs some time after the periods of litter input. Availability of litter as a food resource also varies according to the decomposition rates of the macrophytes present. As discussed previously, the MERP studies and others have shown that dominant macrophytes in prairie wetlands decompose at different rates (Davis and van der Valk 1978a,b; Murkin et al. 1989; van der Valk et al. 1991) and, therefore, are available as food resources for varying lengths of time following submergence.

There has been some work on the effects of changing algal levels on invertebrate densities in prairie wetlands. High algal production also ensures the availability of high-quality nutrient sources. The MERP researchers and others have documented increases in chironomid abundance in direct response to increased algal population levels (Fairchild et al. 1989; Campeau et al. 1994). Campeau et al. (1994) also recorded an increase in cladoceran and snail populations when algal levels were artificially increased in a series of wetland enclosures. There is some evidence that the availability of specific species of algae can influence habitat selection by snails (Lodge 1986). Murkin et al. (1991a) suggested that the increased numbers of cladocerans in response to deep flooding of the MERP experimental cells were linked to changes in phytoplankton levels in the cells.

The relative importance of algae and plant detritus as food resources for aquatic invertebrates has been discussed in several MERP publications. Neckles (1984) reported no reduction in invertebrate abundance when detrital levels were reduced in a whitetop pond. Murkin (1989) suggested that algae may be more important than detritus as a food source for aquatic invertebrates in prairie wetlands. Emergent macrophyte litter may be important as submersed surfaces for algal growth rather than providing a significant, direct source of nutrients (Murkin 1989). Campeau et al. (1994) reported little change in invertebrate abundance and biomass when artificial litter with no nutritive value replaced natural litter in their MERP study (Figure 9.8). Street and Titmus (1982) also found that litter serves primarily as habitat rather than as a direct food source for aquatic invertebrates.

Salinity

Prairie wetlands range from fresh to several times more saline than seawater (Shay and Shay 1986). Water loss in many prairie wetlands is dominated by evapotranspiration rather than surface outflow or groundwater recharge. Thus, dissolved salts become concentrated and have a marked influence on the biota of the system. In addition to evapoconcentration,

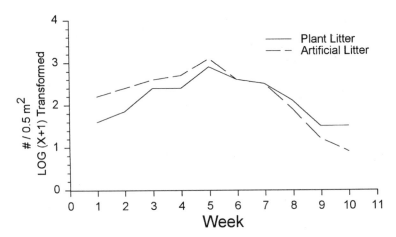

Figure 9.8 Mean number of Chironominae (family Chironomidae) emerging from plant litter and artificial litter (after Campeau et al. 1994).

additional salts originate from groundwater, surface runoff, and the breakdown of glacial till (Shay and Shay 1986). The most obvious impact of salinity on prairie wetlands is its effect on primary productivity (Neill 1993). Distribution and growth of many prairie wetland plant species reflect salinity; the number of plant species decreases as salinity increases. Shay and Shay (1986) list 10 plant species tolerant of saline conditions (15–45 mS/cm), and only one species, the submersed aquatic *Ruppia maritima*, tolerant of hypersaline water conditions (>45 mS/cm). Similarly, Hammer et al. (1983) found that algal production and diversity were limited by salinity in Saskatchewan wetlands. Thus, salinity affects both habitat structure and food resources for aquatic invertebrates.

Salinity also directly affects invertebrate survival; the invertebrate fauna shifts to more salt-tolerant species as salinity increases. A wide range of salt tolerances is evident in the invertebrates of prairie wetlands (Timms et al. 1986; Timms and Hammer 1988; Kantrud et al. (1989); Hammer et al. 1990) (Figure 9.9). As salinity increases, the fauna changes from species with little tolerance such as the molluscs, leeches, and amphipods to salt-tolerant species such as brine shrimp (*Artemia salina*) and shore flies (Ephydridae) (Swanson et al. 1988). The ability of an individual species to survive changes in salinity depends on its ability to regulate the ionic and osmotic concentrations of their tissues (Thorp and Covich 1991). Survival also depends on ability to adapt to alterations in their external environment caused by salinity changes, such as alteration in the complexity of vegetation, shifts in predation, and parasitism. For example, Scudder (1983) found that the corixid *Cenocorixa expleta* occurs naturally only where the salinity tolerance of parasitic mites is exceeded. In contrast, the same study showed that *C. bifida* better resisted parasites and was, therefore, more abundant than *C. expleta* in wetlands with lower salinities.

Oxygen

Dissolved oxygen levels vary spatially and temporally within prairie wetlands (Murkin et al. 1992). Oxygen is often depleted in areas with abundant submersed vegetation and detritus, especially during the summer, resulting in hypoxia particularly at night when photosynthetic activity is low (Suthers and Gee 1986; Davies and Baird 1988; Davies and Gates 1991). During winter, the sediment–water interface is often anoxic for long periods of time, particularly in late winter (Babin and Prepas 1985; Baird et al. 1987). Many invertebrate species possess morphological, physiological, and behavioral adaptations permitting them to survive periods of hypoxia. Species that respire using specialized gills or direct cuticular exchange, such as mayflies, caddisflies, turbellarians, water mites, and odonates are unable to survive long periods of anoxia. Others such as corixids, dytiscids, notonectids, culicids, and syrphids capture air directly from the water surface and can live in completely anoxic waters. Some oligochaetes, chironomids, and cladocerans produce hemoglobin that aids in oxygen uptake and transport. Cladocerans collected in oxygen-poor waters develop a red coloration (Weider and Lampert 1985). Waterman (1961) reported that in poorly oxygenated waters, red *Daphnia* survived longer, fed more, swam faster, produced more eggs, and had eggs with a shorter developmental time than did unpigmented specimens. Landon and Stasiak (1983) found that pigmented specimens

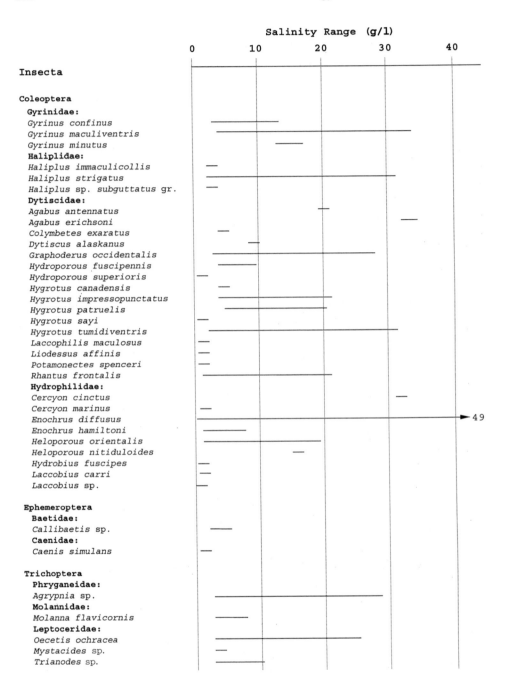

Continued

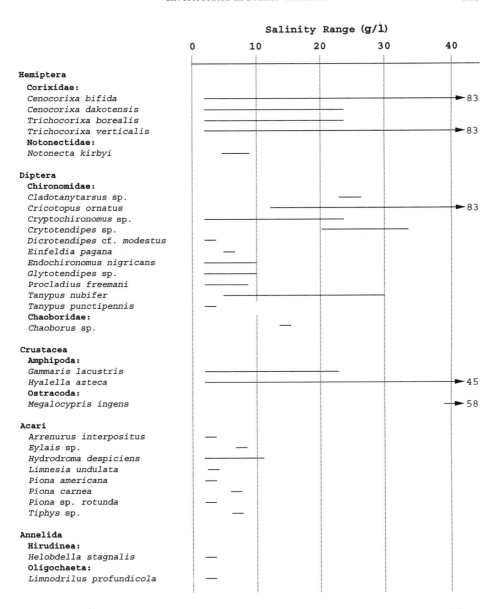

Figure 9.9   Salinity tolerances of invertebrates collected in saline lakes in Saskatchewan and Alberta (after Timms et al. [1986]; Timms and Hammer [1988]; Hammer et al. [1990]).

became clear within a day or two after being placed in well-oxygenated water, suggesting that there is an energetic cost to producing and maintaining hemoglobin. Red pigmentation also may be a strong visual cue to fish and other aquatic predators (Engle 1985). The freshwater leech *Nephelopsis obscura* is able to reduce oxygen depletion by secreting large amounts of mucus over its epidermis (Singhal and Davies 1987). *Chironomus plumosus* can inhabit areas that may be anoxic for several weeks by switching its metabolism to the anaerobic conversion of glycogen to lactic acid (Augenfeld 1967).

Many invertebrates alter their behavior during hypoxic and anoxic periods. Davies and Gates (1991) noted that the feeding rates of *Nephelopsis obscura* decreased as dissolved oxygen levels decreased. Several species of chironomids also reduce their locomotion and feeding activities at low oxygen concentrations (Nagell 1978; Heinis and Crommentuijn 1992). Mobile invertebrates such as leeches (Davies et al. 1992) and some chironomids (Oliver 1971) move to areas with more favorable oxygen conditions.

Structural features of the habitat also result in spatial variations in dissolved oxygen. For example, during MERP and other studies on the Delta Marsh low levels of dissolved oxygen were recorded in stands of dense emergents due to the shading of the water surface and the abundance of litter in the water column (Suthers and Gee 1986; Murkin et al. 1992) (Figure 9.10). Murkin et al. (1992) found that some groups of nektonic invertebrates declined to very low levels in emergent vegetation when oxygen levels were low, whereas cladocerans and ostracods were more abundant in dense cattail stands than in open water during these periods (Figure 9.10). These latter groups are probably more adapted to low oxygen levels and they would benefit from reduced predation by fish and invertebrates that cannot tolerate low oxygen levels (Bennett and Streams 1986; Suthers and Gee 1986).

Temperature

Temperature plays an important role in the development and survival of aquatic invertebrates. Many invertebrates have very specific temperature requirements for reproduction and growth. de March (1981) found that the reproduction of *H. azteca* did not occur in the spring until water temperatures were >20°C, and that growth rates were slow when the

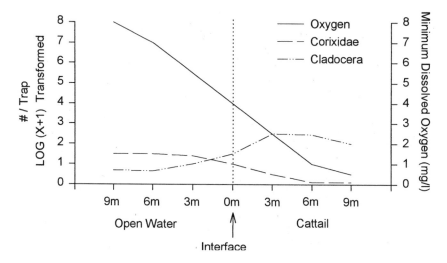

Figure 9.10   Mean number of total Corixidae (order Hemiptera) and Cladocera collected in activity traps and dissolved oxygen levels across the open water–cattail interface in the Delta Marsh, Manitoba. Vertical dotted line indicates edge between open water and cattail (after Murkin et al. 1992).

temperature dropped to <20°C. Leonhard and Lawrence (1981) also found that cladocerans require temperatures >20°C for maximum reproduction. Murkin et al. (1991a) suggested that the low temperatures in the spring and fall in the experimental wetlands of the MERP study may have limited nektonic invertebrate densities during those seasons.

The most severe temperatures in prairie wetlands occur during the winter. Much of the northern prairie region is characterized by temperatures of <0°C for 5 months of the year with minimum air temperatures lower than –40°C in many areas during January (Winter 1989). Extensive freezing of the water column and substrate occurs in most shallow prairie wetlands. During MERP Wrubleski (1984) found the depth to which the water column and substrate froze in the Delta Marsh was directly linked to the presence of emergent vegetation. Up to 48 cm of the substrate froze in open-water areas where no snow cover was present, whereas only the top 10 cm of the substrate froze in areas with bulrush standing litter and 35 cm of snow cover. Ice cover in shallow prairie ponds results in anoxic or near anoxic conditions for extended periods of time (Davies et al. 1987).

Invertebrates subjected to the extreme conditions of a prairie winter must either be adapted to cold conditions, enter a dormant phase, or move to a more favorable microhabitat to survive (Danks 1978). Invertebrates that exhibit cold-hardiness or cold tolerance include two groups: freeze-tolerant invertebrates endure freezing of their body water, whereas freeze-intolerant invertebrates cannot survive ice formation within their tissues (Lee 1991) and must seek out more favorable microhabitats within the substrate or water column or produce cryoprotectant molecules such as glycerol in their bodies (Lee and Delinger 1991). Many freeze-intolerant invertebrates overwinter in dormant phases. For example, some species overwinter in the egg stage with rapid development occurring in the spring (Merritt and Cummins 1984). Many adult insects such as culicids and gerrids overwinter as adults in terrestrial sites. Mobile species such as corixid or dytiscid adults fly to deeper ponds and lakes to overwinter.

Parker (1992) studied an aestival semipermanent wetland in Saskatchewan and found that certain invertebrates were good indicators of substrate freezing. Species such as *Coenagrion angulatum, Lestes disjunctus/unguiculatus, Lestes congener, Corynoneura* cf. *scutellata, Cladopelma viridulus,* and *Parachironomus monochromus* appeared to be less sensitive to substrate freezing. Daborn (1971, 1974) also reported a shift from *Coenagrion* species to *Lestes* species in an aestival pond in Alberta in response to freezing conditions (Dineen 1953).

Predation

Although invertebrates are an important food source for a wide variety of higher-order consumers in prairie wetlands, there is little information on the impact of predation on the distribution and abundance of invertebrates in these systems. Use of invertebrates by waterfowl and other birds is well-documented; however, there are few examples of the effects of birds feeding on invertebrate densities in prairie wetlands. Peterson et al. (1989) observed a significant decrease in benthic chironomid populations due to feeding by mallards and gadwalls on the exposed substrates of the MERP experimental cells soon after

they were drawn down (Figure 9.11). Although birds have been shown to affect invertebrate distribution and abundance in other systems (Quammen 1984; Mercier and McNeil
1994; Szekely and Bamberger 1992), there are no other reports of effects of birds on invertebrate distribution and abundance in natural prairie wetlands.

Fish are important predators on invertebrates in aquatic systems (Bendell and McNicol
1987; Gilliam et al. 1989; Hanson and Riggs 1995; Bouffard and Hanson 1997). Swanson
and Nelson (1970) suggested that there may be competition between fish and waterfowl
for invertebrate foods in prairie wetlands. Most of the fish occurring in prairie wetlands
feed on invertebrates during all or part of their life cycle; however, the impact of fish predation on the distribution and abundance of invertebrates in prairie wetlands has received
little attention. During MERP Murkin et al. (1991a) suggested that the high densities of
cladocerans in stands of emergent vegetation in the Delta Marsh were related in part to the
absence of fish due to low oxygen levels. Suthers and Gee (1986) also showed that during
the summer yellow perch avoided low-oxygen conditions in stands of cattail in the Delta
Marsh, thereby limiting their access to invertebrate foods associated with this habitat type.

Prolonged flooding can result in abundant fish populations developing in ponds with
sufficient depth and suitable environmental conditions. High populations of fish reduce
invertebrate populations in many aquatic systems. Hanson and Butler (1994) found that
extensive fish populations limited invertebrate populations in a shallow waterfowl lake in
Minnesota. Removal of the fish resulted in an increase in invertebrate populations and a
change in the overall trophic structure of the lake system. Cerny and Bytel (1991) found
that populations of *Daphnia* were reduced by high densities of carp and perch. In addition
to reduction in densities over time, one *Daphnia* species underwent a body-size shift to a
smaller size at maturity. A downward shift in body size is supposedly an adaptive mechanism that some invertebrates use for avoiding size-selective predation (Hall et al. 1970).
Peterka (1989) also suggests that tiger salamanders (*Ambystoma tigrinum*) may exert

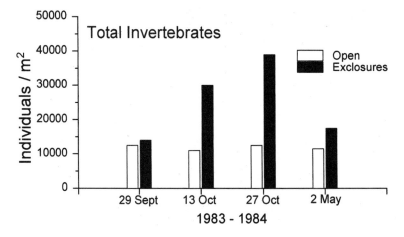

Figure 9.11   Effect of waterfowl feeding on mean total chironomid densities collected from enclosures
and open mudflat sites in the MERP experimental cells (after Peterson et al. 1989).

extensive pressure on available invertebrates in some prairie ponds.

Little information is available on the impact of predation by other invertebrates on overall invertebrate distribution and abundance, specifically in prairie wetland habitats. Like fish, corixids avoid stands of emergent vegetation in the summer when oxygen levels are low, thereby reducing potential predation pressure on cladocerans and other prey species during these periods (Murkin et al. 1991a). Cladocerans also may exhibit a reduction in body size to avoid predation by invertebrate predators such as notonectids (Dodson and Havel 1988). Batzer and Resh (1991) reported that *Berosus* beetles reduced densities of *Cricotopus* larvae in a series of seasonal wetlands in California. Rasmussen (1987) described a reduction in chironomid distribution and abundance due to predation by the leech *Nephelopsis obscura* in a shallow Alberta pond. Further information on invertebrate predation on other invertebrates in aquatic systems can be found in reviews by Bay (1974), Macan (1977), and Cooper (1983).

Several studies on predation in aquatic systems indicate that habitat structure can affect predator–prey interactions (Gilinsky 1984; Bennett and Streams 1986). In general, the more complex the habitat, the more varied are the types of refuges available to prey species (Crowder and Cooper 1982; Hargeby 1990). Foraging efficiency of benthic-feeding perch *(Perca fluviatilis)* was inversely related to the complexity of habitat structure provided by submersed vegetation (Diehl 1993). Gilinsky (1984) suggested that fish in general are much less efficient predators in submersed vegetation. Mattila (1992) found that amphipod densities increased only in areas where submersed vegetation provided refuge from fish.

In summary, predation is probably an important factor in structuring invertebrate communities in prairie wetlands; however, its impact is poorly understood and may vary within years and over the course of the wet–dry cycle, based on the presence and absence of submersed habitats.

Contaminants

Herbicide and insecticide use in agriculture is probably the main source of contaminants entering prairie wetland systems. In 1988 alone, 4.1 million kilograms of broad-leaf weed herbicides and 4.3 million kilograms of grassy weed herbicides were applied to 20 million hectares of cultivated land in Saskatchewan (Waite et al. 1992). Although insecticide use tends to be lower, 3 to 4 million hectares of Saskatchewan farmland were treated with insecticides in 1985 at application rates as high as 1,000 g/ha (Waite et al. 1992). Aerial drift, overspray, postapplication runoff from surrounding land, groundwater, and the treatment of dry basins are all ways in which agricultural chemicals can enter prairie ponds (Buhl et al. 1993). Characteristics of the water affect chemical breakdown rates in wetlands (Muir et al. 1980). For example, pH has a marked effect on the adsorbtion and breakdown of various chemicals (Hamaker and Thompson 1972).

Aquatic invertebrates are exposed to chemicals through feeding and cuticular absorption. The toxicity of agricultural chemicals to invertebrates in their natural environment can be influenced by the following factors: (1) physical–chemical characteristics of the

water such as water temperature, pH, hardness, and the level of dissolved organic material; (2) intrinsic properties of chemicals such as their water solubility and their persistence
in the environment; (3) the species of invertebrate, its size, and life stage (age); and (4)
length of exposure, concentration, and frequency of occurrence of contaminants (Buhl et
al. 1993). These factors result in considerable variability in the impact of chemicals on
invertebrates in prairie wetland systems.

Some chemicals are simply more toxic to invertebrates than others (see Tables 7.15 and
7.17 in Sheehan et al. [1987] for summary of chemical effects on aquatic invertebrates),
and invertebrates often react differently to contamination. For example, some mobile
forms may leave the affected wetland or move to less affected areas within the wetland.
Jones et al. (1991) found that the feeding rate of *Daphnia catawba* decreased in the presence of a herbicide and surfactant, whereas other species of *Daphnia* were unaffected.
Different life stages of the same species may be affected differently. The carbamate insecticide fenoxycarb did not interfere with early development of notonectid embryos; however, fourth and fifth instar nymphs developed morphogenetic abnormalities and were
unable to emerge to the adult stage (Miura and Takahashi 1987). Finally, different timings
of application and rates of exposure modify effects on invertebrates. Parsons and
Surgeoner (1991) found that double-pulse exposures of selected insecticides, particularly
of the pyrethroid insecticide permethrin, were equally or more toxic to mosquito larvae
than a single continuous exposure of equal total duration.

There are few studies on the effects of herbicides on prairie wetland invertebrates (see
Sheehan et al. 1987 for review). Epiphytic insects are particularly susceptible to herbicides. Besides the direct effects mentioned previously, herbicides probably invoke secondary impacts on aquatic invertebrates by eliminating vegetation and predators such as
fish. For example, Robinson (1989) attributed an increase in mosquito emergence to the
death of brook stickleback following an application of the broadleaf, postemergence herbicide bromoxynil to a series of experimental wetlands. Stephenson and Mackie (1986)
found that ponds exposed to phenoxy herbicide 2,4-D had lower species diversity compared with control ponds. They concluded that reduced habitat complexity due to vegetation loss in the treatment ponds caused reduced insect diversity. In a similar study, Smith
and Isom (1967) related decreases in epiphytic insects and other insects such as elmid beetles, dragonflies, and mayflies to loss of vegetation due to herbicide application.

Although the impact of agricultural insecticides has been given more attention, wetland
field studies are still limited. Pyrethroids, carbamates, and organochlorine insecticides are
among the best documented. The carbamate insecticide, carbofuran,, has been one of the
most widely used insecticides on the Canadian prairies, particularly for grasshopper control. Both *Chironomus tentans* larvae and *Gammarus lacustris* can be significantly
reduced in the shallow areas of ponds exposed to the pesticide (Wayland and Boag 1990).
Deltamethrin, a pyrethroid insecticide considered extremely toxic to aquatic invertebrates,
made up 25% of the market share of insecticides used against grasshoppers in
Saskatchewan in 1985 (Sheehan et al. 1987). Day and Maguire (1990) present the acute
toxicities of deltamethrin and its breakdown products for *Daphnia magna*. A study by
Morrill and Neal (1990) reported that deltamethrin decreased the density of chironomid

larvae in Canadian prairie ponds with population recovery taking from 2 months to 1 year. Tidou et al. (1992) reported that deltmethrin, when applied to four experimental meso-cosms in France, killed all species of zooplankton 1 day after application and that the density of *Chaoborus* larvae decreased by 98%. The same authors reported that the organochlorine insecticide lindane was less toxic to nontarget insects and had no significant impact on zooplankton populations.

Use of wetlands in sewage treatment has grown in recent years (Hammer and Bastian 1989; Wetzel 1993); however, there is little information on the effect of nutrient loading on aquatic invertebrates in wetlands. Murkin et al. (1991b) reported elevated invertebrate densities in a Manitoba wetland receiving runoff from a cattle feedlot. Invertebrate abundance increased in response to N and P additions to a shallow wetland in Manitoba (Campeau et al. 1994).

INVERTEBRATES AND THE WET–DRY CYCLE OF PRAIRIE WETLANDS

A wide range of factors affects the distribution and abundance of invertebrates in prairie wetlands and single parameters, in isolation, do not determine invertebrate community characteristics in a wetland or regulate survival of invertebrates in prairie wetlands. In addition, impacts of these factors vary within and among years as wet–dry cycles influence overall wetland productivity. In this section, we examine the wet–dry cycle and its effect on invertebrate distribution and abundance over time (Table 9.5).

Drawdown Stage

Aquatic invertebrates during the dry marsh stage are restricted to dormant forms and flying adults (e.g., mosquitoes and chironomids) from nearby basins holding water. As discussed previously, many species of aquatic invertebrates have mechanisms for surviving dry periods. Dry basins may contain the drought-resistant eggs of a variety of species, including cladocerans and other crustaceans (Wiggins et al. 1980). Mosquito eggs laid by females in the dry basin may be present on the pond litter and substrate (Wood et al. 1979). Snails are buried deep in the available litter and have mucilaginous sheaths over the shell opening to prevent drying. The longevity of the dormant forms of the various species is unknown. It is likely that there is some difference among various species (e.g., the resting eggs of cladocerans versus dormant snails); however, no data are available at the present time. During the MERP study, there was no difference in the invertebrate response to 1- or 2-year drawdowns, indicating that most of the common species can survive dry conditions for at least 2 years.

Regenerating Stage

During the early regenerating stage of the wet–dry cycle, the initial submersed habitat consists of a complex matrix of annual and emergent litter with new emergent vegetation developing later in the spring. The first invertebrate species to appear are those that were

Table 9.5.   General invertebrate response to the wet-dry cycle of northern prairie wetlands

Taxa	Dry stage	Regenerating stage	Degenerating stage	Lake stage
Crustacea				
Cladocera	resistant eggs	abundant early. Some species decline with continuous flooding	distribution affected by distribution of vegetation and related oxygen conditions	lower levels compared with its earlier stages, fewer numbers as diversity decreases
Ostracoda	resistant eggs	similar to Cladocera	similar to Cladocera	similar to Cladocera
Amphipoda	not present	low numbers, distribution affected by subnmerged vegetation	numbers increase over time	present at reduced levels
Gastropoda	dormant in bottom	dormant forms from dry ponds, abundant	previous types replaced by permanent-flooding species	present in moderate numbers
Diptera				
Chironomidae				
Epiphytic	not present	increase as vegetation increases	declines with loss of vegetation	low numbers along the marsh edge
Benthic	may be present in substrate	low numbers present initially	begin to increase	dominant invertebrate group
Coleoptera				
Dytiscidae	not present	move into these areas as invertebrate prey increase	present in high numbers during the early part of the degenerating phase when invertebrate prey are abundant	used as overwintering site and in dry years when other ponds are unavailable
Hemiptera				
Corixidae	not present	similar to dytiscids	similar to dytiscids	similar to dytiscids

present in the pond during the dry stage and very mobile forms that can immigrate from adjacent habitats. Cladocerans, ostracods, and copepods appeared soon after dry basins were flooded in the MERP studies (Neckles et al. 1990, Murkin and Ross 1999). Mosquito larvae appeared at about the same time followed by gastropods shortly thereafter. Corixids, notonectids, and adult dytiscids arrived from adjacent water bodies soon after the other groups appeared. Driver (1977) found that chironomid diversity early in flooding remained low until aquatic plant diversity and abundance increased later in the year. Wrubleski (1991) reported higher densities of chironomids in flooded *Aster* litter than in newly established emergent vegetation shortly after reflooding the MERP cells. *Aster* provided a more complex flooded habitat than the emergent vegetation. Epiphytic chironomid species (e.g., *Corynoneura, Cricotopus, Paratanytarsus*) and the larger mining species (e.g., *Chironomus, Glyptotendipes*) were found on the flooded terrestrial litter.

Flooded plant litter and algae are available to invertebrates as food soon after reflood-ing. Algal growth is probably in response to the abundant submersed surface available and the nutrients leeched from the litter as it decomposes. Algal production is significant dur-ing this period in the absence of an extensive canopy of emergent vegetation. During MERP Gurney and Robinson (1988) found a negative correlation between standing stems of vegetation and metaphytic algal production. Thus, abundant food resources appear to be available to both herbivores and detritivores soon after reflooding. Some reduction in oxygen levels may occur especially at night with the large litter pool present; however, without distinct stands of emergent vegetation, significant differences in oxygen levels across the pond are unlikely.

The first invertebrates to appear from resting eggs in reflooded wetlands with high salin-ity are brine shrimp (*Artemia salina*) (Kantrud et al. 1989). However, gastropod distribu-tion and abundance are restricted in saline systems. Initial flooding reduces salinity; how-ever, it increases over the summer as evapoconcentration occurs.

Continued flooding during the regenerating stage produces changes in the invertebrate community in response to the habitat changes within the wetland basin. The annual plant litter lost through decomposition is replaced by emergent vegetation and litter, causing a shift in invertebrate communities. Wrubleski (1991) found that chironomid communities declined in flooded terrestrial litter as it decomposed and increased in emergent vegetation as stands became more complex. With development of litter within the stands of emergent vegetation, species dependent on this litter such as mining chironomid species increase in abundance. As flooding persists into subsequent years, species that require annual drying events (e.g., *Aedes* mosquitoes, some cladoceran species) are lost from the system and are replaced by others that prefer continuous flooding (e.g., *Culex* mosquitoes, amphipods) (Neckles et al. 1990). Species diversity increases overall as additional species recolonize the basin from adjacent water bodies. The more mobile forms appear first (e.g., predacious species), whereas those with limited dispersal mechanisms appear later in flooding. For example, *G. lacustris* and *H. azteca* are eliminated by drawdown and must reinvade the basin by attaching themselves to aquatic birds and mammals (Daborn 1976; Swanson 1984).

Degenerating Stage

Continued flooding shifts the wetland from the regenerating phase to the degenerating phase. This point of transition in the wet–dry cycle is thought to be the period of maxi-mum invertebrate diversity and abundance in prairie wetlands (Murkin and Ross 1999). The diversity of vegetation types and overall plant biomass are at maximum levels. Overall algal production is high resulting from a diverse community of phytoplankton, epiphyton, and metaphyton. High invertebrate diversity and abundance observed during this period are related to the abundance of plant and detrital material and the diverse matrix of habitat types (Murkin et al. 1992). In the Delta Marsh, some species reached their maximum abundance in cattail stands (e.g., cladocerans), whereas others peaked in submersed vegetation in open-water sites (e.g., corixids). The proximity of different habitat types is also important in fostering maximum diversity. High invertebrate densities are found in emergent vegetation in the spring before submersed vegetation is available. Later in the year, as the submersed vegetation appears in the open areas, some species move to the open-water sites, whereas others remain associated with the emergent vegetation. The differences in oxygen levels among the vegetation types also affects invertebrate distribution at this time (Murkin et al. 1992).

As the cycle continues into the degenerating stage, the emergent vegetation begins to die back and open-water areas begin to expand. There is a reduction in available habitats for epiphytic species with the loss of the emergent vegetation. Litter availability is also reduced due to the decomposition of existing litter and declining production of new litter. This results in reduced habitat structure for invertebrates, reduced submersed surfaces for algae production, and fewer detrital food resources. Algal food resources also are affected by the changes during the degenerating stage. With prolonged above-normal flooding of the MERP experimental cells (Hosseini and van der Valk 1989a, b), phytoplankton levels declined and, although epiphytic algae production did not decline per unit, the overall reduction in submersed surfaces certainly resulted in a lower available epiphyton standing crop.

The invertebrate community responds to these changes in habitat and available food supplies during the regenerating stage in several ways. During MERP, elimination of the emergent and submersed vegetation eliminated the epiphytic chironomid species (Wrubleski 1991) and reduced overall amphipod densities (Murkin et al. 1991a). The chironomid community changed from a variety of epiphytic and mining species to a simpler community of larger benthic forms associated with the bottom substrate (Wrubleski 1991).

The loss of submersed habitat complexity during the degenerating marsh stage is coupled with establishment of a diverse predator community over time. This includes predacious invertebrate species, fish, and other vertebrate predators (e.g., salamanders). The loss of physical structure provided by the submersed plant surfaces reduces refuge from predators. Predation may thus play an increased role in reducing invertebrate populations during the degenerating stage of the wet–dry cycle.

Lake Stage

With continued flooding, the wetland moves to the lake marsh stage. The habitat for invertebrates is reduced to remnant stands of submersed vegetation in sheltered portions of the basin, the thin strip of emergent vegetation at the pond margin, and the unvegetated substrate. The dominant invertebrate taxa associated with the unvegetated substrate are large benthic chironomid species. Both Driver (1977) and Wrubleski (1991) found that benthic chironomids increased with an increase in unvegetated substrate. Benthic chironomids became established after the 2 years of above-normal flooding eliminated the emergent vegetation in the MERP cells. The main support for food webs during the lake marsh stage may be the epipelon associated with the bottom substrate (Murkin 1989). This algal community reaches its maximum production in open-pond situations with soft sediments and few macrophytes to shade the sediment. Although the epipelon does not have a large standing crop at any point in time, its rapid turnover rate results in high annual production, which in turn can support high densities of benthic invertebrates such as chironomids.

Wet–Dry Cycle Summary

The previous discussion is a general outline of the changes in the invertebrate community in response to the wet–dry cycle of prairie wetlands. Variations in this response are common and result from a wide variety of factors. Salinity, for example, can alter the pattern of invertebrate community development. Availability of adjacent permanent-water habitats influences recolonization rates of species that are intolerant of dry conditions. The diversity of the available seed bank determines the vegetation developing in the basin and the resulting habitat structure and food availability. The quality of water reflooding the basin also affects invertebrate populations.

RESEARCH NEEDS

A general lack of basic ecological and taxonomic information on invertebrates in prairie wetlands restricts our understanding of the role of invertebrates in the structure and function of these systems. Species of invertebrates native to prairie wetlands are not well understood (Rosenberg and Danks 1987 and chapters therein). Taxonomic studies and preparation of species lists would aid in the understanding of community diversity, trophic dynamics, competition, and predator–prey relationships. In addition, the lack of basic life-history information, such as simple generation times for common prairie wetland species, limits studies on productivity and trophic dynamics. Functional roles of major taxa must be determined before overall nutrient and energy budgets, and trophic structure of these systems can be worked out in detail. Although information is needed on a wide range of taxa, research should focus on the ecologically and economically important species. For example, in prairie wetlands, the Chironomidae require immediate attention due to their abundance and importance in the general ecology of these systems.

Recent work on MERP and elsewhere has raised several issues related to the trophic structure of prairie wetlands. Further work on the importance of algae in supporting invertebrate production is required. Corollary issues involve the role of invertebrates in litter processing. For example, why are there so many more shredders in streams than in wetlands, given the large amounts of litter found in both systems? The use of living macrophytes as a food source for specific invertebrate groups also requires further study. At higher trophic levels, a better understanding of predator–prey relationships has implications for secondary production, wildlife and fish use of wetland habitats, and possibly biocontrol of pest species.

Little is known of seasonal differences in community composition or population levels across the dominant habitat types in prairie wetlands. The MERP and other studies have provided general information at the level of order or family, but information at the species and population levels is required. Furthermore, the impact of water levels, water chemistry, habitat structure, food resources, or other environmental factors on invertebrate communities and individual species requires further investigation. The widespread use of agricultural chemicals in the prairie landscape requires that studies on the effects of herbicides, pesticides, and fertilizers on prairie wetland invertebrates should be a priority for new research in prairie wetlands. In addition, recent interest in the use of freshwater wetlands for sewage treatment requires an assessment of the ability of wetlands to process domestic and industrial sewage. Information on the impact of nutrient loading and associated environmental impacts on invertebrate production and survival should be part of this assessment.

Study duration is an important consideration in designing wetland invertebrate studies. Most studies to date have been short-term in nature (<1 to 2 years), but there can be considerable interannual variation in invertebrate populations, even among years with seemingly similar environmental conditions. Understanding factors contributing to this variation requires long-term studies on individual study areas.

The development of sampling techniques for wetland habitats is an important step required before research on wetland invertebrates can proceed effectively and efficiently. A wide variety of invertebrate sampling techniques is available for aquatic habitats (Murkin et al. 1994), but the unique characteristics of wetlands make most techniques unsuitable in these habitats. Shallow water with dense submersed or emergent vegetation and unconsolidated sediments often requires modification of existing techniques or the development of completely new procedures. The efficiency of any new procedure should be tested thoroughly before it is used extensively in invertebrate monitoring programs.

ACKNOWLEDGMENTS

We thank the many MERP field assistants who spent countless hours "picking bugs" in the MERP laboratory. Mark Hanson, David Rosenberg, and Dale Wrubleski provided comments on earlier drafts of the manuscript. Tye Gregg and Dale Wrubleski assisted with preparation of the figures. This is paper No. 102 of the Marsh Ecology Research Program, a joint project of Ducks Unlimited Canada and the Delta Waterfowl and Wetlands Research Station.

LITERATURE CITED

Arnott S.E., and M.J. Vanni. 1993. Zooplankton assemblages in fishless bog lakes—influence of biotic and abiotic factors. Ecology 74:2361-2380.
Augenfeld, J.M. 1967. Effects of oxygen deprivation on aquatic midge larvae under natural and laboratory conditions. Physiological Zoology 40:149-158.
Babin, J., and E.E. Prepas. 1985. Modelling winter oxygen depletion rates in ice-covered temperate zone lakes in Canada. Canadian Journal of Fisheries and Aquatic Sciences 42:239-249.
Baird, D.J., T.E. Gates, and R.W. Davies. 1987. Oxygen conditions in two prairie pothole lakes during winter ice cover. Canadian Journal of Fisheries and Aquatic Sciences 44:1092-1095.
Barnes, J.R., J.V. McArthur, and C.E. Cushing. 1986. Effect of excluding shredders on leaf litter decomposition in two streams. Great Basin Naturalist 46:204-207.
Bataille, K.J., and G.A. Baldassarre. 1993. Distribution and abundance of aquatic macroinvertebrates following drought in three prairie pothole wetlands. Wetlands 13:260-269.

Batzer, D.P., and V.H. Resh. 1991. Trophic interactions among a beetle predator, a chironomid grazer, and periphyton in a seasonal wetland. Oikos 60:251-257.

Bay, E.C. 1974. Predator-prey relationships among aquatic insects. Annual Review of Entomology
19:441-453.

Bendell, B.E., and D.K. McNicol. 1987. Fish predation, lake acidity and the composition of aquatic insect assemblages. Hydrobiologia 150:193-202.

Benke, A.C. 1984. Secondary production of aquatic insects. In *The Ecology of Aquatic Insects*.
(Eds.) V.H. Resh and D.M. Rosenberg, pp 289-322. New York: Praeger.

Benke, A.C. 1976. Dragonfly production and prey turnover. Ecology 57:915-927.

Bennett, D.V., and F.A. Streams. 1986. Effects of vegetation on *Notonecta* (Hemiptera) distribution
in ponds with and without fish. Oikos 46:149-158.

Berg, C.O. 1950a. The biology of aquatic caterpillars which feed on Potamogeton. Transactions of
the American Microscopy Society 69:254-266.

Berg, C.O. 1950b. *Hydrellia* (Ephydridae) and some other acalyptrate Diptera reared from
Potamogeton. Annals of the Entomological Society of America 43:374-398.

Berg, C.O. 1950c. Biology of certain Chironomidae reared from *Potamogeton*. Ecological
Monographs 50:85-100.

Berrie, A.D. 1976. Detritus, micro-organisms and animals in freshwater. In *The Role of Terrestrial
and Aquatic Organisms in Decomposition Processes*. (Eds.) J.M. Anderson and A. Macfadyen,
pp. 323-338. London, UK: Blackwell Scientific Publications.

Beule, J.B. 1979. Control and management of cattails in southeastern Wisconsin wetlands.
Wisconsin Department of Natural Resources. Technical Bulletin 110. Madison, Wisconsin.

Bicknese, N.A. 1987. Factors influencing the decomposition of fallen macrophyte litter at the Delta
Marsh, Manitoba, Canada. M.S. thesis, Iowa State University, Ames, Iowa.

Bird, G.A., and N.K. Kaushik. 1992. Invertebrate colonization and processing of maple leaf litter in
a forested and an agricultural reach of a stream. Hydrobiologia 234:65-77.

Botts, P.S. 1993. The impact of small chironomid grazers on epiphytic algal abundance and disper-
sion. Freshwater Biology 30:25-33.

Botts, P.S., and B.C. Cowell. 1992. Feeding selectivity of two epiphytic chironomids in a subtropi-
cal lake. Oecologia (Berl.) 89:331-337.

Bouffard, S.H., and M.A. Hanson. 1997. Fish in waterfowl marshes: Waterfowl managers' perspective. Wildlife Society Bulletin 25:146-157.

Bowen, S.H. 1984. Evidence of a detritus food chain based on consumption of organic precipitates.
Bulletin of Marine Science 35:440-448.

Brönmark, C. 1989. Interaction between epiphytes, macrophytes and freshwater snails: a review.
Journal of Molluscan Studies 55:299-311.

Brönmark, C., and S.E.B. Weisner. 1996. Decoupling of cascading trophic interactions in a freshwater, benthic food chain. Oecologia (Berl.) 108:534-541.

Brönmark, C., S.P. Klosiewski, and R.A. Stein. 1992. Indirect effects of predation in a freshwater,
benthic food chain. Ecology 73:1662-1674.

Buhl, K.J., S.J. Hamilton, and J.C. Schmulbach. 1993. Acute toxicity of the herbicide Bromoxynil
to *Daphnia magna*. Environmental Toxicology and Chemistry 10:1455-1468.

Caldwell, J.P., J.H. Thorp, and T.O. Jervey. 1980. Predator-prey relationships among larval dragonflies, salamanders and frogs. Oecologia (Berl.) 46:285-289.

Campeau, S., H.R. Murkin, and R.D. Titman. 1994. The relative importance of plant detritus and
algae to invertebrates in a northern prairie marsh. Canadian Journal of Fisheries and Aquatic
Sciences 51:681-692.

Cattaneo, A. 1983. Grazing on epiphytes. Limnology and Oceanography 28:104-132.

Cerny, M., and J. Bytel. 1991. Density and size distribution of Daphnia populations at different fish predation levels. Hydrobiologia 225:199-208.

Chambers, P.A., J.M. Hanson, J.M. Burke, and E.E. Prepas. 1990. The impact of the crayfish *Orconectes virilis* on aquatic macrophytes. Freshwater Biology 24:81-91.

Clark, J. 1978. Fresh water wetlands: habitats for aquatic invertebrates, amphibians, reptiles, and fish. In *Wetland Functions and Values: The State of Our Understanding.* (Eds.) P.E. Greeson, J.R. Clark, and J.E. Clark, pp. 330-343. American Water Resources Association Technical Publication TPS79-2.

Cooper, S.D. 1983. Selective predation on cladocerans by common pond insects. Canadian Journal of Zoology 61: 879-886.

Coveney, M.F., G. Cronberg, M. Enell, K. Larsson, and L. Olofsson. 1977. Phytoplankton, zoo-plankton, and bacteria-standing crop and production relationships in an eutrophic lake. Oikos 29:5-21.

Cox, R.R., M.A. Hanson, C.C. Roy, N.E. Euliss, D.H. Johnson, and M.G. Butler. 1998. Mallard duckling growth and survival in relation to aquatic invertebrates. Journal of Wildlife Management 62:104-133.

Crowder, L.B., and W.E. Cooper. 1982. Habitat structural complexity and the interaction between bluegills and their prey. Ecology 63:1802-1813.

Crumpton, W.G. 1989. Algae in northern prairie wetlands. In *Northern Prairie Wetlands.* (Ed.) A.G. van der Valk, pp.188-203. Ames: Iowa State University Press.

Cummins, K.W. 1973. Trophic relations of aquatic insects. Annual Review of Entomology 18:183-206.

Cummins, K.W. 1974. Structure and function of stream ecosystems. Bioscience 24:631-641.

Cummins, K.W., and M.J. Klug. 1979. Feeding ecology of stream invertebrates. Annual Review of Ecology and Systematics 10:147-172.

Daborn, G.R. 1971. Survival and mortality of coenagrionid nymphs (Odonata: Zygoptera) from the ice of an aestival pond. Canadian Journal of Zoology 49:569-571.

Daborn, G.R. 1974. Biological features of an aestival pond in western Canada. Hydrobiologia 44:287-299.

Daborn, G.R. 1976. Colonization of isolated aquatic habitats. Canadian Field-Naturalist 90:56-57.

Danell, K., and K. Sjoberg. 1979. Decomposition of *Carex* and *Equisetum* in a northern Swedish lake: dry-weight loss and colonization by macroinvertebrates. Journal of Ecology 67:191-200.

Danks, H.V. 1978. Modes of seasonal adaption in the insects. 1. Winter survival. Canadian Entomologist 110:1167-1005.

Darby, R.E. 1962. Midges associated with California rice fields, with special reference to their ecology (Diptera: Chironomidae). Hilgardia 32:1-206.

Darnell, R.M. 1976. Organic detritus in relation to the estuarine ecosystem. In *Estuaries.* (Ed.) G.L. Lauff, pp.376-382. Washington: American Association for Advancement of Science Publications 83.

Davies, R.W., and D.J. Baird. 1988. The effects of oxygen regime on the ecology of lentic macroinvertebrates. Verhandlungen International Vereinigung Limnologie 23:2033-2034.

Davies, R.W., and T.E. Gates. 1991. The effects of different oxygen regimes on the feeding and vertical distribution of *Nephelopsis obscura* (Hirudinoidea). Hydrobiologia 211:51-56.

Davies, R.W., D.M.A. Monita, E. Dratnal, and L.R. Linton. 1992. The effects of different dissolved oxygen regimes on the growth of a freshwater leech. Ecography 15:190-194.

Davies, R.W., T. Yang, and F.J. Wrona. 1987. Inter- and intra-specific differences in the effects of anoxia on erpobdellid leeches using static and flow-through systems. Holarctic Ecology 10:149-153.

Davis, C.B., and A.G. van der Valk. 1978a. The decomposition of standing and fallen litter of *Typha glauca* and *Scirpus fluviatilis*. Canadian Journal of Botany 56:662-675.

Davis, C.B., and A.G. van der Valk. 1978b. Litter decomposition in prairie glacial marshes. In *Freshwater Wetlands: Ecological Processes and Management Potential*. (Eds.) R.E. Good, R.L. Simpson, and D.F. Whigham, pp 99-113. New York: Academic Press.

Day, K.E., and R.J. Maguire. 1990. Acute toxicity of isomers of the pyrethroid insecticide deltamethrin and its major degradation products to *Daphnia magna*. Environmental Toxicology and Chemistry 9:1097-1300.

de March, B.G.E. 1981. *Hyalella azteca* (Sassure). In *Manual for the Culture of Selected Freshwater Invertebrates*. (Ed.) S.G. Lawrence, pp.61-77. Canadian Special Publications of Fisheries and Aquatic Sciences 54. Canadian Government Publishing Centre, Hull, Quebec.

Diehl, S. 1993. Effects of habitat structure on resource availability, diet and growth of benthivorous perch, *Perca fluviatilis*. Oikos 67:403-414.

Dineen, C. F. 1953. An ecological study of a Minnesota pond. American Midland Naturalist 50:349-376.

Dodson, S.I. 1987. Animal assemblages in temporary desert rock pools: aspects of the ecology of *Dasyhela subletti* (Diptera: Ceratopogonidae). Journal of the North American Benthological Society 6:65-71.

Dodson, S.I., and J.E. Havel. 1988. Indirect prey effects: some morphological and life history responses of *Daphnia pulex* exposed to *Notonecta undulata*. Limnology and Oceanography 336:1074-1085.

Driver, E.A. 1977. Chironomid communities in small prairie ponds: some characteristics and controls. Freshwater Biology 7:101-133.

Dvörak, J., and E. P. H. Best. 1982. Macro-invertebrate communities within the macrophytes of Lake Vechten: structural and functional relationships. Hydrobiologia 95:115-106.

Elser, M.M., C.N. von Ende, P. Sorrano, and S.R. Carpenter. 1987. *Chaoborus* populations: response to food-web manipulation and potential effects on zooplankton communities. Canadian Journal of Zoology 65:2846-2852.

Engle, D.E. 1985. The production of haemoglobin by small pond *Daphnia pulex*: intraspecific variation and its relation to habitat. Freshwater Biology 15:631-638.

Fairchild, G.W., J.M. Campbell, and R.L. Lowe. 1989. Numerical response of chydorids (*Cladocera*) and chironomids (Diptera) to nutrient-enhanced periphyton growth. Archiv für Hydrobiologie 114:369-382.

Fellton, H.L. 1940. Control of aquatic midges with notes on the biology of certain species. Journal of Economic Entomology 33:252-264.

Fernando, C.H., and D. Galbraith. 1973. Seasonality and dynamics of aquatic insects colonizing small habitats. Verhandlungen International Vereinigung Limnologie 18:1564-1575.

Findlay, S., and K. Tenore. 1982. Nitrogen source for a detritivore: detritus substrate versus associated microbes. Science 218:371-373.

Foote, A.L., J.A. Kadlec, and B.K. Campbell. 1988. Insect herbivory on an inland brackish wetland. Wetlands 8:67-74.

Fukuhara, H., and M. Sakamoto. 1987. Enhancement of inorganic nitrogen and phosphate release from lake sediment by tubificid worms and chironomid larvae. Oikos 48:310-320.

Fuller, R.L., J.L. Roelofs, and T.J. Fry. 1986. The importance of algae to stream invertebrates.

Journal of the North American Benthological Society 5:290-296.

Gallepp, G.W. 1979. Chironomid influence on phosphorus release in sediment-water microcosms. Ecology 60:547-556.

Gee, J.H.R. 1988. Population dynamics and morphometrics of *Gammarus pulex* L.: evidence of seasonal food limitation in a freshwater detritivore. Freshwater Biology 19:333-343.

Gilinsky, E. 1984. The role of fish predation and spatial heterogeneity in determining benthic community structure. Ecology 65:455-468.

Gilliam, J.F., D.F. Fraser, and A.M. Sabat. 1989. Strong effects of foraging minnows on a stream benthic invertebrate community. Ecology 70:445-452.

Graca, M.A.S., L. Maltby, and P. Calow. 1993. Importance of fungi in the diet of *Gammarus pulex* and *Asellus aquaticus*. 1. Feeding strategies. Oecologia (Berl.) 93:139-144.

Gurney, S.E., and G.G.C. Robinson. 1988. VII. Small water bodies and wetlands. The influence of water level manipulation on metaphyton production in a temperate freshwater marsh. Verhandlungen International Vereinigung Limnologie 23:1032-1040.

Hall, D.J., W.E. Cooper, and E.E. Werner. 1970. An experimental approach to the production dynamics and structure of freshwater animal communities. Limnology and Oceanography 15:839-928.

Hamaker, J.W., and J.M. Thompson. 1972. Adsorption. In *Organic Chemicals in the Soil Environment*. (Eds.) C.A.I. Goring and J.W. Hamaker, pp 49-143. New York: Marcel Dekker.

Hammer, D.A., and R.K. Bastian. 1989. Wetlands ecosystems: natural water purifiers? In *Constructed Wetlands for Wastewater Treatment: Municipal, Industrial and Agricultural*. (Ed.) D.A. Hammer, pp.5-20. Chelsea: Lewis Publishers, Inc.

Hammer, U.T., J. Shamess, and R.C. Haynes. 1983. The distribution and abundance of algae in saline lakes of Saskatchewan. Hydrobiologia 105:1-26.

Hammer, U.T., J.S. Sheard, and J. Kranabetter. 1990. Distribution and abundance of littoral benthic fauna in Canadian prairie saline lakes. Hydrobiologia 197: 173-192.

Hann, B.J. 1991. Invertebrate grazer—periphyton interactions in a eutrophic marsh pond. Freshwater Biology 26:87-96.

Hanson, B.A., and G.A. Swanson. 1989. Coleoptera species inhabiting prairie wetlands of the Cottonwood Lake area, Stutsman County, North Dakota. Prairie Naturalist 21:49-57.

Hanson, M.A., and M.G. Butler. 1994. Responses to food web manipulation in a shallow waterfowl lake. Hydrobiologia 280:457-466.

Hanson, M.A., and M.R. Riggs. 1995. Potential effects of fish predation on wetland invertebrates: a comparison of wetlands with and without fathead minnows. Wetlands 15:167-175.

Hargeby, A. 1990. Macrophyte associated invertebrates and the effect of habitat premanence. Oikos 57:338-346.

Hargrave, B.T. 1970. The utilization of benthic microflora by *Hyalella azteca* (Amphipoda). Journal of Animal Ecology 39:427-437.

Heinis, F., and T. Crommentuijn. 1992. Behavioural responses to changing oxygen concentrations of deposit feeding chironomid larvae (Diptera) of littoral and profundal habitats. Archiv für Hydrobiologie 104:173-185.

Henson, E.B. 1988. Macro-invertebrate associations in a marsh ecosystem. Verhandlungen International Vereinigung Limnologie 23:1049-1056.

Hilton, D.F.J. 1987. Odonata of peatlands and marshes in Canada. In *Aquatic Insects of Peatlands and Marshes*. (Eds.) D.M. Rosenberg and H.V. Danks, pp.57-64. Memoirs of the Entomological Society of Canada 140.

Hooper, N., and G.G.C. Robinson. 1976. Primary production of epiphytic algae in a marsh pond. Canadian Journal of Botany 54:2810-2815.

Hosseini, S.M., and A.G. van der Valk. 1989a. Primary productivity and biomass of periphyton and phytoplankton in flooded freshwater marshes. *In Freshwater Wetlands and Wildlife*. (Eds.) R.R. Sharitz and J.W. Gibbons, pp. 303-315. DOE Symposium Series 61, Oak Ridge: USDOE Office of Scientific and Technical Information.

Hosseini, S.M., and A.G. van der Valk. 1989b. The impact of above-normal flooding on metaphyton in a freshwater marsh. In *Freshwater Wetlands and Wildlife*. (Eds.) R.R. Sharitz and J.W. Gibbons, pp. 317-324. DOE Symposium Series 61, Oak Ridge: USDOE Office of Scientific and Technical Information.

Hunter, R.D. 1980. Effects of grazing on the quantity and quality of freshwater aufwuchs. Hydrobiologia 69:251-259.

Jackson, D., S.P. Long, and C.F. Mason. 1986. Net primary production, decomposition, and export of *Spartina anglica* on a Suffolk salt-marsh. Journal of Ecology 74:647-662.

Jones, M., C. Folt, and S. Guarda. 1991. Characterizing individual, population and community effects of sublethal levels of aquatic toxicants: an experimental case study using *Daphnia*. Freshwater Biology 26:35-44.

Joyner, D.E. 1980. Influence of invertebrates on pond selection by ducks in Ontario. Journal of Wildlife Management 44:700-705.

Kadlec, J.A. 1979. Nitrogen and phosphorus dynamics in inland freshwater wetlands. In *Waterfowl and Wetlands—An Integrated Review*. (Ed.) T.A. Bookhout, pp.17-41. Madison: The Wildlife Society.

Kairesalo, T., and I. Koskimies. 1987. Grazing by oligochaetes and snails on epiphytes. Freshwater Biology 17:317-324.

Kajak, Z., and J. Warda. 1968. Feeding of benthic non-predatory Chironomidae in lakes. Annales Zoology Fennici 5:57-64.

Kaminski, R.M., and H.H. Prince. 1981. Dabbling duck and aquatic macroinvertebrate response to manipulated wetland habitat. Journal of Wildlife Management 45: 1-15.

Kankaala, P. 1988. The relative importance of algae and bacteria as food for *Daphnia longispina* (Cladocera) in a polyhumic lake. Freshwater Biology 19:285-296.

Kantrud, H.A., G.L. Krapu, and G.A. Swanson. 1989. Prairie basin wetlands of the Dakotas: a community profile. United States Fish and Wildlife Service Biological Report 85.

Kenow, K.P., and D.H. Rusch. 1989. An evaluation of plant and invertebrate response to water level manipulation of subimpoundments of Horicon Marsh, Wisconsin. In *Freshwater Wetlands and Wildlife*. (Eds.) R.R. Sharitz and J.W. Gibbons, pp.1153-1163. DOE Symposium Series 61, Oak Ridge: USDOE Office of Scientific and Technical Information.

Kesler, D.H. 1981. Grazing rate determination of *Corynoneura scutellata* Winnertz (Chironomidae: Diptera). Hydrobiologia 80:63-66.

Kimerle, R.A., and A.D. Anderson. 1971. Improved traps and techniques for the study of emerging aquatic insects. Entomological News 90:65-78.

Kitchell, J.F., R.V. O'Neill, D. Webb, G.W. Gallepp, S.M. Bartell, J.F. Koonce, and B.S. Ausmus. 1979. Consumer regulation of nutrient cycling. Bioscience 29:28-34.

Kitting, C.L. 1984. Selectivity by dense populations of small invertebrates foraging among seagrass blade surfaces. Estuaries 7:276-288.

Klarer, D.M., and D.F. Millie. 1992. Aquatic macrophytes and algae at Old Woman Creek estuary and other Great Lakes coastal wetlands. Journal of Great Lakes Research 18:622-633.

Krapu, G.L. 1974. Feeding ecology of pintail hens during reproduction. Auk 91:278-290.

Krapu, G.L. 1981. The role of nutrient reserves in mallard reproduction. Auk 98:29-38.

Krapu, G.L., and K.J. Reinecke. 1992. Foraging ecology and nutrition. In *Ecology and Management of Breeding Waterfowl*. (Eds.) B.D. J. Batt, A.D. Afton, M.G. Anderson, C.D. Ankney, D.H. Johnson, J.A. Kadlec, and G.L. Krapu, pp.1-29. Minneapolis: University of Minnesota Press.

Krecker, F.H. 1939. A comparative study of the animal population of certain submerged aquatic plants. Ecology 20:553-562.

LaGrange, T.G., and J.J. Dinsmore. 1989. Habitat use by mallards during spring migration through central Iowa. Journal of Wildlife Management 53:1076-1080.

Lamberti, G.A., and J.W. Moore. 1984. Aquatic insects as primary consumers. In *The Ecology of Aquatic Insects*. (Eds.) V.H. Resh and D.M. Rosenberg, pp.164-195. New York: Praeger Publishers.

Landon, M.S., and R.H. Stasiak. 1983. *Daphnia* hemoglobin concentration as a function of depth and oxygen availability in Arco Lake, Minnesota. Limnology and Oceanography 28:731-737.

Larson, D.J. 1987. Aquatic coleoptera of peatlands and marshes in Canada. In *Aquatic Insects of Peatlands and Marshes*. (Eds.) D.M. Rosenberg and H.V. Danks, pp.99-132. Memoirs of the Entomological Society of Canada 140.

Lawson, D.L., M.J. Klug, and R.W. Merritt. 1984. The influence of the physical, chemical, and microbiological characteristics of decomposing leaves on the growth of the detritivore *Tipula abdominalis* (Diptera: Tipulidae). Canadian Journal of Zoology 62:2339-2343.

Lee, Jr., R.E. 1991. Principles of insect low tolerance temperature. In *Insects at Low Temperature*. (Eds.) R.E. Lee, Jr. and D.L. Delinger, pp.17-46. New York: Chapman & Hall.

Lee, Jr., R.E., and D.L. Delinger. 1991. Insects at Low Temperatures. New York: Chapman & Hall.

Leonhard, S.L., and S.G. Lawrence. 1981. *Daphnia magna, D. pulex*. In *Manual for the Culture of Selected Freshwater Invertebrates*. (Ed.) S.G. Lawrence, pp.31-50. Canadian Special Publications of Fisheries and Aquatic Sciences 54. Canadian Government Publishing Centre, Hull, Quebec.

Levinton, J.S., T.S. Bianchi, and S. Stewart. 1984. What is the role of particulate organic matter in benthic invertebrate nutrition? Bulletin of Marine Sciences 35:270-282.

Linley, E.A.S., and R.C. Newell. 1984. Estimates of bacteria growth yields based on plant detritus. Bulletin of Marine Sciences 35:409-425.

Lodge, D.M. 1986. Selective grazing on periphyton: a determinant of freshwater gastropod microdistributions. Freshwater Biology 16:831-841.

Lodge, D.M. 1991. Herbivory on freshwater macrophytes. Aquatic Botany 41:195-224.

Macan, T.T. 1977. The influence of predation on the composition of fresh-water animal communities. Biological Review 52: 45-70.

Mann, K.H. 1988. Production and use of detritus in various freshwater, estuarine, and coastal marine ecosystems. Limnology and Oceanography 33:910-930.

Mason, C.F., and R.J. Bryant. 1975. Production, nutrient content and decomposition of *Phragmites communis* Trin. and *Typha angustifolia* L. Journal of Ecology 63:71-95.

Matisoff, G., J.B. Fisher, and S. Matis. 1985. Effects of benthic macroinvertebrates on the exchange of solutes between sediments and freshwater. Hydrobiologia 102:19-33.

Mattila, J. 1992. The effect of habitat complexity on predation efficiency of perch *Perca fluviatilis* L. and ruffe *Gymnocephalus cernuus* L. Journal of Experimental Marine Biology and Ecology 157:55-67.

McDonald, M.E. 1955. Cause and effects of a die-off of emergent vegetation. Journal of Wildlife Management 19:24-35.

Menzie, C.A. 1981. Production ecology of *Cricotopus sylvestris* (Fabricius) (Diptera:

Chironomidae) in a shallow estuarine cove. Limnology and Oceanography 26:467-481.

Mercier, F., and R. McNeil. 1994. Seasonal variations in intertidal density of invertebrate prey in a tropical lagoon and effects of shorebird predation. Canadian Journal of Zoology 72:1755-1763.

Merritt, R.W., and K.W. Cummins. 1984. *An Introduction to the Aquatic Insects of North America*. Dubuque: Kendall/Hunt.

Merritt, R.W., and R.S. Wotton. 1988. Effect of grazing by *Physa integra* on periphyton community structure. Journal of the North American Benthological Society 7:29-36.

Miura, T., and R.M. Takahashi. 1987. Impact of fenoxycarb, a carbamate insect growth regulator, on some aquatic invertebrates abundant in mosquito breeding habitats. Journal of the American Mosquito Control Association 3:476-480.

Montague, C.L., S.M. Bunker, E.B. Haines, M.L. Pace, and R.L. Wetzel. 1981. Aquatic macroconsumers. In *The Ecology of a Salt Marsh*. (Eds.) L.R. Pomeroy and R.G. Wiegert, pp.69-85. New York: Springer–Verlag.

Moore, J.W. 1980. Factors influencing the composition, structure and density of a population of benthic invertebrates. Archiv für Hydrobiologia 88:202-218.

Morrill, P.K., and B.R. Neal. 1990. Impact of deltamethrin insecticide on Chironomidae (Diptera) of prairie ponds. Canadian Journal of Zoology 68:289-296.

Motyka, G.L., R.W. Merritt, M.J. Klug, and J.R. Miller. 1985. Food-finding behaviour of selected aquatic detritivores: direct or indirect mechanisms. Canadian Journal of Zoology 63:1388-1394.

Muir, D.C., D.A. Metner, A.P. Blouw, N.P. Grift, and W.L. Lockhart. 1980. Effect of water chemistry on the uptake of organic pollutants by fish in river water. Canadian Technical Report of Fisheries & Aquatic Sciences No. 975.

Munro, J.A. 1941. The grebes: studies of waterfowl in British Columbia. Occasional Papers of the British Columbia Provincial Museum 3.

Murkin, E.J., H.R. Murkin, and R.D. Titman. 1992. Nektonic invertebrate abundance and distribution at the emergent vegetation–open water interface in the Delta Marsh, Manitoba, Canada. Wetlands 10:45-52.

Murkin, H.R. 1983. Responses by aquatic macroinvertebrates to prolonged flooding of marsh habitat. Ph.D. dissertation, Utah State University, Logan, Utah.

Murkin, H.R. 1989. The basis for food chains in prairie wetlands. In *Northern Prairie Wetlands*. (Ed.) A.G. van der Valk, pp.316- 338. Ames: Iowa State University Press.

Murkin, H.R., and B.D.J. Batt. 1987. Interactions of vertebrates and invertebrates in peatlands and marshes. In *Aquatic Insects of Peatlands and Marshes*. (Eds.) D.M. Rosenberg and H.V. Danks, pp.15-30. Memoirs of the Entomological Society of Canada 140.

Murkin, H.R., and J.A. Kadlec. 1986. Responses by benthic macroinvertebrates to prolonged flooding of marsh habitat. Canadian Journal of Zoology 64:65-72.

Murkin, H.R., and L.C.M. Ross. 1999. Northern prairie marshes (Delta Marsh, Manitoba): macroinvertebrate responses to a simulated wet-dry cycle. In *Invertebrates in Freshwater Wetlands of North America: Ecology and Management*. (Eds.) D. Batzer, R.D. Rader, and S.A. Wissinger, pp. 543-569. New York: Wiley.

Murkin, H.R., and D.A. Wrubleski. 1988. Aquatic invertebrates of freshwater wetlands: function and ecology. In *The Ecology and Management of Wetlands Volume 1: Ecology of Wetlands*. (Eds.) D.D. Hook, W.H. McKee, H.K. Smith, J. Gregory, V.G. Burnell, M.R. DeVoe, R.E. Sojka, S. Gilbert, R. Banks, L.H. Stolzy, C. Brooks, T.D. Matthews, and T.H. Shear, pp.239-249. Portland: Timber Press.

Murkin, H.R., R.M. Kaminski, and R.D. Titman. 1982. Responses by dabbling ducks and aquatic invertebrates to an experimentally manipulated cattail marsh. Canadian Journal of Zoology 60:2324-2332.

Murkin, H.R., J.A. Kadlec, and E.J. Murkin. 1991a. Effects of prolonged flooding on nektonic invertebrates in small diked marshes. Canadian Journal of Fisheries and Aquatic Sciences 48:2355-2364.

Murkin, H.R., M.P. Stainton, J.A. Boughen, J.B. Pollard, and R.D. Titman. 1991b. Nutrient status of wetlands in the interlake region of Manitoba, Canada. Wetlands 11:105-102.

Murkin, H.R., A.G. van der Valk, and C.B. Davis. 1989. Decomposition of four dominant macrophytes in the Delta Marsh, Manitoba. Wildlife Society Bulletin 17:215-221.

Murkin, H.R., D.A. Wrubleski, and F.A. Reid. 1994. Sampling invertebrates in aquatic and terrestrial habitats. In *Research and Management Techniques for Wildlife and Habitats, 5th edition.* (Ed.) T.A. Bookhout, pp. 349-369. The Wildlife Society, Bethesda, Maryland.

Nagell, B. 1978. Resistance to anoxia of *Chironomus plumosus* and *Chironomus anthracinus* (Diptera) larvae. Holarctic Ecology 1:333:336.

Neckles, H.A. 1984. Plant and macroinvertebrate responses to water regime in a whitetop marsh. M.S. thesis, University of Minnesota, Minneapolis, Minnesota.

Neckles, H.A., H.R. Murkin, and J.A. Cooper. 1990. Influences of seasonal flooding on macroinvertebrate abundance in wetland habitats. Freshwater Biology 23:311-322.

Neill, C. 1993. Seasonal flooding, soil salinity and primary production in northern prairie marshes. Oecologia (Berl.) 95:499-505.

Nelson, J.W. 1982. Effects of varying detrital nutrient concentration on macroinvertebrate abundance and biomass. M.S. thesis, Utah State University, Logan, Utah.

Nelson, J.W., and J.A. Kadlec. 1984. A conceptual approach to relating habitat structure and macroinvertebrate production in freshwater wetlands. Transactions of the North American Wildlife and Natural Resources Conference 49:262-270.

Nelson, J.W., J.A. Kadlec, and H.R. Murkin. 1990. Responses by macroinvertebrates to cattail litter quality and timing of litter submergence in a northern prairie marsh. Wetlands 10:47-60.

Newell, S.Y., and F. Barlocher. 1993. Removal of fungal and total organic matter from decaying cordgrass leaves by shredder snails. Journal of Experimental Marine Biology and Ecology 171:39-49.

Newman, R.M. 1991. Herbivory and detritivory on freshwater macrophytes by invertebrates: a review. Journal of the North American Benthological Society 10:89-114.

Newman, R.M., and L.M. Maher. 1995. New records and distribution of aquatic insect herbivores of watermilfoils (Haloragaceae: *Myriophyllum* spp.) in Minnesota. Entomological News 106:6-10.

Odum, W.E., and E.J. Heald. 1975. The detritus based food web of an estuarine mangrove community. In *Estuarine Research.* (Ed.) L.E. Cronin, pp.265-286. New York: Academic Press.

Olah, J. 1976. Energy transformation of *Tanypus punctipennis* (Meig.) (Chironomidae) in Lake Balaton. Annales Biology Tihany 43:83-92.

Oliver, D.R. 1971. Life history of the Chironomidae. Annual Review of Entomology 16:211-230.

Otto, C. 1983. Adaptations to benthic freshwater herbivory. In *Periphyton of Freshwater Ecosystems.* (Ed.) R.G. Wetzel, pp.199-205. The Hague, Netherlands: Dr. W. Junk Publishing.

Paasivirta, L. 1974. Insect emergence and output of incorporated energy and nutrients from the oligotrophic lake Paajarvi, southern Finland. Annales Zoology Fennici 10:106-140.

Palmén, E., and L. Aho. 1966. Studies on the ecology and phenology of the Chironomidae (Dipt.)

of the northern Baltic. 2 *Camptochironomus* Kieff. and *Chironomus* Meig. Annales Zoologici Fennici 3:217-244.

Parker, D.W. 1992. Emergence phenologies and patterns of aquatic insects inhabiting a prairie pond. Ph.D. dissertation, University of Saskatchewan, Saskatoon, Saskatchewan.

Parsons, J.T., and G.A. Surgeoner. 1991. Acute toxicities of permethrin, fenitrothion, carbaryl and carbofuran to mosquito larvae during single- or multiple-pulse exposures. Environmental Toxicology and Chemistry 10:1029-1033.

Parsons, K.A., and A.A. de la Cruz. 1980. Energy flow and grazing behavior of conocephaline grasshoppers in a *Juncus roemerianus* marsh. Ecology 61:1045-1050.

Penko, J.M., and D.C. Pratt. 1986. Effects of *Bellura obliqua* on *Typha latifolia* productivity. Journal of Aquatic Plant Management 24:24-28.

Pennak, R.W. 1978. *Freshwater Invertebrates of the United States.* New York: John Wiley & Sons.

Peterka, J.J. 1989. Fishes in northern prairie wetlands. In *Northern Prairie Wetlands.* (Ed.) A.G. van der Valk, pp 302-315. Ames: Iowa State University Press.

Peterson, L.P., H.R. Murkin, and D.A. Wrubleski. 1989. Waterfowl predation on benthic macroin-vertebrates during fall drawdown of a northern prairie marsh. In *Freshwater Wetlands and Wildlife.* (Eds.) R.R. Sharitz and J.W. Gibbons, pp 681-689. USDOE Symposium Series 61, Oak Ridge: USDOE Office of Scientific and Technical Information.

Pip, E. 1978. A survey of the ecology and composition of submerged aquatic snail-plant communi-ties. Canadian Journal of Zoology 56:2263-2279.

Pip, E., and J.M. Stewart. 1976. The dynamics of two aquatic plant-snail associations. Canadian Journal of Zoology 54:1192-1205.

Polunin, N.V.C. 1984. The decomposition of emergent macrophytes in fresh water. Advances in Ecological Research 14:115-166.

Porter, K.G. 1977. The plant-animal interface in freshwater ecosystems. American Scientist 65:159-170.

Quammen, M.L. 1984. Predation by shorebirds, fish and crabs on invertebrates in intertidal mud-flats: an experimental approach. Ecology 65:529-537.

Rasmussen, J.B. 1987. The effect of a predatory leech, *Nephelopsis obscura*, on mortality, growth, and production of chironomid larvae in a small pond. Oecologia (Berl.) 73:133-138.

Robertson, A.I., and K.M. Mann. 1980. The role of isopods and amphipods in the initial fragmen-tation of eelgrass detritus of Nova Scotia, Canada. Marine Biology 59:63-69.

Robinson, R.D. 1989. The effects of a bromoxynil herbicide on experimental prairie wetlands. M.S. thesis, McGill University, Montreal, Quebec.

Rogers, K.H., and C.M. Breen. 1983. An investigation of macrophyte, epiphyte and grazer interac-tions. In *Periphyton of Freshwater Ecosystems.* (Ed.) R.G. Wetzel, pp.217-226. The Hague, Netherlands: Dr. W. Junk Publishers.

Rosenberg, D.M., and H.V. Danks. 1987. Aquatic insects of peatlands and marshes in Canada. Memoirs of the Entomological Society of Canada 140.

Ross, L.C.M., and H.R. Murkin. 1989. Invertebrates. In *Marsh Ecology Research Program Long-Term Monitoring Procedures Manual.* (Eds.) E.J. Murkin and H.R. Murkin, pp.35-38. Delta Waterfowl and Wetlands Research Station Technical Bulletin 2.

Ross, L.C.M., and H.R. Murkin. 1993. The affect of above-normal flooding of a northern prairie marsh on *Agraylea multipunctata* (Trichoptera: Hydroptilidae). Journal of Freshwater Ecology 8:27-35.

Sadler, W.O. 1935. Biology of the midge *Chironomus tentans* Fabricius, and methods for its propagation. Memoirs of the Cornell University Agricultural Experiment Station 173:1-25.

Schindler, D.E., J.F. Kitchell, X. He, S.R. Carpenter, J.R. Hodgson, and K.L. Cottingham. 1993. Food web structure and phosphorous cycling in lakes. Transactions of the American Fisheries Society 102:756-772.

Scudder, G.G.E. 1983. A review of factors governing the distribution of two closely related corixids in the saline lakes of British Columbia. Hydrobiologia 105:143-154.

Scudder, G.G.E. 1987. Aquatic and semiaquatic Hemiptera of peatlands and marshes in Canada. In *Aquatic Insects of Peatlands and Marshes*. (Eds.) D.M. Rosenberg and H.V. Danks, pp.65-98. Memoirs of the Entomological Society of Canada 140.

Shamess, J.J., G.G.C. Robinson, and L.G. Goldsborough. 1985. The structure and comparison of periphytic and planktonic algal communities in two eutrophic prairie lakes. Archiv für Hydrobiologie 103:99-116.

Shay, J.M., and C.T. Shay. 1986. Prairie marshes in western Canada, with specific reference to the ecology of five emergent macrophytes. Canadian Journal of Botany 64:443-454.

Sheehan, P.J., A. Baril, P. Mineau, D.K. Smith, A. Harfenist, and W.K. Marshall. 1987. The impact of pesticides on the ecology of prairie nesting ducks. Canadian Wildlife Service Technical Report, Series Number 19.

Simpson, R.L., D.F. Whigham, and K. Brannigan. 1979. The mid-summer insect communities of freshwater tidal wetland macrophytes. Bulletin of New Jersey Academy of Science 24:22-28.

Simpson, R.L., R.E. Good, M.A. Leck, and D.F. Whigham. 1983. The ecology of freshwater tidal wetlands. Bioscience 33:255-259.

Singhal, R.N., and R.W. Davies. 1987. Histopathology of hyperoxia in *Nephelopsis obscura* (Hirudinoidea: Erpobdellidae): ultrastructural studies. Journal of Invertebrate Pathology 52:409-418.

Sklar, F.H. 1985. Seasonality and community structure of the backswamp invertebrates in a Louisiana cypress-tupelo wetland. Wetlands 5:69-86.

Skuhravy, V. 1978. Invertebrates:destroyers of common reed. In *Pond Littoral Ecosystems: Structure and Functioning*. (Eds.) D. Dykyjova and J. Kvet, pp.376-387. New York: Springer–Verlag.

Smirnov, N.N. 1961. Consumption of emergent vegetation by insects. Verhandlungen International Vereinigung Limnologie 14:232-236.

Smith, G.E., and B.G. Isom. 1967. Investigations on effects of large-scale applications of 2,4-D on aquatic fauna and water quality. Pesticide Monitoring Journal 1:16-22.

Smith, T.G. 1968. Crustacea of the Delta Marsh region, Manitoba. Canadian Field-Naturalist 82:100-139.

Spence, J.R., and N.M. Andersen. 1994. Biology of water striders—interactions between systematics and ecology. Annual Review of Entomology 39:101-108.

Stanhope, M.J., D.W. Powell, and E.B. Hartwick. 1987. Population characteristics of the estuarine isopod *Gnorimosphaeroma insulare* in three contrasting habitats: sedge marsh, algal beds, and wood debris. Canadian Journal of Zoology 65:2097-2104.

Stephenson, M., and G.L. Mackie. 1986. Effects of 2,4-D treatment on natural benthic macroinvertebrate communities in replicate artificial ponds. Aquatic Toxicology 9:243-251.

Stewart, B.A. 1992. The effect of invertebrates on leaf decomposition rates in 2 small woodland streams in southern Africa. Archiv für Hydrobiologie 104:19-33.

Stewart, R.E., and H.A. Kantrud. 1971. Classification of natural ponds and lakes in the glaciated prairie region. Bureau of Sports Fisheries and Wildlife Resource Publication 92.

Street, M. 1977. The food of mallard ducklings in a wet gravel quarry, and its relation to duckling survival. Wildfowl 28:113-105.

Street, M., and G. Titmus. 1982. A field experiment on the value of allochthonous straw as food and substratum for lake macroinvertebrates. Freshwater Biology 10:403-410.

Suthers, I.M., and J.H. Gee. 1986. Role of hypoxia in limiting diel spring and summer distribution of juvenile yellow perch (*Perca flavescens*) in a prairie marsh. Canadian Journal of Fisheries and Aquatic Sciences 43:1562-1570.

Swanson, G.A. 1977. Diel food selection by anatinae on a waste-stabilization system. 1977. Journal of Wildlife Management 41:226-231.

Swanson, G.A. 1984. Dissemination of amphipods by waterfowl. Journal of Wildlife Management 48:988-991.

Swanson, G.A., and M.I. Meyer. 1973. The role of invertebrates in the feeding ecology of Anatinae during the breeding season. In *Waterfowl Habitat Management Symposium*. pp.143-185. Moncton, New Brunswick.

Swanson, G.A., and H.K. Nelson. 1970. Potential influence of fish rearing programs on waterfowl breeding habitat. In *A Symposium of the Management of Mid-Western Winterkill Lakes*. (Ed.) E. Schneberger, pp.65-71. North Central Division, American Fisheries Society Special Publication.

Swanson, G.A., and A.B. Sargeant. 1972. Observation of nighttime feeding behavior of ducks. Journal of Wildlife Management 36:959-961.

Swanson, G.A., T.C. Winter, V.A. Adomaitis, and J.W. LaBaugh. 1988. Chemical characteristics of prairie lakes in south-central North Dakota—their potential for influencing use by fish and wildlife. Fish and Wildlife Technical Report 18

Swanson, S.M., and U.T. Hammer. 1983. Production of *Cricotopus ornatus* (Meigen) (Diptera: Chironomidae) in Waldsea Lake. Saskatchewan. Hydrobiologia 105:155-164.

Szekely, T., and Z. Bamberger. 1992. Predation of waders (*Charadrii*) on prey populations—an exclosure experiment. Journal of Animal Ecology 61:447-456.

Talent, L.G., G.L. Krapu, and R.L. Jarvis. 1982. Habitat use by mallard broods in south central North Dakota. Journal of Wildlife Management 46:629-635.

Taylor, G.T., R. Iturriaga, and C.W. Sullivan. 1985. Interactions of bactivorous grazers and heterotrophic bacteria with dissolved organic matter. Marine Ecology-Progress Series 23:109-141.

Thorp, J.H., and A.P. Covich. 1991. An overview of freshwater habitats. In *Ecology and Classification of North American Freshwater Invertebrates*. (Eds.) J.H. Thorp and A.P. Covich, pp.17-36. Toronto: Academic Press, Inc.

Tidou, A.S., J.-C. Moreteau, and F. Ramade. 1992. Effects of lindane and deltamethrin on zooplankton communities of experimental ponds. Hydrobiologia 232:157-168.

Timms, B.V., and U.T. Hammer. 1988. Water beetles of some saline lakes in Saskatchewan. Canadian Field Naturalist 102:246-250.

Timms, B.V., U.T. Hammer, and J.W. Sheard. 1986. A study of benthic communities in some saline lakes in Saskatchewan and Alberta, Canada. Internationales Revue des gesamten Hydrobiologie 71:759-777.

Titmus, G., and R.M. Badcock. 1981. Distribution and feeding of larval Chironomidae in a gravel-pit lake. Freshwater Biology 11:263-271.

Topping, M.S. 1971. Ecology of larvae of *Chironomus tentans* (Diptera: Chironomidae) in saline lakes in central British Columbia. Canadian Entomologist 103:328-338

Travis, J., W.H. Keen, and J. Juilianna. 1985. The role of relative body size in a predator-prey relationship between dragonfly naiads and larval anurans. Oikos 45:59-65.

Tuck, L.M. 1972. The snipes. Canadian Wildlife Service Monograph Series 5. 429 pp.

Valiela, I., and J.M. Teal. 1978. Nutrient dynamics: summary and recommendations. In *Freshwater Wetlands: Ecological Processes and Management Potential*. (Eds.) R.E. Good, D.F. Whigham, and R.L. Simpson, pp.259-263. New York: Academic Press.

Vallentyne, J.R. 1952. Insect removal of nitrogen and phosphorus compounds from lakes. Ecology 33:573-577.

van der Valk, A.G., J.M. Rhymer, and H.R. Murkin. 1991. Flooding and the decomposition of litter of four emergent plant species in a prairie wetland. Wetlands 11:1-16.

Vannote, R.L., G.W. Minshall, K.W. Cummins, J.R. Sedell, and C.E. Cushing. 1980. The river continuum concept. Canadian Journal of Fisheries and Aquatic Sciences 37:130-137.

Voigts, D.K. 1976. Aquatic invertebrate abundance in relation to changing marsh vegetation. American Midland Naturalist 95:313-322.

Waite, D.T., R. Grover, N.D. Westcott, H. Sommerstad, and L. Kerr. 1992. Pesticides in ground water, surface water and spring runoff in a small Saskatchewan watershed. Environmental Toxicology and Chemistry 11:741-748.

Ward, G.M., and K.W. Cummins. 1979. Effects of food quality on growth of a stream detritivore *Paratendipes albimanus* (Miegen) (Diptera: Chironomidae). Ecology 60:57-64.

Waterman, T. (Ed.). 1961. *The Physiology of Crustacea. Volume 1.* New York: Academic Press.

Wayland, M., and D.A. Boag. 1990. Toxicity of carbofuran to selected macroinvertebrates in prairie ponds. Bulletin of Environmental and Contamination Toxicology 45:74-81.

Weider, L.J., and W. Lampert. 1985. Differential response of *Daphnia* genotypes to oxygen stress: respiration rates, hemoglobin content and low-oxygen tolerance. Oecologia (Berl.) 65:487-491.

Weller, M.W. 1981. *Freshwater Marshes: Ecology and Wildlife Management.* Minneapolis: University of Minnesota Press.

Wen, Y.H. 1992. Life history and production of *Hyalella azteca* (Crustacea: Amphipoda) in a hyper-eutrophic prairie pond in southern Alberta. Canadian Journal of Zoology 70:1417-1424.

Wetzel, R.G. 1993. Constructed wetlands: scientific foundations are critical. In *Constructed Wetlands for Water Quality Improvement.* (Ed.) G.A. Moshiri, pp.3-8. Boca Raton: Lewis Publishers, Inc.

Wiggins, G.B., R.J. MacKay, and I.M. Smith. 1980. Evolutionary and ecological strategies of animals in annual temporary pools. Archiv für Hydrobiologie Supplement 58:97-206.

Williams, W.D. 1985. Biotic adaptations in temporary lentic waters, with special reference to those in semi-arid and arid regions. Hydrobiologia 105:85-110.

Winter, T.C. 1989. Hydrologic studies of wetlands in the northern prairie. In *Northern Prairie Wetlands.* (Ed.) A.G. van der Valk, pp.16-54. Ames: Iowa State University Press.

Wobeser, G.A. 1981. *Diseases of Wild Waterfowl.* New York: Plenum Press.

Wood, D.M., P.T. Dang, and R.A. Ellis. 1979. The mosquitoes of Canada (Diptera: Culicidae). Publication 1686. Research Branch, Agriculture Canada.

Wrona, F.J., R.W. Davies, and L. Linton. 1979. Analysis of the food niche of *Glossiphonia complanata* (Hirudinoidea: Glossiphonidae). Canadian Journal of Zoology 57:2131-2142.

Wrubleski, D.A. 1984. Species composition, emergence phenologies, and relative abundances of Chironomidae (Diptera) from the Delta Marsh, Manitoba, Canada. M.S. thesis, University of Manitoba, Winnipeg, Manitoba.

Wrubleski, D.A. 1987. Chironomidae (Diptera) of peatlands and marshes in Canada. In *Aquatic Insects of Peatlands and Marshes.* (Ed.) D.M. Rosenberg and H.V. Danks, pp.141-161. Memoirs of the Entomological Society of Canada 140.

Wrubleski, D.A. 1989. The effect of waterfowl feeding on a chironomid (Diptera: Chironomidae) community. In *Freshwater Wetlands and Wildlife.* (Eds.) R.R. Sharitz and J.W. Gibbons, pp.691-696. USDOE Symposium Series 61, Oak Ridge: USDOE Office of Scientific and Technical Information.

Wrubleski, D.A. 1991. Chironomidae (Diptera) community development following experimental manipulation of water levels and aquatic vegetation. Ph.D. dissertation, University of Alberta, Edmonton, Alberta.

Wrubleski, D.A., and D.M. Rosenberg. 1984. Overestimates of Chironomidae (Diptera) abundance from emergence traps with polystyrene floats. American Midland Naturalist 111:195-197.

Wrubleski, D.A., and D.M. Rosenberg. 1990. The Chironomidae (Diptera) of Bone Pile Pond, Delta Marsh, Manitoba, Canada. Wetlands 11:1-16.

Wylie, G.D., and J.R. Jones. 1986. Limnology of a wetland complex in the Mississippi alluvial valley of southeast Missouri. Archiv für Hydrobiologie Supplement 74:288-314.

10

Avian Use of Prairie Wetlands

Henry R. Murkin
Patrick J. Caldwell

ABSTRACT

Prairie wetlands support a variety of avian species, although use may vary over time based on the physiological requirements of the birds and resource availability within the basin and adjacent uplands. Some birds are obligate users of wetlands during one or more stages of their annual cycle, whereas others are more opportunistic in their use of these habitats. The objective of this chapter is to review the factors affecting avian use of prairie wetlands using information from the literature and the Marsh Ecology Research Program (MERP). Food and cover resources are two of the most important factors affecting bird use of prairie wetlands. Avian needs for food and cover vary among species and within species during the different stages of the annual cycle (e.g., nesting, migration, etc.). In addition, availability of these resources varies among wetland types and over time as prairie wetlands proceed through the wet–dry cycle characteristic of the variable prairie environment. Other factors affecting bird use of these systems include availability of space (particularly important to territorial birds), presence of conspecifics and other species, water quality in the wetland, and presence of contaminants and excessive nutrients from adjacent land uses. There is some disagreement on the most important factors affecting avian use of wetland habitats, however, habitat selection is probably based on a combination of factors rather than any single factor. At any point in time, complexes of different wetland types provide suitable habitat conditions for the broadest range of bird species. In addition, the wet–dry cycle ensures the highest productivity and diversity of birds over time.

INTRODUCTION

Northern prairie wetlands are dynamic ecosystems supporting abundant and diverse bird communities within the basins and surrounding uplands (Weller and Fredrickson 1974; Burger 1985; Batt et al. 1989; Swanson and Duebbert 1989). Although waterfowl have received much of the attention in the past (Swanson and Duebbert 1989; Kaminski and Weller 1992), factors affecting the distribution and abundance of all birds in these habitats have drawn increasing attention in recent years (Delphey and Dinsmore 1993; Galatowitsch and van der Valk 1994).

Many birds are obligate users of prairie wetlands during one or more stages of their annual cycle, whereas others are more opportunistic in their use of resources available in these systems. About one-third of North American bird species rely on wetlands for some or all of their annual requirements (Kroodsma 1978). For example, Faanes and Stewart (1982) reported that 39% of the 353 bird species observed in North Dakota use wetland habitats at some point in their annual cycle. The differing physiological demands experienced by birds during migration, breeding, molting, staging (in some species), and wintering (Krapu and Reinecke 1992), coupled with the need for thermal cover (Weathers and Sullivan 1993) and avoidance of predators (Sargeant and Raveling 1992) and competitors (Nudds 1992), result in a complex and ever-changing array of nutritional and cover requirements over the course of a single year. An evaluation of the factors affecting the distribution and abundance of birds in prairie wetlands demands an understanding of requirements of the birds during various phases of their life cycles and the ability of the wetlands to meet those requirements within and among years (Swanson and Duebbert 1989).

The dynamic nature of prairie wetlands ensures a great deal of spatial and temporal variation in avian habitat conditions (Weller 1978). At any point in time, the prairie landscape includes a diverse array of wetland types, each with its unique set of physical and biological characteristics (Stewart and Kantrud 1971). The classification of wetland types is primarily a function of the vegetation present (see Figure 9.2 in Chapter 9), a direct result of the frequency and duration of flooding (Kantrud et al. 1989). These wetland classes represent very different habitat conditions for birds and other wetland wildlife. Besides this spatial variation in habitats and available resources, the dynamic nature of the prairie climate affects the ability of an individual wetland or wetland complex to provide avian nutritional and cover requirements over time. The wet–dry cycle described by van der Valk and Davis (1978) (see Figure 1.3 in Chapter 1) also influences productivity of prairie wetlands over time and subsequently their ability to support avian species (Murkin et al. 1997). Birds using prairie wetland habitats must be adapted to this temporal variability in available resources and habitat conditions. This relationship between the environmental changes among and within prairie wetlands and the ability of these wetlands to provide avian food and cover has received only cursory attention to date (Swanson and Duebbert 1989).

The objective of this chapter is to incorporate the MERP avian use information into a review of factors affecting distribution and abundance of birds in prairie wetlands. Changes in these factors among wetland types and during the wet–dry cycle also are examined. An important consideration in addressing this topic is scale (Wiens 1989a). We focus on the food and cover resources for individual species or groups of species provided by individual wetlands or complexes of wetlands at any point in time (second-order and third-order selection as defined by Kaminski and Weller [1992]) rather than on population and overall community responses to habitat availability and conditions on a larger regional scale (first-order selection). The MERP avian use was determined by weekly avian census on the experimental cells through the simulated wet–dry cycle (Murkin et al. 1997; see Appendix 1).

RESOURCE USE BY AVIAN SPECIES IN A VARIABLE ENVIRONMENT

The ability to track and adjust to available resources is critical to survival and reproduction in habitats that fluctuate greatly over time. Changes in bird communities in these environments are the direct result of varying resource levels being tracked by a changing set of consumers (Cody 1981). When resource availability and abundance fall outside the suitable range for an individual species, it must adjust its requirements, move to areas where suitable conditions exist, or perish (Wiens 1989b). Some species are flexible in their resource requirements and can adjust to a fairly wide range of environmental conditions. Others with more fixed requirements must move to areas with suitable conditions. Although it varies among species, birds have the advantage of being highly mobile, which allows them to move within and among habitats to seek out suitable resource availability and quality.

Strategies used to track and exploit available resources vary depending on how closely the individual species is tied to wetland habitat. Species that use wetlands on an opportunistic basis to exploit available resources (e.g., forest passerines moving into wetlands for a few days to feed on a single insect emergence event) (Busby and Sealy 1979), simply avoid wetlands during periods of drought or reduced productivity. Many other species are tied closely to wetlands for their entire life cycle (e.g., ducks and grebes), whereas others use wetlands in a particular season or stage of their life cycle (e.g., blackbirds during the nesting season). Changes in wetland productivity or habitat conditions have serious implications for the species that are most closely tied to wetland habitats. If conditions are not suitable, these species must move to other wetland areas where suitable conditions exist. In a widespread drought, movement out of the prairie region may be required. For example, prairie ducks fly over their normal prairie breeding areas during drought periods and settle in more northern areas with permanent water (Johnson 1986; Batt et al. 1989).

In ecosystems like the prairies where environmental variations are pulsed and extreme, changes in available resources may be so large, and sometimes unpredictable, that it is very difficult to track conditions (Wiens 1989b). Nudds (1983) related the absence of community patterns in prairie dabbling ducks to the extreme environmental variation in their habitats. In the more northern parkland areas where habitat conditions are more stable (particularly for annual precipitation), there is a much more defined community structure among the waterfowl species.

Prairie wetland birds also must assess the spatial variation in resources. The ability to exploit resources across different wetland types within a wetland complex can be an advantage and as such the home range of many species often includes one or more wetland types (Swanson and Duebbert 1989). For example, as reported in an early MERP publication, red-winged blackbirds (*Agelaius phoeniceus*) nesting in cattail (*Typha* spp.) stands in the central areas of the Delta Marsh flew to nearby seasonally flooded whitetop (*Scholochloa festucacea*) meadows to feed (Murkin et al. 1989). The cattail provided nesting habitat, whereas the whitetop areas with their abundant invertebrate populations in the spring (Neckles et al. 1990) provided more profitable foraging opportunities.

The abundance and diversity of birds found in prairie wetlands indicate that many species have adapted to the variable conditions characteristic of these habitats. Understanding of the overall impacts of environmental changes on resource use by birds requires an examination of how available resources vary in time and space and an assessment of the birds' ability to adjust to those changes.

FOOD RESOURCES

Northern prairie wetlands provide a diverse array of food resources for avian consumers. Plant foods consist of seeds, tubers, and structural components including leaves, stems, roots, and rhizomes. Algal production is very high in prairie wetlands (see Chapter 8) and provides a source of nutrients for several bird species (Murkin 1989). Potential animal foods for higher-order consumers include invertebrates and all classes of vertebrates: fish, amphibians, reptiles, mammals, and other birds. The abundance and diversity of animal life associated with various wetland types and stages of the wet–dry cycle ensure a variety of potential prey when the basins are flooded. However, as the MERP studies on vegetation, algae, and invertebrates indicated, availability and abundance of food resources change over time due to variations in wetland conditions (see Chapters 7–9).

Avian Food Habits

In prairie wetlands, avian food habits range from strict herbivory through varying degrees of omnivory to carnivory. Often, the high variability in food resource availability favors opportunistic feeding behavior by species closely tied to prairie wetlands. Potential foods are determined in part by coevolved patterns of habitat use and morphology (Krapu and Reinecke 1992). Individual species are adapted to exploit resources within a specific habitat type. Long bills and legs allow shorebirds to probe for benthic invertebrates in shallow wetland edges. Herons are adapted for feeding on fish, frogs, and invertebrates by wading in deeper water. Canada geese (*Branta canadensis*) are adapted to graze on vegetation in adjacent uplands and shallow ponds, whereas lesser snow geese (*Chen caerulescens*) have strong mandibles that can be used for grazing or for digging for tubers and rhizomes in mudflats or shallowly flooded stands of emergent vegetation (Cargill and Jefferies 1984). Swans also have powerful mandibles for feeding on belowground plant parts; however, their large bodies and long necks allow them to exploit deeper-water areas than geese. Dabbling ducks feed near the surface and on bottom in shallow wetland areas by dabbling or tipping up. Within dabbling ducks, types of foods or prey are further determined by length of the neck (Pöysä 1983) and morphological characteristics of their bills (Nudds and Kaminski 1984). For example, northern shovelers (*Anas clypeata*) with very fine lamellar spacing are able to feed on smaller invertebrates than mallards (*A. platyrhynchos*) with their wider spacing (Nudds and Bowlby 1984). Diving ducks and grebes with their adaptations for swimming underwater are able to feed on submersed plants, invertebrates, and other prey in deep water. Canvasbacks (*Aythya valisineria*) combine this diving ability with a bill adapted to dig out tubers in deep water (Kehoe and Thomas 1987; Lagerquist and Ankney 1989). American coots (*Fulica americana*) have the ability to feed at the surface of the water and dive to feed at slightly greater depths (Desrochers and Ankney 1986). Passerines with their perching ability and agility are adapted to feed on adult insects in the stands of emergent vegetation. In summary, each species of bird is adapted to exploit a subset of the resources available within the prairie wetland system.

Few species of birds are strictly herbivorous in northern prairie wetlands. Examples are geese and swans. Tundra (*Cygnus columbianus*) and trumpeter swans (*C. buccinator*) feed on roots, tubers, stems, and leaves of emergent and submersed vegetation (Banko 1960; Bailey and Batt 1974; Castelli and Applegate 1989). In the prairie region in recent years, there has been increased use of waste grains in agricultural fields by migrating tundra swans (Krapu and Reinecke 1992). Most species of geese migrating through mid-continent North America feed on waste grain. These include Canada geese (Raveling and Lumsden 1977), white-fronted geese (*Anser albifrons*) (U.S. Fish and Wildlife Service 1981), and lesser snow geese (Alisauskas and Ankney 1992a). As green vegetation becomes available in the spring, these species switch to browsing on new-growth vegetation. Canada geese that remain on the prairies to breed feed entirely on green vegetation during the breeding season (Krapu and Reinecke 1992). Unlike many other bird species, the diet of geese appears to follow food availability rather than reproductive status of the birds (Buchsbaum and Valiela 1987; Krapu and Reinecke 1992). There is some evidence to suggest that laying females and both adult geese and broods during brood-rearing select shoots of grasses and sedges based on protein content (Gauthier and Bédard 1990) and digestibility (Boudewijn 1984). The timing of seasonal migration and local foraging patterns must track vegetation development for geese to maximize intake of required nutrients from green vegetation.

Many wetland bird species are omnivores, although the proportion of plant and animal matter may change according to availability and stage of the annual cycle. Many species of ducks feed on plants during most times of the year, however, they switch to invertebrates during the breeding season to meet the high protein demands of reproduction (Murkin and Batt 1987; Krapu and Reinecke 1992). Unlike geese and swans that depend largely on body reserves for the nutrient requirements of reproduction, duck species rely to varying degrees on resources obtained on the breeding grounds (Alisauskas and Ankney 1992b). Fat reserves are often acquired by feeding on plant foods on the wintering grounds and in migration areas (Krapu and Reinecke 1992). These reserves are used for the energy requirements of migration and breeding. The protein and calcium requirements of gonadal development and egg laying are largely obtained on the breeding grounds by feeding on invertebrates (Swanson et al. 1979). Lipid reserves carried to the breeding grounds also allow females to forage for aquatic invertebrates for long periods in the early spring when invertebrate densities are still low and protein requirements of the birds are high. For example, mallards and northern pintails (*Anas acuta*) feed extensively on plant foods during the winter (Jorde 1981; Heitmeyer 1985). The use of invertebrates begins to increase in late winter, and once they return to the prairie breeding grounds becomes a dominant part of their diet during the breeding season. Canvasbacks are often classified as herbivores (Barnes and Thomas 1987) and feed almost exclusively on the tubers and rootstocks of submersed vegetation throughout much of the year; however, they increase their use of invertebrates during spring migration and throughout the egg-laying period (Austin et al. 1990). Waterfowl broods feed almost exclusively on invertebrates to meet the protein demands of rapid growth (Krapu and Reinecke 1992).

The MERP avian-use monitoring indicated the importance of invertebrates to waterfowl during early spring. During periods of normal water levels within the experimental cells, the highest waterfowl densities during May and June were found on cells with the highest invertebrate densities (Table 10.1). These correlations were supported by behavioral observations showing that waterfowl using the cells during this period spent a large proportion of time feeding (Murkin and Kadlec 1986a). Later in the fall when most species are feeding on plant foods, there was no relationship found between waterfowl and invertebrate densities.

Red-winged and yellow-headed (*Xanthocephalus xanthocephalus*) blackbirds also switch from plant foods to a largely invertebrate diet during breeding and brood-rearing periods (Bird and Smith 1964). Considered agricultural pests during fall and winter, blackbirds feed almost exclusively on insects in and around their wetland breeding territories in the spring (Orians 1980). American coots feed on a combination of plant and animal material throughout the year, and although during the MERP study there was no relationship found between coot and invertebrate densities in the experimental cells (Murkin and Kadlec 1986a), their intake of invertebrates increases during the breeding season (Jones 1940). This lack of relationship may be because coots are territorial in the spring thereby limiting the number of birds that can use a specific site. Coots feed almost exclu-

Table 10.1. Kendall rank correlation coefficients for waterfowl and American coot densities (numbers per hectare) versus invertebrate densities during the spring (May and June, 1980–1982) in the MERP experimental cells, Delta Marsh, Manitoba

Year	Avian group	Nekton density (no./liter)	Benthos density (no./m^2)
1980	Blue-winged teal	0.51[a]	0.55[b]
	Mallards	0.08	−0.16
	Total dabblers	0.44[b]	0.45[b]
	Total ducks	0.44[b]	0.49[b]
	American coots	−0.14	−0.10
1981	Blue-winged teal	0.78[b]	0.65[b]
	Mallards	0.17	0.42
	Total dabblers	0.67[b]	0.70[b]
	Total ducks	0.67[b]	0.70[a]
	American coots	0.32	0.24

Source: After Murkin and Kadlec 1986a
[a]$P < 0.05$.
[b]$P < 0.01$.

sively within their breeding territory and the abundance of both plant and animal food within the territory can influence reproductive output (Ryan and Dinsmore 1980; Hill 1988).

Other species are strictly predators feeding on a variety of invertebrate and vertebrate prey. Herons feed primarily on fish and amphibians (Kantrud et al. 1989). Grebes prey on fish and large invertebrates (Boe 1992), whereas white pelicans (*Pelecanus erythrorhynchos*) feed primarily on fish. Passerine species associated with prairie wetlands feed almost exclusively on emerging insects (Busby and Sealy 1979; Sealy 1980; Guinan and Sealy 1987). Shorebirds using temporary and seasonal ponds in North Dakota consume primarily crustaceans and insects during spring migration (Kantrud et al. 1989). Other avian predators include hawks, owls, and falcons that feed on birds and mammals in these systems, however, few, if any, of these species feed exclusively in wetland habitats. Wetlands normally form a portion of a larger feeding territory. For example, although voles in upland areas are the principle component of its diet, the northern harrier (*Circus cyaneus*) feeds on a wide range of birds and small mammals in wetlands (Errington and Breckenridge 1936; Hamerstrom 1979). Swainson's hawks (*Buteo swainsoni*) make extensive use of prey in prairie wetland habitats (Houston 1990).

Wetland Type and Food Resources

Ephemeral wetlands (Class I in Figure 9.2 in Chapter 9), often termed sheet water, are simply shallow depressions that hold meltwater and runoff for brief periods in early spring (Stewart and Kantrud 1971). Water is not present long enough to allow development of aquatic invertebrate communities. Avian feeding opportunities are limited primarily to seeds of terrestrial plants from the previous growing season and earthworms that are forced to the surface by saturated soil conditions (Krapu 1974; Swanson et al. 1985). Waterfowl and shorebirds make extensive use of ephemeral ponds during migration (Wishart et al. 1981; LaGrange and Dinsmore 1989). Resources provided by these wetlands are important because they are often the only available open-water feeding areas available in early spring. In sheet water on agricultural fields, many birds, especially migrating waterfowl, make extensive use of waste grain (Kantrud et al. 1989; Krapu and Reinecke 1992). LaGrange and Dinsmore (1989) noted that most field feeding by mallards in the spring was in ephemeral ponds rather than in dry uplands as is common in the fall. Field treatment the previous fall (e.g., disking or plowing) may affect availability of waste grain or other seeds (Froud-Williams et al. 1983). In most years, the first green vegetation of the spring occurs in ephemeral ponds providing new growth forage for migrant geese and other grazers.

The longer duration of flooding in temporary wetlands (Class II in Figure 9.2) results in the development of a central zone of wet-meadow vegetation in undisturbed basins. Plants in this zone consist of moist soil species that produce abundant seed supplies, an important food of migrating waterfowl (Fredrickson and Taylor 1982). Invertebrate fauna is limited to species that complete their life cycles in a short period of time in the cold water that often exists in early spring. These include several crustacean species within the orders

Anostraca (fairy shrimp), *Conchostraca* (clam shrimp), and *Notostraca* (tadpole shrimp) (Kantrud et al. 1989). Northern pintails make extensive use of fairy shrimp in temporary wetlands, moving only to other invertebrate groups once the temporary ponds dry up (Krapu 1974). The shallow nature of temporary wetlands makes them particularly suitable for dabbling ducks and wading birds (Kantrud et al. 1989). Temporary ponds are used by large numbers of arctic and subarctic nesting shorebirds that migrate through the prairie region each spring (Wishart et al. 1981; Kantrud et al. 1989). Like the more ephemerally flooded areas, temporary ponds also "green up" early in spring and serve as grazing areas for migrant geese.

Seasonal wetlands (Class III in Figure 9.2) are flooded long enough that an extensive wetland community develops. Water depths are greater than those in ephemeral or temporary ponds thereby providing additional habitats. Besides establishment of species common to shallow wetland zones such as whitetop, sedges (*Carex* spp.), and spikerush (*Eleocharis* spp.), there is some development of submersed vegetation and algae in ponds with extended flooding (Stewart and Kantrud 1971). The emergent species present provide seeds and serve as browse to swans and geese. American wigeon (*Anas americana*) is the only North American duck adapted for grazing and they were observed on many occasions clipping the newly emerged shoots of whitetop in the MERP experimental cells. These plant foods provided by the shallow marsh zone are in addition to the seeds and forage produced in the low prairie and wet-meadow zones at the periphery of the basin (Figure 9.2).

An important difference between seasonal ponds and those with shorter flooding durations is the development of an extensive and varied aquatic invertebrate community (Kantrud et al. 1989; Neckles et al. 1990). Although the ponds normally dry out on an annual basis, they do remain flooded long enough for a wide range of invertebrate species to complete their life cycles. The aquatic invertebrates in these ponds have evolved structural, behavioral, and physiological adaptations to survive periods of dry conditions (see Chapter 9). Common invertebrates in seasonal prairie ponds include cladocerans (Cladocera), ostracods (Ostracoda), Culicidae (Diptera), Chironomidae (Diptera), and a variety of snails (Gastropoda). Cladocerans and ostracods survive the dry period as drought-resistant eggs (Wiggins et al. 1980). Adult females of some mosquito (Culicidae) species (e.g., *Aedes*) oviposit on the substrate of dry basins and eggs hatch upon reflooding (Merritt and Cummins 1984). Chironomidae survive through the dry period by the larvae burrowing into the substrate. Snails secrete a mucous covering over the shell opening to avoid desiccation during the dry period. Snails adapted to the seasonal flooding regime include those in the *Aplexa hypnorum–Gyraulus circumstriatus* association (Kantrud et al. 1989).

Water-manipulation experiments in association with MERP showed that eliminating the dry period had a negative effect on many of the important invertebrate species in seasonal whitetop ponds (Neckles et al. 1990). Cladocerans were much reduced in ponds that were flooded >1 year and ostracods were virtually eliminated. Eventually, species adapted to continuous flooding begin to dominate the system.

A wide range of bird species is attracted to the abundant invertebrate foods provided by seasonal ponds. Dabbling ducks make extensive use of seasonal ponds and the greater depth and duration of flooding make use by diving ducks possible (Kantrud et al. 1989). Kaminski and Prince (1981a) found that the foraging rate of mallards, blue-winged teal (*Anas discors*), northern shovelers, gadwalls (*Anas strepera*), and pintails was related to the invertebrate abundance on whitetop ponds in the Delta Marsh. Talent et al. (1982) found that seasonal ponds were the preferred habitat for brood-rearing mallards. As mentioned previously, Murkin et al. (1989) reported that red-winged blackbirds made use of seasonally flooded whitetop meadows to feed on emerging insects.

Development of a deep-marsh zone with its associated emergent and submersed vegetation in semipermanent wetlands (Class IV in Figure 9.2) adds a new dimension to food availability. Although some grazing by geese and other waterfowl occurs in the peripheral areas of the wetland, few bird species feed extensively on the vegetative tissues of the emergent vegetation associated with the deep-marsh zone (Murkin 1989). There is some use of seeds (e.g., bulrush [*Scirpus* spp.]), at various times of the year by some species, including ducks (Woodin and Swanson 1989) and rails (Melvin and Gibbs 1994). Submersed vegetation beds that develop in the deep-marsh zone are fed on by several bird species. During the MERP studies, feeding by mallards and blue-winged teal completely eliminated sago pondweed (*Potamogeton pectinatus*) aboveground biomass from a shallow bay in the Delta Marsh (Wrubleski 1989). Canvasbacks feed on tubers of sago pondweed (Anderson and Low 1976). Coots also feed on pondweed and associated filamentous algae (Jones 1940).

Several MERP studies have shown that an important function of submersed vegetation in the deep-marsh zone is the habitat provided for aquatic invertebrates (Murkin 1989, Wrubleski 1989; Murkin et al. 1992). Many species of invertebrates are attracted to the submersed habitat structure and food supply within the beds of submersed vegetation. These invertebrates in turn are fed upon by foraging waterfowl and other bird species. Elimination of submersed vegetation has a major effect on the invertebrate community and subsequently potential avian foods in the wetland (Wrubleski 1989).

Fish are found in semipermanent wetlands with extended flooding (Hanson and Riggs 1995) and serve as food for herons, grebes, several duck species, and other birds. Fish may compete with birds for available invertebrate resources, thereby reducing habitat suitability for some bird species (Hanson and Butler 1994). Frogs and salamanders are also more common in these wetlands than in the shallower wetland types and are preyed on by a variety of bird species (Kantrud et al. 1989).

Permanent wetlands (Class V in Figure 9.2) add a permanent open-water zone to the diversity of habitats associated with these usually deep basins. The permanent open-water zone and the long-term presence of water result in several differences from other wetland types. Continued flooding allows development of more extensive fish communities (Hanson and Butler 1990). The open-water zone provides an often extensive benthic invertebrate community (Murkin and Kadlec 1986b). The invertebrate community switches from species adapted to periodic drying to those requiring permanent water (Neckles et al. 1990). The greater water depth associated with these basins excludes most wading birds and dabbling ducks other than at the shallow margins. Many of the diving ducks are able to exploit the resources associated with the deep water of permanent ponds. For example, lesser scaup (*Aythya affinis*) feed extensively on the often abundant amphipod (Amphipoda) populations associated with deep permanent ponds (Rogers and Korschgen 1966; Afton et al. 1991).

The Wet–Dry Cycle and Food Resources

Available food resources vary widely through the various stages of the wet–dry cycle (Table 10.2). Absence of water during the dry stage of the wet–dry cycle (see Figure 1.3 in Chapter 1) results in the eventual elimination of aquatic invertebrates and submersed vegetation from the wetland basin. However, as the water levels recede during the early stages of drawdown, surviving aquatic invertebrates and fish are concentrated in the shrinking pools, creating an accessible food supply for a wide range of bird species. Shorebirds also are attracted to newly drawndown areas to feed on invertebrates in the newly exposed substrates (Skagen and Knopf 1994). Wading birds such as herons congregate at drawdown pools to prey on fish and amphibians stranded by receding water. During the MERP study, Peterson et al. (1989) noted increased bird use of the experimental cells during the first 2 months of the initial drawdown in the fall of 1984. Mallards, which normally feed on plant foods and grain in the fall, were observed feeding in high numbers during this period in the MERP cells. Densities of chironomid larvae in the newly exposed substrate were reduced >70% by the feeding waterfowl (see Figure 9.11 in Chapter 9).

As the substrate dries out completely during the later stages of the drawdown, conditions within the basin change markedly. The vegetation resources on the drawdown surface are a combination of terrestrial mudflat species and newly germinated emergent species (van der Valk and Davis 1978; Welling et al. 1988). There may be some grazing by geese in spring on new-growth vegetation; however, use of dry marsh basins by feeding waterfowl is limited. Available invertebrate foods include terrestrial and semiterrestrial invertebrates and the dormant stages of aquatic invertebrates. Birds using these resources are primarily those species adapted to feeding in uplands (Schitoskey and Linder 1978), including bobolinks (*Dolichonyx oryzivorus*), meadowlarks (*Sturnella neglecta*), and mourning doves (*Zenaida macroura*) (Weller and Spatcher 1965). A large concentration of terrestrial spiders was noted in the MERP experimental cells soon after drawdown, presumably feeding on small invertebrates (primarily dipterans) attracted to the decomposing vegetation and algae. Large numbers of common grackles (*Quiscalus quiscalus*) were observed feeding on spiders in the MERP cells at that time (Murkin, personal observation).

The prey base for higher-order predators also changes during dry conditions. Water birds, their young, and eggs are much reduced or no longer available; however, they are replaced to some degree by upland birds. Small mammals such as deer mice (*Peromyscus maniculatus*), usually abundant in adjacent uplands, increase in wetlands during dry periods (Weller 1981) and serve as prey for several types of avian predators. Some raptors commonly associated with wetlands during breeding may have reduced reproductive effort during drought conditions, For example, the extended drought in southern

Table 10.2. Changes in avian food resources during the various stages of the wet–dry cycle in prairie wetlands

	Dry marsh	Regenerating marsh	Hemi-marsh	Degenerating marsh	Lake marsh
Plant foods					
Annual plant biomass	Abundant, increasing growing season	Eliminated in flooded basin, some present at damp edges	Some present at damp edges	Some present at damp edges	Some present at damp edges
Annual plant seeds	Abundant following initial growing season	Some production at damp edges, some persist in flooded substrate	Some production at damp edges, some persist in flooded substrate	Some production at damp edges, some persist in flooded substrate Decreasing over time	Some production at damp edges, viable seeds in flooded substrate decline over time
Emergent plant biomass: above-ground	Little present initially, increasing over time	Increasing over time	Stabilized, may be slightly increasing or decreasing over time depending on water depth		Decreasing over time
Emergent plant biomass: below-ground	Little or none present initially, increasing over time	Increasing over time	Slight increase or decrease	Decreasing over time	Rapid decrease over time
Emergent plant seeds	Little present, increasing over time	Increasing over time	Increasing over time	Decreasing over time	Decreasing over time
Submersed plant biomass	None	Increasing over time	Increasing over time	Increasing in shallow areas, begin to decline in deeper areas	Stable in shallow areas, eliminated in deep areas
Submersed plant seeds, tubers	Some present in dry substrate, little value as food resource	Little initially, increasing over time	Increasing over time	Increasing over time in shallow areas, stable in deeper areas	Stable, some decrease over time in deep areas
Algae	Some present on surface in damp areas, not important as food	Increasing, epiphytic forms dominant	Increasing, epiphytic, metaphytic forms dominant	Stabilizing, epipelic, and planktonic forms increasing	Epipelic forms dominate
Animal foods					

Continued

Table 10.2. (Continued)

	Dry marsh	Regenerating marsh	Hemi-marsh	Degenerating marsh	Lake marsh
Terrestrial and semiterrestrial insects	Present throughout basin	Present at margins only	Present at margins only	Present at margins only	Present at margins only
Aquatic invertebrates: water column	No active forms present, eggs and other resting stages not important food resource	Abundant, crustaceans peak early, aquatic insects increase over time	Abundant, insects well established in water column, high diversity	Densities beginning to decline as vegetation is lost from the basin	Densities low, except possibly in association with basin margin
Aquatic invertebrates: benthic	Some forms (e.g., Chironomid larvae) survive for period following drawdown, decline as substrate dries, eggs and other resting forms not important foods	Low densities early, begin to increase in open-water areas	Densities increasing in open-water areas, high diversity due to different communities developing in areas with no vegetation, submersed aquatics, and emergents	Densities increasing in open water areas due to algae production and litter inputs	Densities high in open-water areas
Aquatic invertebrates: epiphytic	None	Low densities early, increase as submersed plant surfaces become available	High densities and diversity associated with emergent and submersed vegetation	Densities declining as vegetation lost from basin	Densities low, except in shallow areas where vegetation persists
Fish	None	None present initially, increase over time once established	Increasing over time, excluded from stands of emergent vegetation by low oxygen levels	Increasing over time	May be abundant
Amphibians	None	None present initially, increase over time once established	Increasing over time	Stabilized, begin to decline	Decreasing over time with elimination of vegetation
Mammals	Small-mammal densities increase with vegetation development	Muskrat densities increase over time	Muskrat densities may be high	Muskrat densities will be declining	Low muskrat densities
Birds	Shorebirds on newly exposed mud flats, increase in upland bird use with increase in vegetation	Increase in overall bird use	Abundant and diverse bird use	Decreasing diversity in bird use	Low abundance (except possibly during migration) and diversity

Saskatchewan and its associated impact on wetlands during the 1980s are considered the main reasons for the decline of the Swainson's hawk in the region during that period (Kirk and Houston 1994).

With the return of water during the regenerating stage of the wet–dry cycle (Figure 1.3), habitat becomes available to birds requiring flooded conditions. Terrestrial mudflat plants that established during the drawdown are eliminated by the flooding; however, the abundant seeds produced by these species are available to feeding birds. Seeds are important to blue-winged teal and other dabbling ducks as a source of protein and carbohydrates in early spring before invertebrates are available (Murkin et al. 1982). Green-winged teal (*Anas crecca*) are attracted to shallowly flooded habitats in fall where they feed primarily on seeds (Murkin et al. 1997). Other plant foods available soon after drawdown include the newly germinated emergent vegetation, particularly grasses such as whitetop. Geese and wigeon make extensive use of the newly emerging shoots of whitetop and sedges especially in the spring (Krapu and Reinecke 1992). Submersed vegetation also appears soon after flooding, providing both vegetative material and tubers for herbivores such as canvasbacks.

Aquatic invertebrates reappear quickly following reflooding, especially those species that remained dormant in the basin during the dry period (Neckles et al. 1990; see Chapter 9). Abundant food and habitat created by the flooded litter, high algal production, and an initial lack of predators result in high densities of these invertebrate groups. These early peaks of invertebrates are followed by those mobile species that are able to colonize new habitats quickly (e.g., Hemiptera and Coleoptera) (Hanson and Swanson 1989). Rapid increase in invertebrates following reflooding has been documented in the MERP experimental cells (Wrubleski 1991; Murkin and Ross 1999) and in other studies (Whitman 1974). Amphipods are reintroduced to the ponds by clinging to birds and mammals that are moving from pond to pond (Swanson 1984). Birds that feed on aquatic invertebrates respond quickly to these abundant food resources. Algal production is also very high during this period. Although the abundant forms at this time, phytoplankton and epiphyton, are not eaten directly by birds, they do support invertebrate herbivores (Murkin 1989). With reflooding, amphibians and fish establish from adjacent permanent-water habitats (Kantrud et al. 1989; Hanson and Butler 1990). They gain access to the ponds during periods of overland flow or in association with other species (e.g., eggs caught in the feathers of birds). Once established, they remain part of the wetland fauna until the next drawdown period.

With continued flooding, the emergent vegetation begins to die back as the wetland enters the degenerating stage of the wet–dry cycle (see Figure 1.3 in Chapter 1). Submersed vegetation expands as more light is made available within the water column by the thinning stands of emergent vegetation. The invertebrate community changes as the prolonged flooding eliminates taxa requiring a dry period to complete their life cycles (Neckles et al. 1990). Invertebrates more representative of permanently flooded wetlands appear (e.g., amphipods and benthic chironomids) and dominate the invertebrate community. During the MERP invertebrate monitoring, both amphipods and chironomids became established during the first year of reflooding and increased in abundance during the subsequent years of flooding (Murkin, unpublished data; Wrubleski 1991).

Eventually, the wetland enters the lake marsh stage (Figure 1.3) and food resources provided for birds resemble those found in permanent, shallow lake systems. With the elimination of much of the emergent vegetation, there is little shelter available from the wind. The resulting turbulence restricts the distribution of the submersed macrophytes (Anderson 1978) thereby reducing their abundance and availability to herbivores. Fish communities often are well established by this time and are suggested to compete with birds for the available invertebrate resources in the ponds (Hanson and Butler 1994; Hanson and Riggs 1995). Piscivorous birds may increase during these periods. The invertebrate fauna is dominated by benthic insect species that inhabit the unvegetated bottom substrates of the ponds (Murkin and Kadlec 1986b; Wrubleski 1989). In general, although overall availability and productivity of many food resources are at a low point during the lake marsh stage, there is a subset of food resources (e.g., fish and benthic invertebrates) that may be abundant at this time (Table 10.3).

COVER RESOURCES

Cover resources important to birds in wetlands consist primarily of emergent vegetation within the basin and vegetation in the surrounding uplands. Plant litter from the previous growing season is also an important component of the available cover especially during early spring, fall, and winter. Besides cover provided by living and dead vegetation, variations in topography (e.g., convoluted shorelines, presence of islands) also play a role in providing cover for wildlife species in these habitats. Although open water is not generally considered a form of cover, the availability of deep water may have some value as cover for diving birds. The distribution and abundance of these various forms of cover play a significant role in avian use of these habitats.

The primary objective of the MERP avian monitoring was to relate the avian use of the experimental cells to the changes in emergent vegetation during the various stages of the simulated wet–dry cycle (Murkin et al. 1997). Each year of the study, vegetation maps of the cells were developed using geographic information system (GIS) techniques (see Appendix 1). Avian counts were related to this vegetation information and to water depths within the individual cells by using principal components analyses (PCA) procedures (Murkin et al. 1997). The PCA identified two principal components (one related primarily to water depth and the other to vegetation cover) that could be used to diagrammatically represent the variety of habitat conditions within the experimental cells during the simulated wet–dry cycle (Figure 10.1). The PCA procedures assigned scores to the various bird groups at different times of the year based on habitat variables in the cells where they were observed. These scores could then be used to graphically depict habitat use at any particular time during the study (Figures 10.2–10.5) These results are described in detail in Murkin et al. (1997) and are summarized in the following review of avian cover requirements.

Table 10.3. Changes in avian cover resources during the various stages of the wet–dry cycle in prairie wetlands

	Dry marsh	Regenerating marsh	Hemi-marsh	Degenerating marsh	Lake marsh
Nesting cover	Little cover available initially, increases over time, some use by upland passerines, some use by dabbling ducks if permanent water available nearby	Over-water nesting cover sparse initially, increases over time	Over-water nesting cover abundant, lots of vegetation–open water edges	Over-water nesting cover declines	Little over-water nesting cover except at basin margin
General escape cover	Little available initially, increases over time for upland birds	Increases over time as emergent vegetation develops	Abundant escape cover interspersed with open water	Decreases over time as emergent vegetation stands thin out	Little available except at basin margin
Visual isolation	Low early, increases over time as vegetation develops	Increases over time	High visual isolation for water birds	Decreases over time	Low
Deep water	None	Some available depending on initial flooding depths	Some available interspersed with vegetation	Open-water areas increase over time	High
Thermal cover	Little available initially, increases over time for upland birds	Increases as vegetation stands develop	High, especially for water birds that prefer open water in association with vegetation	Decrease over time	Low

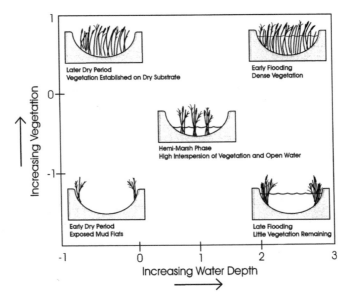

Figure 10.1 Diagrammatic representation of the two significant PCA components related to avian use of the MERP experimental cells (1980–1989). The PC1 is related generally to factors associated with flooding extent and depth, whereas PC2 is related to extent and mixture of vegetative cover types (after Murkin et al. 1997).

Avian Cover Requirements

Cover resources serve a variety of functions for avian species, including provision of nest sites, predator and nest parasite avoidance, visual isolation from conspecifics, and shelter from adverse weather (thermal cover) (Kaminski and Prince 1984; Kaminski and Weller 1992). Cover requirements vary according to the stage of the annual cycle (e.g., migrating waterfowl flocks in the spring have different cover requirements than hens with broods during the summer). In addition, climatic conditions vary seasonally resulting in different needs for thermal cover throughout the year.

Migrants to the northern prairie region in early spring often arrive before the snow and freezing conditions of the prairie winter have left the landscape. Many Canada geese, mallards, and northern pintails arrive as soon as sheet water is available. At this time the deeper, permanent wetland areas are still frozen and may remain so for several weeks. Although conditions may be harsh, cover requirements during these periods are minimal because these birds are relatively large bodied and well insulated against low temperatures. Although in some situations they may stay on the sheet water throughout the day and night, they generally feed and loaf on or near sheet water during the day and move to the larger wetland areas to roost at night (LaGrange and Dinsmore 1989). Although still frozen in many cases, the open-water areas used for roosting also provide good visibility for predator detection. Emergent vegetation stands in the large wetland areas may provide some thermal cover at this time of year.

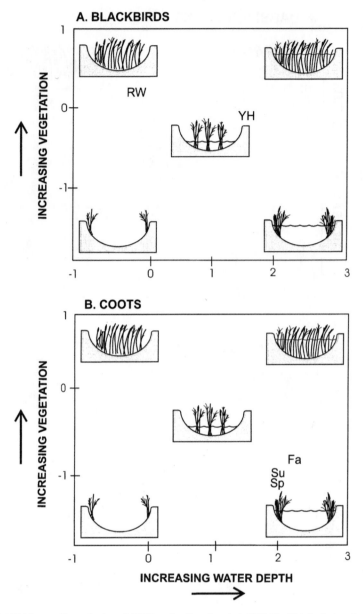

Figure 10.2   Habitat use by red-winged (RW) and yellow-headed (YH) blackbirds during the spring and American coots during the spring (Sp), summer (Su), and fall (Fa) in the MERP experimental cells (after Murkin et al. 1997).

A. DABBLING DUCKS

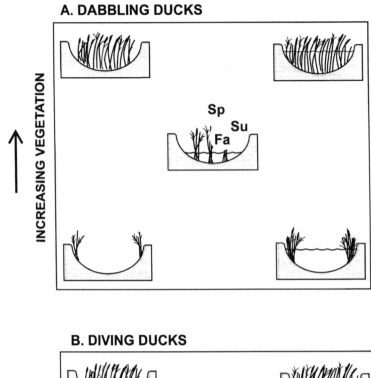

B. DIVING DUCKS

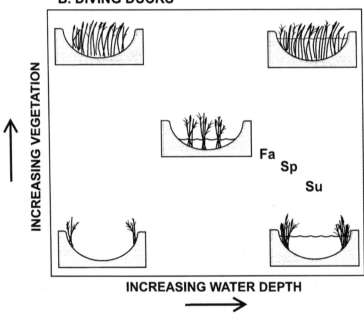

Figure 10.3 Habitat use by dabbling and diving ducks during spring (Sp), summer (Su), and fall (Fa) within the MERP experimental cells (after Murkin et al. 1997).

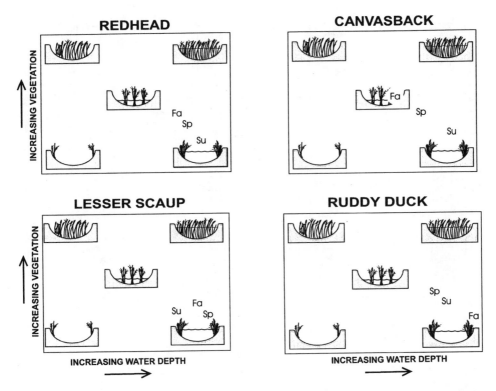

Figure 10.4   Habitat use by diving duck species during spring (Sp), summer (Su), and fall (Fa) within the MERP experimental cells (after Murkin et al. 1997).

Many of the smaller-bodied spring migrants are cold sensitive (e.g., rails) and do require thermal cover during periods of reduced temperatures in the spring (Melvin and Gibbs 1994). Some birds such as blackbirds arrive in flocks and require blocks of cover (e.g., cattail stands) for roosting (Orians 1961; Bird and Smith 1964). These sites serve as thermal cover and protection from predators. Roosting blackbirds do not feed within the roost, therefore, provision of cover is the primary factor in roost-site selection (Hayes and Caslick 1984).

Once migrants move through an area in the spring, the cover requirements of those birds remaining change as the breeding process begins. Although cover for thermal protection and predator avoidance remain a consideration, other needs for cover vary based on the breeding strategy of the species involved. Territorial species such as blackbirds require expanses of cover to provide singing perches for territorial males and nest sites within the territory (Orians 1961). Colonial species (e.g., eared grebes [*Podiceps nigricollis*] [Boe 1992] and black terns [*Chlidonias niger*]) [Burger 1985]) require blocks of cover large enough to provide thermal cover, predator avoidance, and nest sites for the entire colony. Some solitary territorial species such as the sora rail (*Porzana carolina*) require blocks of cover to meet all breeding needs through brood rearing, and at the same time minimize inter- and intraspecific interactions (Melvin and Gibbs 1994). Murkin et al. (1982) sug-

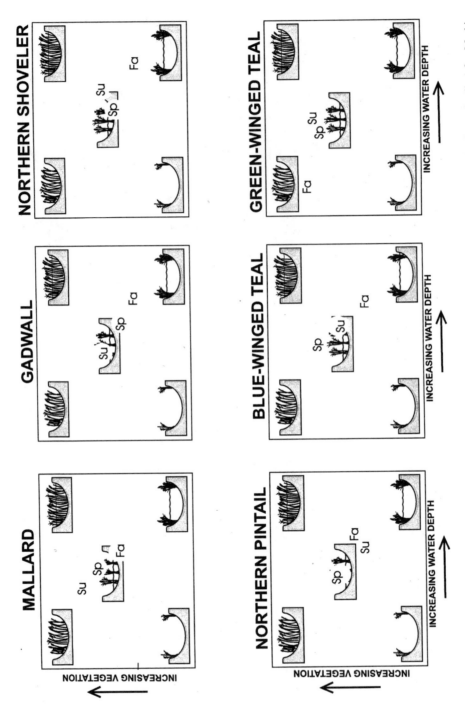

Figure 10.5 Habitat use by dabbling duck species during spring (Sp), summer (Su), and fall (Fa) within the MERP experimental cells (after Murkin et al. 1997).

268

gested that cover during early spring provides visual isolation for territorial dabbling duck pairs, thereby reducing antagonistic interactions and allowing more pairs to settle in areas with an interspersion of cover and open water.

As the breeding season progresses and nesting begins, provision of suitable nesting sites becomes an important issue related to available cover. Birds nest in a variety of cover types within and around prairie wetlands. Nest sites range from upland sites used by dabbling ducks that are located several kilometers from the actual basin to floating vegetation mats used by grebes in open-water areas of the wetland (Burger 1985). For many bird species, suitable cover for nest sites may be more important than food resources in overall avian productivity (Martin 1988). Cover requirements related to nest-site selection include material for nest construction and support as well as concealment from predators and avian nest parasites. Egg and nest mortality are directly affected by the quality and quantity of the available cover and the resulting nest characteristics (Greenwood et al. 1995). Suitability of nesting cover also is related to security from inundation due to changing water levels (Burger 1985) and to proximity to other important habitats (e.g., feeding areas).

Upland nesters, such as many dabbling duck species, require suitable areas of dense grass, forbs, and shrubs close to flooded wetland basins (Duebbert and Lokemoen 1976; Lokemoen et al. 1984; Kantrud 1993; Greenwood et al. 1995). Many dabbling ducks nest within 100 m of wetland edges (Bellrose 1976), although nests of mallards, gadwalls, and pintails may be up to 3 km from water (Swanson and Duebbert 1989). Greenwood et al. (1987) found that most dabbling ducks preferred to nest in undisturbed dense cover or odd sites (e.g., rock piles, shelterbelts). Upland cover on islands within wetlands is particularly attractive to nesting dabbling ducks (Duebbert et al. 1986; Lokemoen et al. 1984). Mallards and gadwalls prefer cover that is tall and dense and may be grass, herbaceous, or shrub. Patches of shrub are highly preferred (Duebbert et al. 1986). Some of the smaller dabbling ducks (e.g., blue-winged teal and northern shovelers) prefer nesting cover composed of short- to medium-height grasses. Northern pintails are an exception and can be found in sparse upland cover 1 to 2 km from water (Duncan 1987).

Cover resources within the wetland are vital to species that nest over water. Over-water nesters are composed of three groups: species that build elevated nests attached to emergent vegetation, species that build nests at the water surface but attached to emergent vegetation, and species that build floating nests or nest on floating mats of plant litter. Perching species such as blackbirds and marsh wrens require erect stems of emergent vegetation to attach their nests (Orians 1961). Although vegetation structure is probably more important than the actual plant species, cattail and bulrush appear to be preferred for elevated nests (Orians 1961; Burger 1985). Many species avoid large unbroken stands of emergent vegetation even though nest-building materials are available (Murkin et al. 1982; Murkin et al. 1989).

Often, preferred nesting cover for elevated nests is located in close proximity to open water (Burger 1985). In a previous MERP publication, we reported that red-winged blackbirds appeared to have a preferred zone for nesting along the emergent vegetation–open water interface and that nesting densities were much reduced farther into the emergent vegetation (Murkin et al. 1989). In the MERP experimental cells, yellow-headed blackbird densities during the spring nesting period were highest in areas of emergent vegetation interspersed with open water (Figure 10.2).

Many species build nests attached to emergent vegetation at the water surface. For example, diving ducks build over-water nests that are mounds of vegetation either supported by emergent vegetation or anchored directly to it (Maxson and Riggs 1996). The availability of flooded nesting cover influences wetland use by diving ducks (Kaminski and Weller 1992). Suitable cover for diving duck nests must include access to open water and some residual cover around the nest site. Large monotypic stands of dense vegetation with few breaks are avoided (Swanson and Duebbert 1989). In the MERP experimental cells, canvasbacks, redheads (*Aythya americana*), and ruddy ducks (*Oxyura jamaicensis*) were located most frequently in open-water areas with clumps of vegetation during the spring nesting season (Figure 10.4). This is similar to observations of these species in other areas (Swanson and Duebbert 1989), although in some areas ruddy ducks nest in shallow wetlands or wetland margins in clumps of sedge or grasses. In the MERP experimental cells, lesser scaup were found in very open areas in spring (Figure 10.4), however, they commonly nest in upland situations and therefore are not tied to emergent vegetation within the wetland basin as nesting cover (Hammel 1973; Afton 1983). Nest success of diving ducks has been positively correlated with the amount of vegetative concealment, although this depends to a large extent on the makeup of the predator community (Clark and Nudds 1991).

An important consideration for nests at or near the water surface is shelter from wave action. Although relatively large nests such as those built by diving ducks may be resistant to wave action, smaller, floating nests (e.g., the small, flat nests built by black terns on floating plant litter) (Dunn 1979) are very susceptible to damage by wave action (McNichol 1982). Nests of these species must be built well back in the stand of emergent vegetation or in sheltered areas of open water with little potential for wave action. Black terns became very abundant on the MERP experimental cells during the 2 years of deep flooding (1981–1982) when the water surface in many parts of the cells was covered by a thick mat of floating emergent plant litter. Although standing emergent vegetation was eliminated by flooding and much of the area was open water, the thick mat of litter dampened or eliminated wave action.

Although cover from predators is a primary consideration during nest-site selection, protection from parasitic avian species also may play a role (Sayler 1992). Brown-headed cowbirds (*Molothrus ater*) are a common parasitic layer in the nests of marsh passerine birds (Delphey and Dinsmore 1993). Cowbirds search for potential host nests from elevated perches (Norman and Robertson 1975). Thus, cowbird parasitism rates are higher near shrubs or trees. It is also higher in restored wetlands and disturbed habitats, presumably due to lower vegetation densities and greater visibility of nests in these habitats (Delphey and Dinsmore 1993). Some waterfowl species lay eggs parasitically in the nests of conspecifics and other species (Sayler 1992). Redheads routinely lay eggs in the nests of other diving duck species, especially canvasbacks (Weller 1959; Bouffard 1983). Nests that are more visible and easy to locate are more likely to be parasitized (Sayler 1992).

Following hatch, altricial birds remain in the nest, consequently cover requirements do not change during the brood-rearing period. For precocial birds, however, there are specific requirements for brood cover that may or may not necessitate movement away from the nesting habitat. Young precocial birds are very susceptible to predation and cold, wet weather conditions. Most precocial birds are not cold hardy at hatch and must be brooded periodically by a parent until they are able to thermoregulate on their own (Afton and Paulus 1992). Although it probably plays a role in brood habitat use, the importance of thermal cover to young broods has not been addressed in detail. In contrast, the critical role of vegetative cover in predator avoidance for broods has been well documented (Swanson and Duebbert 1989; Sedinger 1992).

Some species of precocial birds remain in the vicinity of the nesting site or territory until fledging. This is particularly true for species such as rails and coots, where the adults play a role in feeding the young. Sora rails remain within the densely vegetated areas of the breeding territory during this period (Melvin and Gibbs 1994). They are adapted for wading in shallow water and walking on litter mats and, therefore, require fairly dense emergent vegetation with a significant litter component. Coots also remain on the breeding territory to brood and feed their young (Alisaukas and Arnold 1994). The mixture of flooded vegetation and open water of the breeding territory provides the necessary food and cover to meet the needs of juvenile coots. On the MERP experimental cells, there was very little change in habitat use between the spring nesting period and summer brood-rearing period (Figure 10.2) because they stay on their territories throughout these periods. As the young birds become less dependent on the adults, the territory begins to break down. In other species, the cover requirements for nesting are very different than for brood-rearing (note the difference in habitat use between spring and summer in the experimental cells [Figure 10.3]) and often require moves to distant habitats.

Dabbling duck broods have very different cover requirements than diving duck broods. Generally, dabbling ducks prefer a mixture of emergent vegetation and cover, whereas diving ducks are found in deeper, more open areas of the wetland (Sedinger 1992). This trend was evident in the summer waterfowl use of the MERP experimental cells (Figure 10.3). Mallard broods were found in shallow densely vegetated areas (Figure 10.5). Other dabbling duck broods were found in slightly more open areas than mallards; however, emergent vegetation remained an important component of the habitats used during the summer brood-rearing period. Diving duck broods, in contrast, were found in the more open areas of the MERP experimental cells during the summer (Figure 10.4). These differences in habitat use are probably related to predator response behaviors. Dabbling duck broods scatter and hide in the emergent vegetation in response to predators, whereas diving duck broods dive to avoid predators. These deeper areas away from emergent vegetation that are preferred by diving ducks also have less dense submersed vegetation due to water-depth intolerance and wave action (Anderson 1978). Hochbaum (1944) and Monda and Ratti (1988) suggest that young diving ducks avoid dense, submersed vegetation beds due to problems with entanglement below the surface.

Waterfowl and coots undergo a wing moult and flightless period during the summer and the reduced mobility during this period has important implications for avoiding predators.

Mallards use the Delta Marsh extensively during the moulting period particularly areas of dense vegetation (Gordon 1985). Other dabbling ducks make extensive use of stands of emergent vegetation during the moult and flightless period. In the MERP experimental cells, most dabbling duck species used areas of dense cover during the summer, however, gadwalls and northern shovelers were often found in more open areas during this period (Figure 10.5). Moulting diving ducks prefer open-water areas during the wing molt (Oring 1964); in fact, many diving duck species abandon prairie breeding marshes to moult in northern lakes (Bergman 1973; Bailey and Titman 1984). Diving ducks and coots remaining in the MERP experimental cells during late summer were found in the most open areas of the available habitats (Figures 10.4 and 10.2).

Cover requirements change once again as late summer and fall brings the onset of migration. Territorial boundaries and activities break down and many species become highly gregarious and form migratory flocks. Cover requirements are very similar to the spring migration and early arrival periods. Blackbirds form flocks and roost in dense vegetation in wetlands, making regular feeding forays to agricultural fields (Hayes and Caslick 1984). Many waterfowl form flocks and use large open areas of the wetlands for staging. On the MERP experimental cells, many of the dabbling duck species moved to more open habitats during the fall compared with spring and summer (Figure 10.5). Mallards and pintails feed primarily in agricultural fields at this time of the year (Clark et al. 1986), therefore, wetland use is related mostly to protection from predators and possibly some thermal cover as the season progresses. Some dabbling ducks such as gadwalls and northern shovelers move to open habitats in the fall (Figure 10.5) and are often found in the company of diving ducks during this time of year (Oring 1964; Poston 1974)). Diving ducks also use open-water areas during the fall. Ruddy ducks in the MERP experimental cells used the most open habitats of any waterfowl species present during the fall (Figure 10.4).

As the fall season advances, most migratory birds leave the northern prairie region. Some waterfowl, primarily mallards, remain in northern areas along rivers where there is open water associated with flowing water (Pattenden and Boag 1989; Pawlina et al. 1993). These birds normally roost on the water at night and feed in adjacent agricultural fields during the day. Because prairie wetlands freeze during the winter, they are not generally a factor in the habitat use of these wintering mallards. Some nonmigratory birds do make use of prairie wetlands during the winter. The residual standing litter of the emergent vegetation is available as cover as are the snowdrifts that form within and around the wetland. Ring-necked pheasants (*Phasianus colchicus*) and sharp-tailed grouse (*Pedioecetes phasianellus*) use the emergent vegetation as cover from predators and weather (Kantrud et al. 1989; Moyles 1981). Sharp-tailed grouse burrow into the accumulated snow both for protection from weather and concealment from predators (Gratson 1988).

Wetland Type and Cover Resources

Ephemeral or sheet-water ponds are used extensively by migrating waterfowl, shorebirds, and other species during spring; however, this use is probably linked to food resources rather than to available cover. There is little cover associated with these basins other than residual agricultural crop or grass litter. Most species use these wet areas during the day and then roost in larger wetlands with emergent cover at night. Temporary wetlands with their longer flooding duration, greater water depths, and more diverse cover are used extensively by early spring migrants. Although they are generally too shallow for use by diving ducks, the available cover in these basins during wet years provides important pair space and visual isolation for a wide range of dabbling duck species (Kantrud and Stewart 1977; Kantrud et al. 1989), most notably blue-winged teal, pintails, and mallards (Ruwaldt et al. 1979). Although the available cover appears important to pair use, little nesting takes place in temporary wetland basins when they are flooded. Kantrud and Stewart (1984) reported high use of these habitats by upland species during dry years or once the basins dry up later in spring.

Seasonal wetlands add another zone of vegetation and associated cover to that provided by the ephemeral and temporary wetlands (Figure 9.2). Besides providing visual isolation for waterfowl breeding pairs, the shallow marsh zone also supports over-water nests. Thus, during years that these basins are flooded, dabbling duck pairs and several diving duck species make extensive use of these wetlands (Ruwaldt et al. 1979). In wet years, when flooding of these basins continues well into summer, they are also attractive as waterfowl brood habitat (Talent et al. 1982) and for moulting due to the dense cover in the central portion of these ponds. Eared grebes also respond to the deeper water and the flooded cover provided by seasonal wetlands (Kantrud et al. 1989). Faanes (1982) observed breeding pairs of bobolink, savannah sparrows (*Passerculus sandwichensis*), and Nelson's sharp-tailed sparrows (*Ammodramus nelsoni*) in seasonal wetlands. In summary, seasonal wetlands add a nesting component to the available cover and although the vegetation does not include the large-stemmed species such as cattail, those species present (primarily sedge and whitetop) support a variety of nesting birds.

Kantrud et al. (1989) state that semipermanent wetlands supply most of the needs of prairie-nesting waterfowl and their broods. Cover in these basins provides visual isolation for pairs nest sites and nesting cover, brood cover, cover for moulting birds, and cover and roosting space for migrating birds in the fall. Semipermanent ponds are the principle breeding habitat for diving ducks in North Dakota (Kantrud and Stewart 1977). Faanes (1982) found that all common nonwaterfowl bird species characteristic of wetland habitats in North Dakota were found in semipermanent wetlands. He also found the highest densities of breeding pairs of most species on these habitats. These included nesting Forster's terns (*Sterna forsteri*), red-winged blackbirds, coots, and black terns. Importantly, the available habitat and cover conditions can vary widely among years in semipermanent ponds based on climatic conditions and the stage of the wet–dry cycle.

In permanent wetlands, available cover is normally limited to the emergent fringe and the deep water for diving birds. The vegetation diversity in the emergent fringe is usually very limited (e.g., cattail, bulrush, or both). The actual importance of the available cover is determined in part by the topography of the basin. If there is a gradual sloping shoreline, an extensive fringe of flooded, emergent vegetation may develop providing some nesting cover and flooded cover for broods and moulting birds. A steep shoreline results in a narrow band of emergent vegetation and therefore much reduced cover. The open nature of these ponds supports diving-duck broods and flocks of both dabbling and diving ducks in the fall. Other migrating birds such as coots, geese, swans and gulls also make use of these open ponds.

The Wet–Dry Cycle and Cover Resources

As with food, there are marked changes in available cover during the various stages of the wet–dry cycle (Table 10.3). As water levels recede with the onset of the dry stage (Figure 1.3), loss of flooded emergent vegetation has implications for several bird species. Loss of over-water nesting sites, flooded brood cover, and flooded escape cover reduces the water-bird use of these systems. Once the marsh is completely dry, upland birds use the cover provided by the dry stands of emergent vegetation and associated litter. Dabbling ducks nest in the residual emergent cover in dry wetlands (Swanson and Duebbert 1989). Mudflats exposed during the early dry marsh stage have little cover value; however, they provide open feeding areas and roost sites suitable to a variety of shorebirds (Skagen and Knopf 1994). As emergent vegetation and mudflat annuals germinate and grow on the exposed substrate, the availability of new cover attracts a variety of upland birds, including bobolinks and meadowlarks (Weller and Spatcher 1965). Dabbling ducks also nest in these newly established stands of vegetation if the drought persists for more than one growing season, thereby providing substantial residual cover from the previous year's growth. During the drawdown period in the MERP experimental cells, dabbling ducks were not observed nesting in the cells until the second year of drawdown. Ducks nesting in dry wetland basins require permanent water nearby as pair and brood habitats.

Soon after reflooding and the onset of the regenerating stage of the wet–dry cycle (Figure 1.3), available cover within the basins consists of standing litter of the emergent and annual vegetation that had established on the dry marsh surface. Winter and early-spring storms topple some of the standing litter (Davis and van der Valk 1978), therefore, there are usually numerous open-water areas associated with this litter. If the initial flooding is deep enough to cover the existing litter, there will be large areas of open water. The annual plant litter persists, often well through the first year of flooding and therefore provides some cover during this period. However, in the MERP experimental cells, there were no birds observed nesting in the annual plant litter following reflooding. The emergent vegetation also may be somewhat sparse at first, but eventually the stands expand through vegetative development to provide cover within the basin.

Available cover in newly flooded basins provides some isolation for dabbling duck pairs later in the season. Diving duck use may be delayed until submersed vegetation becomes established in open-water areas and emergent vegetation stands develop to a point that they provide sufficient nesting cover (Murkin et al. 1997). Brood use depends on development of emergent vegetation stands following flooding. Species that depend on emergent vegetation for nest sites (e.g., marsh wrens), territorial singing perches (e.g., blackbirds), or foraging (e.g., rails) increase in density as emergent vegetation stands develop during the regenerating stage of the wet–dry cycle.

Water depth determines the rate of change in vegetation and associated cover during the wet–dry cycle (van der Valk and Davis 1978; see Chapter 7). Shallow flooding encourages the development of emergent vegetation. If the wetland is flooded below the long-term tolerance of the emergent species, the vegetation stands expand and eventually form unbroken (often monotypic) stands with little or no open water. These stands typically receive little bird use other than possibly some winter use by pheasants and sharp-tailed grouse. In basins where the flooding is somewhat deeper, the basin enters the degenerating stage of the marsh cycle as the continued presence of standing water eventually eliminates emergent vegetation. Flooding intolerance, muskrats (*Ondrata zibethicus*), insects, and disease all contribute to loss of emergent vegetation from the basin (van der Valk and Davis 1978; see Chapter 11). The transition from the regenerating to the degenerating stages of the wet–dry cycle includes a period of hemi-marsh conditions when the emergent cover and open water are available in equal amounts in an interspersed pattern (Weller and Spatcher 1965; Weller and Fredrickson 1974). This is generally considered the period of maximum avian use and production in prairie wetlands (Weller and Spatcher 1965). Experimental studies on the Delta Marsh showed maximum dabbling duck pair densities on study plots with a 50:50 cover:water ratio (Kaminski and Prince 1981b; Murkin et al. 1982). Murkin et al. (1989) also found maximum blackbird nest densities on experimental hemi-marsh plots. Dabbling ducks on the MERP experimental cells showed a preference for hemi-marsh conditions during the spring and summer (Figure 10.5). Some of the diving ducks such as canvasbacks also were found in hemi-marsh conditions in the spring, probably in response to the need for over-water nesting sites (Figure 10.4). The mixture of cover and water during the hemi-marsh stage also results in a large edge (open water–cover interface) component to the habitat. Many species prefer edges or access to edges, which may account for the high abundance of some species during this period.

As the emergent vegetation stands die back during the degenerating phase, the available over-water cover is reduced (see Figure 1.3 in Chapter 1). This eliminates species tied to emergent vegetation for cover. As the open-water areas expand, they are used during the summer by diving duck broods and moulting adults (Murkin et al. 1997). Diving-duck use of the MERP experimental cells was limited to the deeply flooded open-water sites characteristic of the degenerating stage of the wet–dry cycle (Figure 10.4). These open areas also attracted dabbling ducks in the fall (Figure 10.5). Open marsh areas also are used by migrating waterfowl, coots, grebes, and other birds in the spring and fall.

With the complete elimination of emergent vegetation and its associated cover from the central area of the basin during the lake marsh stage (see Figure 1.3 in Chapter 1), avian use is limited to birds requiring open water. These include migrating waterfowl and coots, and to some degree diving-duck broods. With loss of the sheltering effects of the emergent vegetation, the submersed vegetation also is lost from much of the basin. Use by some birds depends on the extent of the emergent fringe that remains at the periphery of the basin. A flooded fringe provides some nesting cover for birds such as coots, blackbirds, and diving ducks; however, this use is minor compared with that in earlier stages of the wet–dry cycle.

OTHER RESOURCES

Although food and cover are often the primary factors affecting the distribution and abundance of birds, there are several other resources that require consideration especially in wetland habitats. The provision of suitable space is important when addressing avian use of individual wetlands. Many wetland birds are territorial during part of the time they spend on prairie wetlands. Territorial behavior ranges from defending a small, mobile area immediately around the female (e.g., northern pintails) to large areas defended throughout the breeding and brood-rearing seasons (e.g., American coots) (Anderson and Titman 1992). This requirement for space limits the number of territories and therefore individuals that can occupy the available habitat (Davies 1978).

The presence of other birds also may be a factor in determining the distribution and abundance of some bird species in prairie wetland habitats. Within a species, the need for mates during the breeding season is obvious. The presence of bird prey species affects the distribution of birds that feed on other birds. Red-winged blackbirds avoid the territories of marsh wrens because wrens are nest predators on other marsh passerines (Picman 1980). Nest parasites are attracted to habitats with an abundance of potential host species. Some species nest in association with other species to avoid predators. Black-crowned night herons (*Nycticorax nycticorax*) and western grebes (*Aechmophorus occidentalis*) often nest in association with terns or gulls (Nuechterlein 1981; Burger 1985). Some studies have noted that breeding waterfowl are attracted to nesting on islands with nesting gulls (Vermeer 1968; Evans 1970). Gulls are much better adapted for driving off predators than most ducks. Ducks also have been shown to nest in association with Canada geese (Long 1970; Giroux 1981) for much the same reason.

Competition with other species for available resources also plays a role in habitat use. For example, red-winged blackbird habitat use on the Delta Marsh is determined in large part by competition with yellow-headed blackbirds (Murkin et al. 1989). Red-winged males arrive in the spring before yellow-headed males and establish territories in the central, more open areas of the wetland. Once the yellow-headed males arrive, they force the red-winged males out of these preferred habitats and into the drier, more heavily vegetated wetland edge. Consequently, most red-winged blackbird territories and nests are found in the shallow edges of the Delta Marsh. In the MERP cells the red-winged blackbirds also were found in the peripheral areas of the cells, whereas the yellowheads were found in the more preferred interspersed areas of the cells (Figure 10.2). This same competitive exclusion has been observed in other areas (Willson 1966; Krapu 1978). Thus, any discussion of habitat use by red-winged blackbirds must include yellow-headed blackbird abundance in the same area (Murkin et al. 1997).

Water quality also influences avian use of prairie wetlands. There is considerable variability in the chemistry of the water in prairie wetlands (LaBaugh 1989). The chemical characteristics of the water influence the structure and distribution of plant and invertebrate communities within the wetland basin (Swanson and Duebbert 1989), and thereby determine the available food and cover resources for birds and other animals within these habitats. Prairie wetlands tend to be shallow, highly eutrophic, and often high in dissolved salts. Ponds that receive regional or local groundwater inputs and do not have outputs equal to these inputs have elevated salt concentrations. Stewart and Kantrud (1971) described the influence of varying salt concentrations on vegetation development in prairie wetlands. Increasing salt concentrations results in species shifts in the dominant emergent vegetation and therefore available cover. For example, cattail is replaced by emergent species such as alkali bulrush (*Scirpus maritimus*) at higher salt levels. Although alkali bulrush may not provide suitable cover for birds such as nesting blackbirds, it does produce seeds that are eaten by a wide range of wetland birds. Salt concentrations also affect invertebrate populations. Increasing salt concentrations reduces diversity favoring species that are tolerant of saline conditions (see Chapter 9). Increased salinity also affects the prey base for piscivorous birds. Fathead minnows (*Pimephalus promelas*) and brook sticklebacks (*Culaea inconstans*) are more resistant to increased salt than other fish species in the prairie region (Peterka 1989). Few amphibians are tolerant of saline conditions (Swanson and Duebbert 1989).

Salt concentrations also affect the quality of water for drinking. Although this does not appear to be a problem for most birds species, Swanson et al. (1984) reported duckling mortality at elevated salt levels. The type of salts found in some prairie wetlands does not appear to be processed effectively by the salt glands of young ducks (Schmidt-Nielson 1960). Thus, hens that hatch broods on saline wetlands must move them to wetlands that have suitable water quality (Mitcham and Wobeser 1988; Wobeser 1988, Swanson and Duebbert 1989).

Another factor to be considered in prairie wetland water quality is the presence of agricultural chemicals or other contaminants. Prairie wetlands are susceptible to agricultural chemical inputs because they are often interspersed within agricultural fields, which increases the probability for contamination due to direct spraying, aerial drift, and runoff. These chemicals may affect wetland birds directly through lethal or sublethal effects, or indirectly by affecting plant and animal foods (Grue et al. 1986). Many commonly used agricultural chemicals, and particularly insecticides, are highly toxic to birds species; however, we know very little regarding their impact on birds or resources important to these birds (e.g., invertebrates). The impact of nutrient loading due to fertilizer application on adjacent agricultural fields or runoff from feedlot operations has not be determined. Although there may be some increased productivity and waterfowl use of ponds due to agricultural nutrient inputs (Murkin et al. 1991), impacts of varying nutrient inputs on wetlands are unknown.

CONCLUSIONS

We have presented a discussion of the individual factors affecting the distribution and abundance of birds in prairie wetland habitats. Many studies have examined habitat selection or use by wetland birds (Hildén 1965; Courcelles and Bedard 1979), but often disagree on which factor is most important in overall habitat selection. For example, Gilmer et al. (1975) and Godin and Joyner (1981) suggested that mallards choose habitats depending on the pond or opening size. Other studies showed that mallard distribution is affected by food (Murkin et al. 1982; Krapu et al. 1983), or that there is no relationship with food (Godin and Joyner 1981; Kaminski and Prince 1981a). Some of the lack of agreement is caused by the intercorrelations among many factors determining habitat use (Murkin et al. 1997). Overall habitat selection, however, is probably based on a combination of factors rather than any single factor.

In summary, prairie wetlands are dynamic, offering a wide array of habitats to birds. Habitat conditions vary among the various wetland types at any point in time and over time as climatic conditions force the basins through the wet–dry cycle. Certain wetland types or stages of the wet–dry cycle are more suitable for some species than others. For example, although hemi-marsh conditions result in maximum densities for a variety of species, they do not meet the needs of all species for all seasons. Few, if any, species are found in equal abundance during all stages of the wet–dry cycle. This fact coupled with the importance of the wet–dry cycle in maintaining overall wetland productivity has important implications for understanding the avian use of these systems. At any point in time, wetland complexes of varying depths (and therefore wetland classes) provide the widest array of conditions for avian species. The wet–dry cycle provides different habitat conditions over time and ensures the highest productivity and greatest diversity of birds over the long term.

ACKNOWLEDGMENTS

This chapter is dedicated to the many MERP field assistants who participated in the "morning bird counts" on the MERP experimental cells from 1980 to 1989. James Dinsmore, Richard Kaminski, and Mickey Heitmeyer reviewed earlier drafts of this manuscript. Lisette Ross prepared the figures for this chapter. This is Paper No. 103 of the Marsh Ecology Research Program, a joint project of Ducks Unlimited Canada and the Delta Waterfowl and Wetlands Research Station.

LITERATURE CITED

Afton, A.D. 1983. Male and female strategies for reproduction in lesser scaup. Ph.D. dissertation, University of North Dakota, Grand Forks, North Dakota.

Afton, A.D., R.H. Hier, and S.L. Paulus. 1991. Lesser scaup diets during migration and winter in the Mississippi flyway. Canadian Journal of Zoology 69:328-333.

Afton, A.D., and S.L. Paulus. 1992. Incubation and brood care. In *Ecology and Management of Breeding Waterfowl*. (Eds.) B.D.J. Batt, A.D. Afton, M.G. Anderson, C.D. Ankney, D.H. Johnson, J.A. Kadlec, and G.L. Krapu, pp.62-108. Minneapolis: University of Minnesota Press.

Alisauskas, R.T., and C.D. Ankney. 1992a. Spring habitat use and diets of midcontinent adult lesser snow geese. Journal of Wildlife Management 56:43-54.

Alisauskas, R.T., and C.D. Ankney. 1992b. The cost of egg laying and its relationship to nutrient reserves in waterfowl. In *Ecology and Management of Breeding Waterfowl*. (Eds.) B.D.J. Batt, A.D. Afton, M.G. Anderson, C.D. Ankney, D.H. Johnson, J.A. Kadlec, and G.L. Krapu, pp.30-61. Minneapolis: University of Minnesota Press.

Alisauskas, R.T., and T.W. Arnold. 1994. American coot. In *Migratory Shore and Upland Game Bird Management in North America*. (Eds.) T.C. Tacha and C.E. Braun, pp.127-143. Washington: International Association of Fish and Game Agencies.

Anderson, M.G. 1978. Distribution and production of sago pondweed *Potamogeton pectinatus* L. on a northern prairie marsh. Ecology 59:154-160.

Anderson, M.G., and J.B. Low. 1976. The use of sago pondweed by waterfowl on the Delta Marsh, Manitoba. Journal of Wildlife Management 40:233-242.

Anderson, M.G., and R.D. Titman. 1992. Spacing patterns. In *Ecology and Management of Breeding Waterfowl.* (Eds.) B.D.J. Batt, A.D. Afton, M.G. Anderson, C.D. Ankney, D.H. Johnson, J.A. Kadlec, and G.L. Krapu, pp.251-289. Minneapolis: University of Minnesota Press.

Austin, J.E., J.R. Serie, and J.H. Noyes. 1990. Diets of canvasbacks during breeding. Prairie Naturalist 22:171-176.

Bailey, R.O., and B.D.J. Batt. 1974. Hierarchy of waterfowl feeding with whistling swans. Auk 91:488-493.

Bailey, R.O., and R.D. Titman. 1984. Habitat use and feeding ecology of postbreeding redheads. Journal of Wildlife Management 48:1144-1155.

Banko, W.E. 1960. The Trumpeter Swan: Its History, Habits, and Population in the United States. North American Fauna 63, Washington: Bureau of Sport Fisheries and Wildlife.

Barnes, G.G., and V.G. Thomas. 1987. Digestive organ morphology, diet, and guild structure of North American Anatidae. Canadian Journal of Zoology 65:1812-1817.

Batt, B.D.J., M.G. Anderson, C.D. Anderson, and F.D. Caswell. 1989. The use of prairie potholes by North American ducks. In *Northern Prairie Wetlands.* (Ed.) A.G. van der Valk, pp.204-227. Ames: Iowa State University Press.

Bellrose, F.C. 1976. Ducks, Geese, and Swans of North America. Harrisburg: Stackpole Books.

Bergman, R.D. 1973. Use of southern boreal lakes by postbreeding canvasbacks and redheads. Journal of Wildlife Management 37:160-170.

Bird, R.D., and L.B. Smith. 1964. The food habits of the red-winged blackbird, *Agelaius phoeniceus,* in Manitoba. Canadian Field Naturalist 78:179-186.

Boe, J.S. 1992. Wetland selection by eared grebes, *Podiceps nigricollis,* in Minnesota. Canadian Field-Naturalist 106:480-488.

Boudewijn, T. 1984. The role of digestibility in the selection of spring feeding sites by brent geese. Wildfowl 35:97-105.

Bouffard, S.H. 1983. Redhead egg parasitism of canvasback nests. Journal of Wildlife Management 47:213-216.

Buchsbaum, R., and I. Valiela. 1987. Variability in the chemistry of estuarine plants and its effect on feeding by Canada geese. Oecologia (Berl.) 73:146-153.

Burger, J. 1985. Habitat selection in temperate marsh-nesting birds. In *Habitat Selection in Birds.* (Ed.) M.L. Cody, pp.253-281. New York: Academic Press.

Busby, D.G., and S.G. Sealy. 1979. Feeding ecology of a population of nesting yellow warblers. Canadian Journal of Zoology 57:1670-1681.

Cargill, S.M., and R.L. Jefferies. 1984. The effects of grazing by lesser snow geese on the vegetation of a sub-arctic salt marsh. Journal of Applied Ecology 21:669-686.

Castelli, P.M., and J.E. Applegate. 1989. Economic loss caused by tundra swans feeding in cranberry bogs. Transactions of the Northeast Section of the Wildlife Society 46:17-23.

Clark, R.G., H. Greenwood, and L.G. Sugden. 1986. Estimation of grain wasted by field-feeding ducks in Saskatchewan. Journal of Wildlife Management 50:184-189.

Clark, R.G., and T.D. Nudds. 1991. Habitat patch size and duck nesting success: the crucial experiments have not been performed. Wildlife Society Bulletin 19:534-543.

Cody, M.L. 1981. Habitat selection in birds: the roles of vegetation structure, competitors, and productivity. Bioscience 31:107-111.

Courcelles, R., and J. Bedard. 1979. Habitat selection by dabbling ducks in the Baie Noire marsh, southwestern Quebec. Canadian Journal of Zoology 57:2230-2238.

Davies, N.B. 1978. Ecological questions about territorial behaviour. In *Behavioural Ecology - An Evolutionary Approach.* (Eds.) J.R. Krebs and N.B. Davies, pp.317-350. Oxford, UK: Blackwell Scientific.

Davis, C.B., and A.G. van der Valk. 1978. The decomposition of standing and fallen litter of *Typha glauca* and *Scirpus fluviatilis.* Canadian Journal of Botany 56:662-675.

Delphey, P.J., and J.J. Dinsmore. 1993. Breeding bird communities of recently restored and natural prairie potholes. Wetlands 13:200-206.

Desrochers, B.A., and C.D. Ankney. 1986. Effect of brood size and age on the feeding behavior of adult and juvenile American coots (*Fulica americana*). Canadian Journal of Zoology 64:1400-1406.

Duebbert, H.F., and J.T. Lokemoen. 1976. Duck nesting in fields of undisturbed grass-legume cover. Journal of Wildlife Management 40:39-49.

Duebbert, H.F., J.T. Lokemoen, and D.E. Sharp. 1986. Nest sites of ducks in grazed mixed-grass prairie in North Dakota. Prairie Naturalist 18:99-108.

Duncan, D.C. 1987. Nest-site distribution and overland brood movements of northern pintails in Alberta. Journal of Wildlife Management 51:716-723.

Dunn, E.H. 1979. Nesting biology and development of young in Ontario black terns. Canadian Field-Naturalist 93:276-281.

Errington, P.L., and W.J. Breckenridge. 1936. Food habits of marsh hawks in the glaciated prairie region of north-central United States. American Midland Naturalist 17:831-848.

Evans, R.M. 1970. Oldsquaws nesting in association with arctic terns at Churchill, Manitoba. Wilson Bulletin 82:383-390.

Faanes, C.A. 1982. Avian use of Sheyenne Lake and associated habitats in central North Dakota. U.S. Fish and Wildlife Service Resource Publication 144.

Faanes, C.A., and R.E. Stewart. 1982. Revised checklist of North Dakota birds. Prairie Naturalist 14:81-92.

Fredrickson, L.H., and T.S. Taylor. 1982. Management of seasonally flooded impoundments for wildlife. U.S. Department of Interior, Fish and Wildlife Service, Resources Publication 148, Washington, DC.

Froud-Williams, R.J., R.J. Chancellor, and D.S.H. Drennan. 1983. Influence of cultivation regime upon buried weed seeds in arable cropping systems. Journal of Applied Ecology 20:199-208.

Galatowitsch, S.M., and A.G. van der Valk. 1994. *Restoring Prairie Wetlands.* Ames: Iowa State University Press.

Gauthier, G., and J. Bédard. 1990. The role of phenolic compounds and nutrients in determining food preference in greater snow geese. Oecologia (Berl.) 84:553-558.

Gilmer, D.S., I.J. Ball, L.M. Cowardin, J.H. Riechmann, and J.R. Tester. 1975. Habitat use and home range of mallards breeding in Minnesota. Journal of Wildlife Management 39:781-789.

Giroux, J.P. 1981. Use of artificial islands by nesting waterfowl in southeastern Alberta. Journal of Wildlife Management 45:669-679.

Godin, P.R., and D.E. Joyner. 1981. Pond ecology and its influence on mallard use in Ontario. Wildfowl 32:28-34.

Gordon, D.H. 1985. Postbreeding ecology of adult male mallards on the Delta Marsh, Manitoba. Ph.D. dissertation, Michigan State University, East Lansing, Michigan.

Gratson, M.W. 1988. Spatial patterns, movements, and cover selection by sharp-tailed grouse. In

Adaptive Strategies and Population Ecology of Northern Grouse. (Eds.) A.T. Bergerud and M.W. Gratson, pp.158-192. Minneapolis: University of Minnesota Press.

Greenwood, R.J., A.B. Sargeant, D.H. Johnson, L.M. Cowardin, and T.L. Shaffer. 1987. Mallard nest success and recruitment in prairie Canada. Transactions of the North American Wildlife and Natural Resources Conference 52:298-309.

Greenwood, R.J., A.B. Sargeant, D.H. Johnson, L.M. Cowardin, and T.L. Shaffer. 1995. Factors associated with the duck nest success in the prairie pothole region of Canada. Wildlife Monographs 128:1-57.

Grue, C.E., L.R. DeWeese, P. Mineau, G.A. Swanson, J.R. Foster, P.M. Arnold, J.N. Huckins, P.J. Sheehan, W.K. Marshall, and A.P. Ludden. 1986. Potential impacts of agricultural chemicals on waterfowl and other wildlife inhabiting prairie wetlands: an evaluation of research needs and approaches. Transactions of the North American Wildlife and Natural Resources Conference 51:357-383.

Guinan, D.M., and S.G. Sealy. 1987. Diet of house wrens (*Troglodytes aedon*) and abundance of the invertebrate prey in the dune-ridge forest, Delta Marsh, Manitoba. Canadian Journal of Zoology 65:1587-1596.

Hamerstrom, F. 1979. Effect of prey on predator: voles and harriers. Auk 96:370-374.

Hammel, G.S. 1973. The ecology of lesser scaup (*Aythya affinis Eyton*) in southwestern Manitoba. M.S. thesis, University of Guelph, Guelph, Ontario.

Hanson, B.A., and G.A. Swanson. 1989. Coleoptera species inhabiting the Cottonwood Lake area, Stutsman County, North Dakota. Prairie Naturalist 21:49-57.

Hanson, M.A., and M.G. Butler. 1990. Early responses of plankton and turbidity to biomanipulation in a shallow prairie lake. Hydrobiologia 200/201:317-327.

Hanson, M.A., and M.G. Butler. 1994. Responses to food web manipulation in a shallow waterfowl lake. Hydrobiologia 280:457-466.

Hanson, M.A., and M.R. Riggs. 1995. Potential effects of fish predation on wetland invertebrates: a comparison of wetlands with and without fathead minnows. Wetlands 15:167-175.

Hayes, J.P., and J.W. Caslick. 1984. Nutrient deposition in cattail stands by communally roosting blackbirds and starlings. American Midland Naturalist 112:320-331.

Heitmeyer, M.E. 1985. Wintering strategies of female mallards related to dynamics of lowland hardwood wetlands in the upper Mississippi Delta. Ph.D. dissertation, University of Missouri, Columbia, Missouri.

Hildén, O. 1965. Habitat selection in birds. Annales Zoologici Fennici 2:53-75.

Hill, W.L. 1988. The effect of food abundance on the reproductive patterns of coots. Condor 90:324-331.

Hochbaum, H.A. 1944. *The Canvasback on a Prairie Marsh.* Washington: American Wildlife Institute.

Houston, C.S. 1990. Saskatchewan's Swainson's hawks. American Birds 44:215-220.

Johnson, D.H. 1986. Determinants of the distributions of ducks. Ph.D. dissertation, North Dakota State University, Fargo, North Dakota.

Jones, J.C. 1940. Food habits of the American coot with notes on distribution. U.S. Fish and Wildlife Service Research Bulletin 2.

Jorde, D.G. 1981. Winter and spring staging ecology of mallards in south central Nebraska. M.S. thesis, University of North Dakota, Grand Forks, North Dakota.

Kaminski, R.M., and H.H. Prince. 1981a. Dabbling duck activity and foraging responses to aquatic invertebrates. Auk 98:115-126.

Kaminski, R.M., and H.H. Prince. 1981b. Dabbling duck and aquatic macroinvertebrate responses to manipulated wetland habitat. Journal of Wildlife Management 45:1-15.

Kaminski, R.M., and H.H. Prince. 1984. Dabbling duck-habitat associations during spring in Delta Marsh, Manitoba. Journal of Wildlife Management 48:37-50.

Kaminski, R.M., and M.W. Weller. 1992. Breeding habitats of nearctic waterfowl. In *Ecology and Management of Breeding Waterfowl.* (Eds.) B.D.J. Batt, A.D. Afton, M.G. Anderson, C.D. Ankney, D.H. Johnson, J.A. Kadlec, and G.L. Krapu, pp.568-589. Minneapolis: University of Minnesota Press.

Kantrud, H.A. 1993. Duck nest success on conservation reserve program land in the prairie pothole region. Journal of Soil and Water Conservation 48:238-242.

Kantrud, H.A., and R.E. Stewart. 1977. Use of natural basin wetlands by breeding waterfowl in North Dakota. Journal of Wildlife Management 41:243-253.

Kantrud, H.A., and R.E. Stewart. 1984. Ecological distribution and crude density of breeding birds on prairie wetlands. Journal of Wildlife Management 48:426-437.

Kantrud, H.A., G.L. Krapu, and G.A. Swanson. 1989. Prairie basin wetlands of the Dakotas: a community profile. U.S. Fish and Wildlife Service Biological Report 85(7.28).

Kehoe, F.P., and V.G. Thomas. 1987. A comparison of interspecific differences in the morphology of external and internal feeding apparatus among North American Anatidae. Canadian Journal of Zoology 65:1818-1822.

Kirk, D.A., and C.S. Houston. 1994. Productivity declines in Swainson's hawks and their significance to population trends. Bird Trends 4:19-20.

Krapu, G.L. 1974. Foods of breeding pintails in North Dakota. Journal of Wildlife Management 38:408-417.

Krapu, G.L. 1978. Productivity of red-winged blackbirds in prairie pothole habitat. Iowa Bird Life 48:1-9.

Krapu, G.L., and K.J. Reinecke. 1992. Foraging ecology and nutrition. In *Ecology and Management of Breeding Waterfowl.* (Eds.) B.D.J. Batt, A.D. Afton, M.G. Anderson, C.D. Ankney, D.H. Johnson, J.A. Kadlec, and G.L. Krapu, pp.1-29. Minneapolis: University of Minnesota Press.

Krapu, G.L., A.T. Klett, and D.G. Jorde. 1983. The effect of variable spring water conditions on mallard reproduction. Auk 100:689-698.

Kroodsma, D.E. 1978. Habitat values for nongame wetland birds. In *Wetland Functions and Values: The State of Our Understanding.* (Eds.) P.E. Gresson, J.R. Clark, and J.E. Clark, pp.320-329. American Water Resources Association Technical Publication TPS79-2.

LaBaugh, J.W. 1989. Chemical characteristics of water in northern prairie wetlands. In *Northern Prairie Wetlands.* (Ed.) A.G. van der Valk, pp.56-90. Ames: Iowa State University Press.

Lagerquist, B.A., and C.D. Ankney. 1989. Interspecific differences in bill and tongue morphology among diving ducks (*Aythya* spp., *Oxyura jamaicensis*). Canadian Journal of Zoology 67:2694-2699.

LaGrange, T.G., and J.J. Dinsmore. 1989. Habitat use by mallards during spring migration through central Iowa. Journal of Wildlife Management 53:1076-1080.

Lokemoen, J.T., H.F. Duebbert, and D.E. Sharp. 1984. Nest spacing, habitat selection, and behavior of waterfowl on Miller Lake Island, North Dakota. Journal of Wildlife Management 48: 309-321.

Long, R.J. 1970. A study of nest-site selection by island-nesting anatids in central Alberta. M.S. thesis, University of Alberta, Edmonton, Alberta.

Martin, T.E. 1988. Habitat and area effects on forest bird assemblages: is nest predation an influence? Ecology 69:74-84.

Maxson, S.J., and M.R. Riggs. 1996. Habitat use and nest success of overwater nesting ducks in west central Minnesota. Journal of Wildlife Management 60: 108-119.

McNichol, M.K. 1982. Factors affecting reproductive success of Forster's Terns at Delta Marsh, Manitoba. Colonial Waterbirds 5:32-38.

Melvin, S.M., and J.P. Gibbs. 1994. Sora. In *Migratory Shore and Upland Game Bird Management in North America.* (Eds.) T.C. Tacha and C. E. Braun, pp.209-217. Washington: International Association of Fish and Wildlife Agencies.

Merritt, R.W., and K.W. Cummins. 1984. An Introduction to the Aquatic Insects of North America. Dubuque: Kendall/Hunt.

Mitcham, S.A., and G. Wobeser. 1988. Toxic effects of natural saline waters on mallard ducklings. Journal of Wildlife Diseases 24:45-50.

Monda, M.J., and J.T. Ratti. 1988. Niche overlap and habitat use by sympatric duck broods in eastern Washington. Journal of Wildlife Management 52:95-102.

Moyles, D.L.J. 1981. Seasonal and daily use of plant communities by sharp-tailed grouse (*Pedioecetes phasianellus*) in the parklands of Alberta. Canadian Field-Naturalist 95:287-291.

Murkin, H.R. 1989. The basis for food chains in prairie wetlands. In *Northern Prairie Wetlands.* (Ed.) A.G. van der Valk, pp.316-338. Ames: Iowa State University Press.

Murkin, H.R., and B.D.J. Batt. 1987. Interactions of vertebrates and invertebrates in peatlands and marshes. In *Aquatic Insects of Peatlands and Marshes.* (Eds.) D.M. Rosenberg and H.V. Danks, pp.15-30. Memoirs of the Entomological Society of Canada 140.

Murkin, H.R., and J.A. Kadlec. 1986a. Relationships between waterfowl and macroinvertebrates in a northern prairie marsh. Journal of Wildlife Management 50:212-217.

Murkin, H.R., and J.A. Kadlec. 1986b. Responses by benthic macroinvertebrates to prolonged flooding of marsh habitat. Canadian Journal of Zoology 64:65-72.

Murkin, H.R., and L.C.M. Ross. 1999. Northern prairie marshes (Delta Marsh, Manitoba): macroinvertebrate responses to a simulated wet-dry cycle. *In Invertebrates in Freshwater Wetlands of North America: Ecology and Management.* (Eds.) D. Batzer, R.D. Rader, and S.A. Wissinger, pp. 543-569. New York: Wiley.

Murkin, H.R., R.M. Kaminski, and R.D. Titman. 1982. Responses by dabbling ducks and aquatic invertebrates to an experimentally manipulated cattail marsh. Canadian Journal of Zoology 60:2324-2332.

Murkin, H.R., R.M. Kaminski, and R.D. Titman. 1989. Responses by nesting red-winged blackbirds to manipulated cattail habitat. In *Freshwater Wetlands and Wildlife.* (Eds.) R.R. Sharitz and J.W. Gibbons, pp.673-680. USDOE Symposium Series 61. Oak Ridge: USDOE Office of Scientific and Technical Information.

Murkin, H.R., E.J. Murkin, and J.P. Ball. 1997. Avian habitat selection and prairie wetland dynamics. Ecological Applications 7:1144-1159.

Murkin, E.J., H.R. Murkin, and R.D. Titman. 1992. Nektonic invertebrate abundance and distribution at the emergent vegetation - open water interface in the Delta Marsh, Manitoba, Canada. Wetlands 12:45-52.

Murkin, H.R., M.P. Stainton, J.A. Boughen, J.B. Pollard, and R.D. Titman. 1991. Nutrient status of wetlands in the Interlake Region of Manitoba, Canada. Wetlands 11:105-122.

Neckles, H.A., H.R. Murkin, and J.A. Cooper. 1990. Influences of seasonal flooding on macroinvertebrate abundance in wetland habitats. Freshwater Biology 23:311-322.

Norman, R.F., and R.J. Robertson. 1975. Nest-searching behavior in the brown-headed cowbird. Auk 92:610-611.

Nudds, T.D. 1983. Niche dynamics and organization of duck guilds in variable environments. Ecology 64:319-330.

Nudds, T.D. 1992. Patterns in breeding waterfowl communities. In *Ecology and Management of Breeding Waterfowl*. (Eds.) B.D.J. Batt, A.D. Afton, M.G. Anderson, C.D. Ankney, D.H. Johnson, J.A. Kadlec, and G.L. Krapu, pp.540-567. Minneapolis: University of Minnesota Press.

Nudds, T.D., and J.N. Bowlby. 1984. Predator-prey relationships in North American dabbling ducks. Canadian Journal of Zoology 62:2002-2008.

Nudds, T.D., and R.M. Kaminski. 1984. Sexual size dimorphism in relation to resource partitioning in North American dabbling ducks. Canadian Journal of Zoology 62:2009-2012.

Nuechterlein, G.L. 1981. 'Information parasitism' in mixed colonies of western grebes and Forster's terns. Animal Behaviour 29:985-989.

Orians, G.H. 1961. The ecology of blackbird (*Agelaius*) social systems. Ecological Monographs 31:285-312.

Orians, G.H. 1980. *Some Adaptations of Marsh-nesting Blackbirds*. Princeton: Princeton University Press.

Oring, L.W. 1964. Behaviour and biology of certain ducks during the post-breeding period. Journal of Wildlife Management 28:223-233.

Pattenden, R.K., and D.A. Boag. 1989. Skewed sex ratio in a northern wintering population of mallards. Canadian Journal of Zoology 67:1084-1087.

Pawlina, I.A., D.A. Boag, and F.E. Robinson. 1993. Population structure and changes in body mass and composition of mallards (*Anas platyrhynchos*) wintering in Edmonton, Alberta. Canadian Journal of Zoology 71:2275-2281.

Peterka, J.J. 1989. Fishes in northern prairie wetlands. In *Northern Prairie Wetlands*. (Ed.) A.G. van der Valk, pp.302-315. Ames: Iowa State University Press.

Peterson, L.P., H.R. Murkin, and D.A. Wrubleski. 1989. Waterfowl predation on benthic macroinvertebrates during fall drawdown of a northern prairie marsh. In *Freshwater Wetlands and Wildlife*. (Eds.) R.R. Sharitz and J.W. Gibbons, pp.681-689. USDOE Symposium Series 61. Oak Ridge: USDOE Office of Scientific and Technical Information.

Picman, J. 1980. Impacts of marsh wrens on reproductive strategy of red-winged blackbirds. Canadian Journal of Zoology 58:337-350.

Poston, H.J. 1974. Home range and breeding biology of the shoveler. Canadian Wildlife Service Report Series 25.

Pöysä, H. 1983. Morphology-mediated niche organization in a guild of dabbling ducks. Ornis Scandinavica 14:317-326.

Raveling, D.G., and H.G. Lumsden. 1977. Nesting ecology of Canada Geese in the Hudson Bay Lowlands of Ontario: evolution and population regulation. Ontario Ministry of Natural Resources, Fish and Wildlife Research Report No. 98.

Rogers, J.P., and L.J. Korschgen. 1966. Foods of lesser scaup on breeding, migration, and wintering areas. Journal of Wildlife Management 30:258-264.

Ruwaldt. J.J., L.D. Flake, and J.M. Gates. 1979. Waterfowl pair use of natural and man-made wetlands in South Dakota. Journal of Wildlife Management 43:375-383.

Ryan, M.R., and J.J. Dinsmore. 1980. The behavioral ecology of breeding American coots in relation to age. Condor 82:320-327.

Sargeant, A.B., and D.G. Raveling. 1992. Mortality during the breeding season. In *Ecology and Management of Breeding Waterfowl*. (Eds.) B.D.J. Batt, A.D. Afton, M.G. Anderson, C.D. Ankney, D.H. Johnson, J.A. Kadlec, and G.L. Krapu, pp.396-422. Minneapolis: University of Minnesota Press.

Sayler, R.D. 1992. Ecology and evolution of brood parasitism. In *Ecology and Management of Breeding Waterfowl*. (Eds.) B.D.J. Batt, A.D. Afton, M.G. Anderson, C.D. Ankney, D.H. Johnson, J.A. Kadlec, and G.L. Krapu, pp.290-322. Minneapolis: University of Minnesota Press.

Schitoskey, Jr., F., and R.L. Linder. 1978. Use of wetlands by upland wildlife. In *Wetland Functions and Values: The State of our Understanding*. (Eds.) P.E. Greeson, J.R. Clark, and J.E. Clark, pp.307-311. American Water Resources Association Technical Publication TPS79-2.

Schmidt-Nielson, K. 1960. The salt-secreting gland of marine birds. Circulation 21:955-967.

Sealy, S.G. 1980. Reproductive responses of northern orioles to a changing food supply. Canadian Journal of Zoology 58:221-227.

Sedinger, J.S. 1992. Ecology of pre-fledging waterfowl. In *Ecology and Management of Breeding Waterfowl*. (Eds.) B.D.J. Batt, A.D. Afton, M.G. Anderson, C.D. Ankney, D.H. Johnson, J.A. Kadlec, and G.L. Krapu, pp.109-127. Minneapolis: University of Minnesota Press.

Skagen, S.K., and F.L. Knopf. 1994. Migrating shorebirds and habitat dynamics at a prairie wetland complex. Wilson Bulletin 106:91-105.

Stewart, R.E., and H.A. Kantrud. 1971. Classification of natural ponds and lakes in the glaciated prairie region. Bureau of Sports Fisheries and Wildlife Resource Publication No. 92.

Swanson, G.A. 1984. Dissemination of amphipods by waterfowl. Journal of Wildlife Management 48:988-991.

Swanson, G.A., and H.F. Duebbert. 1989. Wetland habitats of waterfowl in the prairie pothole region. In *Northern Prairie Wetlands*. (Ed.) A.G. van der Valk, pp.228-267. Ames: Iowa State University Press.

Swanson, G.A., V.A. Adomaitis, F.B. Lee, J.R. Serie, and J.A. Shoesmith. 1984. Limnological conditions influencing duckling use of saline lakes in south-central North Dakota. Journal of Wildlife Management 48:340-349.

Swanson, G.A., G.L. Krapu, and J.R. Serie. 1979. Foods of laying female dabbling ducks on the breeding grounds. In *Waterfowl and Wetlands*. (Ed.) T.A. Bookhout, pp.47-57. Madison: The Wildlife Society.

Swanson, G.A., M.I. Meyer, and V.A. Adomaitis. 1985. Foods consumed by breeding mallards on wetlands of south central North Dakota. Journal of Wildlife Management 49:197-202.

Talent, L.G., G.L. Krapu, and R.L. Jarvis. 1982. Habitat use by mallard broods in south central North Dakota. Journal of Wildlife Management 46:629-635.

U.S. Fish and Wildlife Service. 1981. The Platte River ecology study. U.S. Fish and Wildlife Service Special Research Report, Northern Prairie Wildlife Research Center, Jamestown, North Dakota.

van der Valk, A.G., and C.B. Davis. 1978. The role of seed banks in the vegetation dynamics of prairie glacial marshes. Ecology 59:322-335.

Vermeer, K. 1968. Ecological aspects of ducks nesting in high densities among larids. Wilson Bulletin 80:78-83.

Weathers, W.W., and K.A. Sullivan. 1993. Seasonal patterns of time and energy allocation in birds. Physiological Ecology 66: 511-536.

Weller, M.W. 1959. Parasitic egg-laying in the redhead (*Aythya americana*) and other North American Anatidae. Ecological Monographs 29:333-365.

Weller, M.W. 1978. Wetland habitats. In *Wetland Functions and Values: The State of Our Understanding*. (Eds.) P.E. Greeson, J.R. Clark, and J.E. Clark, pp.210-234. American Water Resources Association Technical Publication TPS79-2.

Weller, M.W. 1981. *Freshwater Marshes: Ecology and Wildlife Management*. Minneapolis: University of Minnesota Press.

Weller, M.W., and L.H. Fredrickson. 1974. Avian ecology of a managed glacial marsh. Living Bird 12:269-291.

Weller, M.W., and C.E. Spatcher. 1965. Role of habitat in the distribution and abundance of marsh birds. Department of Zoology and Entomology Special Report 43. Agricultural and Home Economics Experiment Station, Iowa State University, Ames, Iowa.

Welling, C.H., R.L. Pederson, and A.G. van der Valk. 1988. Recruitment from the seed bank and the development of zonation of emergent vegetation during drawdown in a prairie marsh. Journal of Ecology 76:483-496.

Whitman, W.R. 1974. The response of macro-invertebrates to experimental marsh management. Ph.D. dissertation, University of Maine, Orono, Maine.

Wiens, J.A. 1989a. Spatial scaling in ecology. Functional Ecology 3:385-397.

Wiens, J.A. 1989b. The *Ecology of Bird Communities: Volume 2. Processes and Variations.* Cambridge: Cambridge University Press, UK.

Wiggins, G.B., R.J. MacKay, and I.M. Smith. 1980. Evolutionary and ecological strategies of animals in annual temporary pools. Archiv Hydrobiologie Supplement 58:97-206.

Willson, M.F. 1966. Breeding ecology of the yellow-headed blackbird. Ecological Monographs 36:51-77.

Wishart, R.A., P.J. Caldwell, and S.G. Sealy. 1981. Feeding and social behavior of some migrant shorebirds in southern Manitoba. Canadian Field-Naturalist 95:183-185.

Wobeser, G. 1988. Effects of sodium and magnesium sulfate on drinking water on mallard ducklings. Journal of Wildlife Diseases 24:30-44.

Woodin, M.C., and G.A. Swanson. 1989. Foods and dietary strategies of prairie-nesting ruddy ducks and redheads. Condor 91:280-287.

Wrubleski, D.A. 1989. The effect of waterfowl feeding on a chironomid (*Diptera: Chironomidae*) community. In *Freshwater Wetlands and Wildlife.* (Eds.) R.R. Sharitz and J.W. Gibbons, pp.691-696. DOE Symposium Series 61. Oak Ridge: USDOE Office of Scientific and Technical Information.

Wrubleski, D.A. 1991. Chironomid recolonization of marsh drawdown surfaces following reflooding. Ph.D dissertation, University of Alberta, Edmonton, Alberta.

11

Ecology of Muskrats in Prairie Wetlands

William R. Clark

ABSTRACT

I review the literature on the ecology and management of the muskrat (*Ondatra zibethicus*), particularly in prairie wetlands, and integrate data from studies on population dynamics and habitat relationships conducted in the MERP experimental cells. In northern prairie wetlands, muskrats produce two or three litters, averaging from four to eight young per litter, although breeding may be nearly continuous in regions farther south. Annual survival is estimated to range from 13 to 22%. Density in spring averages 1 or 2 muskrats per hectare and autumn density may reach 50 individuals per hectare. Reproductive rates and winter survival are inversely density dependent. The largest fraction of mortality is caused by trapping, and the maximum sustainable rate of harvest is ≈67% of the fall population. Changes in other mortality factors compensate for trapping mortality so that it is difficult to overharvest a muskrat population. Muskrats have dispersed as much as 25–30 km under drought conditions, but movement is normally confined to a small home range of up to 30 m in diameter, centered on the main dwelling. Muskrats consume the above- and belowground parts of a variety of wetland emergent macrophytes, and also eat submersed macrophytes, algae, and animal matter. Cattail and bulrush are preferred vegetation types for foraging and lodge building in northern prairie wetlands. Water depths of 30–40 cm are preferred for lodge sites, and selection is influenced by freezing and microclimatic conditions, access to food resources, and risk of predation. At densities of 20–30 muskrats per hectare, these herbivores were estimated to remove between 1% and 11% of the total standing crop of emergents in the MERP cells between August and May. The MERP study supports the explanation that declines of emergent macrophytes in prairie wetlands are caused by flooding, hastened by the effects of herbivory by abundant muskrats. Substantial removal of senescent aboveground parts of emergents short-circuits wetland decomposition pathways. Data on muskrat–vegetation interactions from exclosure experiments would not only be valuable in determining their role in wetland ecosystems but also would elucidate the basis for density-dependent effects of deteriorating habitat on populations. The interaction of declining abundance of food and patchy habitat quality causes predictable changes in mortality and reproduction, similar to the patterns that explain the population dynamics of other small mammals.

INTRODUCTION

The muskrat (*Ondatra zibethicus*) is one of most familiar and ecologically important vertebrates inhabiting prairie wetlands. Studies of the species have been influential in shaping our basic understanding of population dynamics (Errington 1963) and wetland ecology. Observations of the visible effects of muskrat feeding and lodge-building activities have influenced views on dynamics and management of wetland vegetation (Weller 1981; Perry 1982). The species is distributed throughout North America, from north of tree line in Alaska and Canada to the coastal estuaries of the southeastern United States (Perry 1982), and it is an important furbearer throughout its range (Boutin and Birkenholz 1987). The niche-gestalt of the species can be visualized as that of a large aquatic vole (Figure 11.1) and many aspects of its ecology are similar to smaller microtines (Boutin and Birkenholz 1987).

Muskrats have been of great interest to wildlife managers and ecologists and a voluminous amount of literature has been published about the species (Errington 1963; Perry 1982; Boutin and Birkenholz 1987; Fritzell 1989). However, for such a frequently studied species, much of what we know about its population dynamics and role in wetland dynamics is surprisingly descriptive. In this chapter, I selectively review what is known about the life history, population dynamics, habitat relationships, and management of muskrats by integrating current information with the contribution of the Marsh Ecology Research Program (Clark and Kroeker 1993; Clark 1994). Information in this chapter applies pri-

Figure 11.1   A muskrat preening its fur on a feeding platform (Photo by D. Kroeker).

marily to prairie wetlands, including potholes, delta marshes, and large riverine wetlands throughout the north central United States and central Canada. However, I reference studies conducted outside this area when they have made particularly important contributions to our understanding of muskrat ecology.

LIFE HISTORY

Reproduction

Reproductive parameters have been derived from studies in a variety of wetland habitats throughout North America (Willner et al. 1980; Perry 1982). The sex ratio in adult populations is often reported to be >50% male (Perry 1982), although nearly all ratios have been derived from trapped samples that are male biased. The sex ratio also may be female biased, especially under conditions of low population size (Marinelli and Messier 1993). The basic social unit seems to be the sexual pair and a monogamous mating system (Errington 1963; Caley 1987; Marinelli and Messier 1993, 1995). Polygyny is common when there are more females than males in the population, providing flexibility in the social system (Marinelli and Messier 1993, 1995). Breeding is generally initiated when wetlands become ice-free and may extend through September, but peak activity is from April to July. Normal gestation is 28–30 days and females come into estrus immediately after parturition (Errington 1963). In the northern prairie region, muskrats have two or three litters per season, averaging between four and eight young per litter, although breeding may be continuous farther south in the range (Perry 1982; Boutin and Birkenholz 1987). Litter size as large as 14 embryos have been reported (Chubbs and Phillips 1993). The number of young per litter is positively correlated with latitude, although the relationship reaches a plateau at $\approx 50°N$ (Boyce 1978; Simpson and Boutin 1993). The number of litters per year is inversely correlated with latitude (Boyce 1978), so that total production may actually be similar in northern and southern populations (Simpson and Boutin 1993).

Neonates weigh 15–20 g and develop and grow rapidly (Errington 1939; Perry 1982; Virgl and Messier 1992b, 1995) until they are weaned at ≈ 28 days and 150–180 g (Perry 1982; Virgl and Messier 1992b, 1995). The most reliable growth curves have been developed from studies in Saskatchewan (Virgl and Messier 1995), although comparative data are available from studies in Iowa (Errington 1939), Wisconsin (Dorney and Rusch 1953), Delta Marsh Manitoba (Olsen 1959), Nebraska (Sather 1958), New Brunswick (Parker and Maxwell 1980), and Alberta (Welch 1980). Detailed results on developmental changes in body composition have been published by Virgl and Messier (1992b) as well as an evaluation of changes in adult composition (Virgl and Messier 1992a) and body size and condition indices (Virgl and Messier 1993). In wetlands of the far north, juveniles may only reach ≈ 325 g by the end of the first growing season (Simpson and Boutin 1993). Only 1 to 1.5% of muskrats in northern prairie wetlands reproduce in the calendar year of their birth (Errington 1963) and precocial breeding at this rate has no measurable consequence on population dynamics.

Mortality

Estimates of annual survival of muskrats from birth to 1 year range from 13 to 22% (Perry 1982) although the most reliable methods place the mean in the range of 13–16% (Clark 1987; Clark and Kroeker 1993). Survival of adults older than 1 year has been estimated to be ≈6% (Clark 1987). Survival of young-of-the-year to age 5 months is more variable than it is during other life-history periods; reported values range from 13 to 84% (Perry 1982; Boutin and Birkenholz 1987), although estimates are not always based on similar methods and consistent definitions of time periods. Mortality before weaning is often likely to occur to whole litters (Dorney and Rusch 1953; Boutin et al. 1988), but survival after weaning is more randomly distributed among individuals and is not related to litter size (Boutin et al. 1988). LeBoulenge and LeBoulenge-Nguyen (1981) concluded that survival of later-born litters was less than earlier litters, although Boutin et al. (1988) concluded that birthdate was not correlated with survival of young. Because mean life span is <1 year, survival in the winter immediately after birth has a major influence on demographics. In MERP, winter survival during the first two winters after reflooding was slightly higher than the averages of 31–41% reported elsewhere (Errington 1963; Proulx and Gilbert 1983; Clay and Clark 1985), contributing to the population buildup. The lower survival in winters subsequent to 1987 corresponded to a decline in population level.

Although a major cause of mortality in most muskrat populations is legal harvest (Perry 1982; Clark 1987), trapping was not a factor in the MERP cells because they were not open to trapping. Predators are an important cause of mortality in all populations, and the relationship between muskrats and mink (*Mustela vison*) has been particularly well-studied (Errington 1943, 1954b, 1963). Under low-water conditions mortality from predators such as mink, red fox (*Vulpes vulpes*), and raccoons (*Procyon lotor*) may be increased, apparently because of increased access to a larger fraction of the wetland (Errington 1943, 1963; Wilson 1953; Danell 1978b; Proulx et al. 1987). Tyzzer's disease (also called hemorrhagic or Errington's disease, Wobeser [1981]) is a highly infectious bacterial disease that has been credited with substantial declines in populations (Errington 1954b, 1963; Sather 1958; Woebeser et al. 1978).

During droughts, loss of food and shelter may account for direct mortality of muskrats during the growing season but most of these losses are attributable to exposure to predators (Errington 1938, 1963; Proulx et al. 1987). The effects of drought are most noticeable in winter, when freezeout may occur or when animals die of exposure when dispersing over ice (Errington 1963; Seabloom and Beer 1964). Flooding has been cited as a direct cause of death, especially of young in burrows and lodges (Errington 1937; Bellrose and Low 1943). However, females can move young by carrying them while swimming (Perry 1982) and young can swim and dive by about age 14 days (Errington 1939). The behavioral versatility of this aquatic rodent suggests that such occasional mortality is not likely to have major demographic consequences.

Movements and Dispersal

Muskrats are generally sedentary during most of the year, moving within relatively small home ranges averaging 7–30 m in radius centered on the main dwelling (Perry 1982;

Boutin and Birkenholz 1987; Marinelli and Messier 1993). Home ranges tend to increase somewhat in autumn when animals construct lodges and secure suitable winter habitat (Proulx and Gilbert 1983; Boutin and Birkenholz 1987). However, MacArthur (1978) found that in winter most foraging movements of radio-tagged muskrats were within 5–10 m of a lodge or pushup (breathing holes with vegetation piled above holes in the ice).

Spring dispersal is associated with ice-thaw and commencement of breeding (Errington 1939; Sprugel 1951; Sather 1958). Using settling-distance criteria based on center of activity, Caley (1987) concluded that natal dispersal was male biased and that natal philopatry was female biased; females settled an average of 41 m from their natal site, whereas males settled an average of 65 m away. DNA fingerprinting has been used to determine parentage in muskrats (Marinelli et al. 1992; Marinelli et al. 1997) so it is now possible to fully assess the role of dispersal in reproductive success of the species.

Recapture data indicate that only a small proportion of the population moves long distances, especially in continuous wetland habitat. For example, <5% of marked muskrats at Horicon Marsh Wisconsin was recaptured in an adjacent trapping unit (Mathiak 1966) and <1% of muskrats marked along the upper Mississippi River was recovered in areas other than where they were marked (Clay and Clark 1985; Clark 1986). In MERP, Clark and Kroeker (1993) found that movements from one cell to another were fivefold greater during October and May than during the intervening breeding season. Local movement between habitat types is most likely to occur in late winter or early spring (Parker and Maxwell 1980).

Despite their generally sedentary behavior, habitat changes and population pressures on muskrats are thought to induce dispersals, some as far as 25–30 km (Errington 1940; Sprugel 1951). Occasionally muskrats are observed far from any wetland, particularly under drought conditions in the prairie pothole region, and such dispersers are subject to mortality factors (Errington 1963). Because of these population implications, Clark and Kroeker (1993) attempted to relate estimated dispersal rates to population levels in the MERP study. But they were unable to conclude that movements among cells were related to population density.

POPULATION DYNAMICS

Fluctuations in Numbers

Density of muskrats varies considerably depending on location and wetland type. Rapid invasion of newly flooded wetlands is typical of muskrat populations (Kroll and Meeks 1985; Clark and Kroeker 1993). The magnitude of response and seasonal fluctuations in the MERP cells is similar to dynamics reported by Errington (1954a) for Wall Lake in central Iowa (Figure 11.2; Clark and Kroeker 1993). Spring densities ranging from less than one to five muskrats per hectare and autumn densities ranging from eight to 50 muskrats per hectare have been reported from *Phragmites*, *Sagittaria*, *Scirpus*, and *Typha* habitats (Errington 1963; Perry 1982; Proulx and Gilbert 1983; Clay and Clark 1985). Densities are usually somewhat less along streams, ditches, and around ponds (Errington 1963; Stewart and Bider 1974; Proulx and Buckland 1986). Muskrats are reported to undergo cyclic fluctuations over much of their range (Errington 1954a; Butler 1962; Danell 1978b;

Counts at Wall Lake, Iowa

Errington (1954)

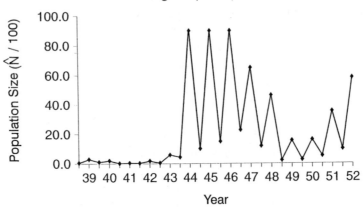

Average Population in MERP

Clark and Kroeker (1993)

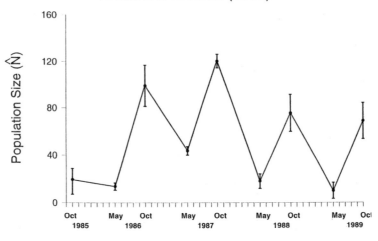

Figure 11.2 Muskrat populations fluctuate widely, both seasonally and annually (Adapted from Errington [1954a] and Clark and Kroeker [1993]).

Finerty 1980). Errington (1954a, 1963) contended that such dynamics were intrinsically produced by changes in social tolerance and physiology interacting with disease and predation (Fritzell 1989), whereas others have argued that the mechanisms are largely extrinsic, induced by causes such as depletion of food supply (O'Neil 1949; Weller 1981) and changing predation pressure (Danell 1978b, 1985).

Density Dependence

Reproductive rates.   Errington (1954a) presented the first quantitative description of density-dependent reproductive rates in muskrats when he showed that the spring-to-fall increase in population, which he equated to net recruitment, was inversely related to the prebreeding population size. This principle of inversity appears widely in the literature. The MERP muskrat data are an improvement over previously published accounts because per capita recruitment ($\hat{B}_i$) and May population size ($\hat{N}_i$) were independently estimated. The relationship between $\hat{B}_i/\hat{N}_i$ and ln ($\hat{N}_i$) (Figure 11.3; Clark and Kroeker 1993) is very similar to that observed by others among a variety of species (Clark 1990). Thus, the MERP studies add convincing evidence to the prevalence of density dependence in per capita recruitment.

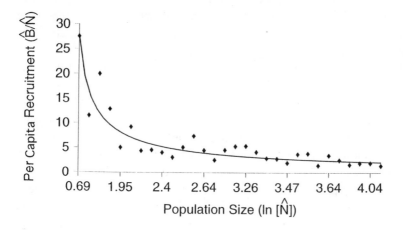

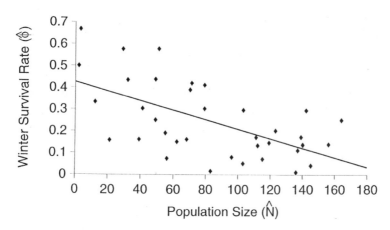

Figure 11.3   Per capita birth rate ($\hat{B}/\hat{N}$) and winter survival ($\hat{\varnothing}$) in MERP muskrat populations were both density dependent (reprinted by permission from Clark and Kroeker 1993).

Mortality.   Estimates of winter survival $(\hat{\phi}_i)$ in MERP were also conclusively density dependent (Figure 11.3), although the relationship was substantially more variable than the one regarding recruitment (Clark and Kroeker 1993). Considering the difficulties with independently estimating the survival and population size parameters (Clark 1990; Clark and Kroeker 1993), this result is especially significant ecologically. Although Errington (1946, 1954a) implied density-dependent mortality was an important population regulation mechanism based on his observations of muskrats, he never actually formulated a test similar to that provided by the MERP data. Based on differences in survival of birth-year and older muskrats, Clark (1987, 1990) suggested that the previously described density-dependent relationship might differ between age groups. However, we were not able to confirm this hypothesis with the MERP data (Clark and Kroeker 1993).

Compensation.   Although individual muskrats may be exposed to a variety of mortality factors, the net effect may not significantly alter population demographics. Heavy losses of animals from one cause of mortality may be offset by decreased mortality due to another cause. The idea of compensatory or replacive mortality was largely derived by Errington (1946, 1954b, 1963) based on his observations of losses of muskrats to mink and disease. Specifically, he wrote (Errington 1946): "We may see that a great deal of predation is without depressive influence. In the sense that victims of one agency miss becoming victims of another, many types of loss—including loss from predation—are at least partly intercompensatory in net population effect." By this statement Errington reduced the role of predators to taking individuals that would have died anyway from disease or exposure, especially those that were in insecure habitat (Errington 1946, 1954b).

The evidence supporting this concept has been correctly criticized because of the circular logic that the value of habitat was largely determined by security from predators, but predators did not in turn limit prey populations (Taylor 1984). Because harvest management is traditionally based on understanding the nature of the compensatory interactions of natural and harvest mortality (Allen 1962), I have made a particular effort to quantify the relationship between mortality rates in this review.

Habitat Relationships

Food Habits

Muskrats are chiefly herbivorous, consuming the above- and belowground parts of a variety of aquatic and terrestrial plants (Perry 1982). In prairie wetlands, meristematic bases of shoots and belowground storage organs of emergent macrophytes, especially *Scirpus* and *Typha*, are important food items when present, although submersed macrophytes and algae may be predominant parts of the diet, depending on season and availability of other foods (Willner et al. 1975). In winter, muskrats depend on access to belowground parts of plants under the ice (Errington 1963; Jelinski 1989; Virgl and Messier 1992a), although they eat parts of lodges and feeding platforms when stressed (Errington 1963). Along streams and ditches where emergent vegetation is less prominent, muskrats may eat a wider variety of foods, including grasses, woody browse, crops, and animal matter

(Errington 1963; Stewart and Bider 1974).

Trapping data from MERP indicated that muskrats foraged in a variety of vegetation types, from areas with no emergents to areas that were nearly dry (Clark 1994). However, based on the observation that >60% of the muskrats were first captured in emergent stands of *Scolochloa*, *Typha*, and *Scirpus*, it is clear that these habitats are most important for foraging and other activities. Muskrats avoided vegetated areas where the water was <1 cm in depth.

Lodge and Burrow Site Selection

Lodges and burrows may be used for periods exceeding a single year (Errington 1963; Pelikán et al. 1970; Proulx and Gilbert 1983; Messier and Virgl 1992) and the characteristics of burrows, lodges, feeding platforms, and pushups have been studied under a wide range of circumstances (Perry 1982). Selection of the location for a lodge is influenced by the distribution of building materials, food resources, physiological requirements, water depth, and risk of predation (MacArthur and Aleksiuk 1979; Messier et al. 1990; Messier and Virgl 1992). Burrows are often preferred in summer when they may be cooler than lodges (MacArthur and Aleksiuk 1979) and because they require less maintenance (Messier and Virgl 1992). However, locations for burrows may be limited by bank steepness and by suitable substrate (Earhart 1969). Muskrats build lodges any time during the ice-free season, although there is an increase in activity about the time of first frost (Danell 1978a; Jelinski 1989). Lodges are frequently begun on places that were used for sitting, such as a stump or feeding platform (Errington 1963; Danell 1978a). They are often located at the boundary between emergent vegetation types or within a few meters of the edge of open water (Pelikán et al. 1970; Danell 1978a; Proulx and Gilbert 1983; Figure 11.4).

Vegetation characteristics. In the MERP cells, muskrats statistically preferred to build lodges in *Typha* and *Scirpus*. They avoided *Scolochloa* and open water for lodge locations, and were neutral toward *Phragmites* (Clark 1994). This pattern is consistent with that reported from similar wetlands (Pelikán et al. 1970; Perry 1982; Proulx and Gilbert 1983; Messier et al. 1990), although *Equisetum fluviatile* and *Sagittaria latifolia* are more important than the previously mentioned emergents in more northerly areas (Danell 1978a; Jelinski 1989) and along rivers (Clay and Clark 1985), respectively.

Messier et al. (1990), Clark (1994), and Virgl and Messier (1996, 1997) connected these observations of habitat selection by muskrats to contemporary theory on source–sink population dynamics (Morris 1987; Pulliam and Danielson 1991). Theory predicts that relative density in preferred habitats should increase more slowly than in less suitable habitats as overall population density increases because of density-dependent competition within habitats with the highest value for reproduction and survival. The MERP habitat selection data (Clark 1994) and those of Danell (1978a) and Virgl and Messier (1996, 1997) are consistent with this prediction. For example, although *Scolochloa* was a predominant vegetation type in MERP during 1985 and 1986 and many muskrats were initially captured there, this habitat was avoided for lodge construction. Winter weight gain

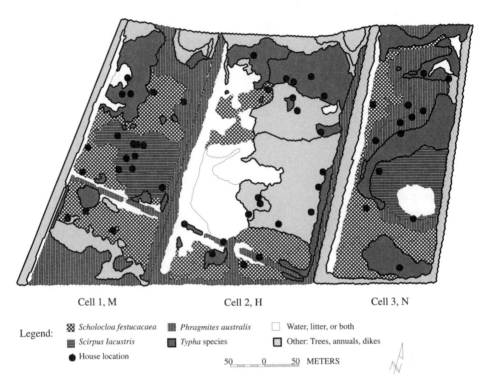

Figure 11.4   Muskrats preferred locating lodges in *Typha* and *Scirpus* in water averaging 38 cm in depth (reprinted by permission from Clark 1994).

and survival of muskrats intially captured in *Scolochloa* was significantly less than in *Scirpus* and *Typha*. The temporal and spatial variability of vegetation composition and productivity in prairie wetlands, coupled with density-dependent habitat selection by muskrats, causes variation in condition, survival, and recruitment (Welch 1980; Messier et al. 1990; Clark and Kroeker 1993), contributing to temporal variation in demography.

Water levels. Depth of water is another critical factor that interacts strongly with vegetation type in determining placement and success of occupancy of lodges. High water levels or catastrophic flooding in spring is usually of brief duration, although it may temporarily displace muskrats (Bellrose and Low 1943; Sather 1958; Errington 1963; Olsen 1959) and can have demographic consequences (Bellrose and Low 1943; Errington 1963). The prolonged flooding reconstructed by the MERP treatments was designed to examine long-term effects on habitat selection rather than rapid changes in water levels. Even the flooding treatment of 60 cm above normal did not affect habitat selection (Clark 1994). Average water depth at lodges constructed in the MERP cells was 38 cm, similar to the range of averages of 30–40 cm reported for the species in a variety of wetland conditions

(Bellrose and Brown 1941; Errington 1963; Danell 1978a; Proulx and Gilbert 1983).

Low water levels, such as under drought conditions, have a more frequent and direct impact on habitat selection and population dynamics than high water levels. Lodge-site selection appears to be more predictably influenced by potential low water in winter, which is related to freezing, access to below-ground food resources, predation, and exposure to harsh microclimatic conditions (Friend et al. 1964; Seabloom and Beer 1964; MacArthur 1978; Messier et al. 1990; Clark 1994). The use of winter forage resources and predator escape are facilitated by closely spaced lodges and pushups in water of sufficient depth to prevent freezing to the bottom (MacArthur 1978).

Robust emergent vegetation traps snow, which insulates the water column, thereby reducing the chance that the water will freeze to the bottom (Messier et al. 1990; Clark 1994). Seabloom and Beer (1964) associated lack of snow cover and heavy ice formation to high mortality of muskrats in North Dakota. In MERP, vegetation types such as *Phragmites* and *Typha* trapped the most snow, whereas it had blown away in open-water areas. The combination of moderate water depth, snow accumulation, and quality food resources made *Typha* the best habitat for muskrats to select in the MERP cells to enhance winter survival.

Role in Wetland Succession

Muskrats are an important functional component of wetlands, directly influencing vegetation structure, distribution, and primary production, and indirectly influencing decomposition, invertebrate communities, and habitat use by birds. Virtually every review of the ecology of muskrats includes some statements about the destruction of wetland emergents by muskrats (Errington 1963; Perry 1982; Weller 1981; Fritzell 1989; Murkin 1989). The observations that muskrats become overpopulated and "eatout" marsh vegetation were initially described in southern estuarine wetlands (Lynch et al. 1947; O'Neil 1949; Wilson 1968; Sipple 1979) although Errington (1963) and contemporaries (Sather 1958; Weller and Spatcher 1965) described a similar interaction in prairie wetlands. Boutin (Boutin and Birkenholz 1987) stated that he knew of no documented cases of eatouts in Canada. Danell (1978b) observed that peak densities of muskrats in a Swedish lake did not cause any heavy decline in available food resources.

Although most accounts of eatouts acknowledge the importance of both flooding and muskrat activities on emergent vegetation changes (Errington et al. 1963; Weller and Spatcher 1965; Weller 1981), few data have been presented to distinguish the relative importance of these factors. The often-cited examples of eatouts described in Iowa prairie marshes (Errington et al. 1963; Weller and Spatcher 1965; Weller 1981) include statements such as "water level had been rising since spring 1957" and "during summer 1960 vegetation steadily deteriorated, apparently from a combination of high water and excessive exploitation by the muskrats" (Errington et al. 1963). In similar Iowa wetlands with relatively stable water conditions, cover of emergents has been documented to remain essentially unchanged while muskrat densities increased and decreased 6- to 10-fold (Bishop et al. 1979). It is difficult to separate cause from effect based on the descriptions.

Consumption.   Although there is great interest in the impact of muskrats on wetland vegetation, there are very few quantitative estimates of consumption of emergents by
muskrats (Fritzell 1989; Murkin 1989). Pelikán et al. (1970) calculated that muskrats
destroyed 20% of the net production within a 14-m radius of lodges, which translated to
5–10% of *Typha* production within the wetland. I (unpublished data) calculated that
muskrats consumed or wasted 75% of aboveground standing crop of *Sagittaria* within a
4- to 5-m radius of lodges along the Mississippi River and that this estimate was <1% of
the total standing crop. Other than the data of Smith and Kadlec (1985), there are no published estimates of consumption of emergents by muskrats that are based on paired exclosures. They concluded that the combined herbivory of geese and muskrats significantly
reduced total annual production of *Typha*, *Scirpus lacustris*, and *Scirpus maritimus* in
burned, but not in unburned, areas of a marsh in Utah. Under these circumstances, they
estimated that grazing reduced the total annual production by 48% for *Typha*, 25% for *S.
lacustris*, and 9% for *S. maritimus*. But on a trophic basis, energy flow through mammals,
including muskrats is only a small fraction of estimated net production in wetlands
(Mitsch and Gosselink 1986). Campbell and MacArthur (1994) have estimated assimilation efficiency of >60%, which is somewhat greater than that estimated for other micro-
tine rodents under a variety of circumstances (Grodzinski and Wunder 1975). There has
been some excellent work done on energy requirements, nitrogen balance, and consumption rates of muskrats (McKwan et al. 1974; MacArthur 1986; Campbell and MacArthur
1994, 1996; Campbell et al. 1998). Energy consumption estimates of muskrats range from
331 kJ/kg/d (79 kcal/kg/d) under summer and fall conditions (Campbell and MacArthur
1994) to as much as 710 kJ/kg/d (169 kcal/kg/d) under simulated winter microhabitat conditions (MacArthur 1986). Based on these requirements and population levels in the
MERP cells, consumption (exclusive of wastage) would average 0.08 g/m²/d, equivalent
to 1.35 kJ/m²/d. If muskrats consumed at this constant rate over a year, average populations would consume ≈492 kJ/m²/yr (118 kcal/m²/yr), a small fraction of the net production.

   It is apparent that indirect control of plant production is likely to be much more significant than the direct flow of energy through muskrats. Muskrats clearly cut and discard
much plant material, and although the amount of wastage is not well known, their activity clearly increases spatial diversity in emergent vegetation. Fortuitously, energy requirements and consumption of vegetation by individual muskrats have been studied under laboratory and natural conditions in Manitoba during both summer and winter (Campbell and
MacArthur 1994, 1996; Campbell et al. 1998). Most recently, Campbell et al. (1998) estimated daily intake of fresh vegetation, dry matter, and assimilated energy of free-ranging
muskrats by using deuterated water.

   By using estimates from Campbell and MacArthur (1996) and Campbell et al. (1998)
and the demographic data from MERP, I calculated estimates of removal of standing crop
of emergents in each of the cells. I compared my estimates of consumption to the aboveground standing crop (AGSC), belowground standing crop (BGSC) and total standing
crop (TSC) of emergents in the MERP cells estimated in August 1986 and 1987 (see
Chapter 7). These calculations were made from data reflecting standing crops at the end

of each growing season when muskrats had reached peak densities (see Figure 12.2 in Clark and Kroeker [1993]) but do not estimate the effect of consumption by muskrats on net primary production (NPP) during the growing season. I calculated a weighted mean of 81.6 g dry vegetation/kg body mass/d consumption by using Campbell et al. (1998) estimates for September, December and February. I assumed that muskrats cut and waste twice the amount they consume (Pelikán et al. 1970; i.e., 3.0 times baseline consumption) and that wastage is at a peak during lodge construction. I also assumed that wastage declined linearly to 1.5 times baseline consumption in midwinter because of increased digestive efficiency and reduced activity (Campbell and MacArthur 1996). With these assumptions muskrats would cut or consume as much as 245 g dry mass/kg/d.

I integrated the effect on the vegetation in each MERP cell over the period from 1 October to 1 May (212 days) by using my estimate of 81.6 g/kg/d consumption adjusted for wastage, and Clark and Kroeker's (1993) estimates of October populations, accounting for survival and body mass changes. Estimated total removal ranged from as little as 1% to as much as 11% of TSC (Table 11.1). Removal of TSC was not significantly different between 1986 and 1987 ($F = 1.41$; df = 1, 14; $P = 0.254$) nor among water-level treatments ($F = 0.32$; df = 2, 14; $P = 0.73$).

Recognizing that removal of senescing aboveground vegetation during the nongrowing season may not reflect true impact on the vegetation, I also calculated an estimate of removal assuming that muskrats met all their requirements by consuming only belowground rhizomes of emergents. This potentially could be a major impact on the reserves that the plant community would draw upon during the next growing season, affecting the ability of these species to respond to changing water levels. Calculating this maximum effect on the BGSC, muskrats would have removed an average of 10% of BGSC during

Table 11.1. Estimated removal (kg/ha), consumption, and wastage of total standing crop (TSC, kg/ha) by initial populations of muskrats ($\hat{N}_i$) during October through April (212 days) following the growing seasons in which muskrat populations peaked in the MERP cells at Delta Marsh Manitoba in 1986 and 1987

Cell	1986				1987			
	$\hat{N}_i$	TSC	Removal	%	$\hat{N}_i$	TSC	Removal	%299
1	103	5,983	426	7	136	9,880	366	4
2	164	6,579	499	8	145	3,343	322	10
3	79	7,154	337	5	112	5,796	319	6
4	51	5,550	201	4	83	10,022	218	2
5	71	6,391	280	4	115	5,236	296	6
6	70	7,744	244	3	140	10,868	356	3
7	55	8,663	202	2	62	12,653	152	1
8	123	4,401	495	11	103	5,577	277	5
9	142	5,834	594	10	139	11,812	379	3
10	111	9,896	427	4	137	11,755	385	3
Mean	97	6,819	371	6	117	8,694	307	4

both winters, and there was no effect of water level treatment on the estimated removal of BGSC ($F = 0.40$; df $= 2$, 14; $P = 0.68$). Without question my calculations represent a maximum effect because I have not accounted for adaptations such as conservative lodge temperatures, huddling behavior, reduced foraging (MacArthur and Aleksiuk 1979), or reduced metabolic rate (McKwan et al. 1974; Campbell and MacArthur 1996).

Certainly, I would have liked to estimate consumption during the growing season so that I could consider the effects of herbivory on annual net primary production (NPP) of emergents. Several lines of evidence suggest that the effect of muskrat herbivory during the growing season would be <5%. First, the harvest method used for estimating TSC of emergents does not account for removal of gross primary production occurring during the growing season (Odum 1971). Furthermore, consumption of submergents and algae, which can be substantial at some times of the year (Willner et al. 1975; Welch 1980; Jelinski 1989), would account for some of the food used by the muskrats during the growing season. And muskrats are known to have reduced daily consumption during the growing season (Campbell and MacArthur 1996; Campbell et al. 1998).

Based on my calculations of removal, I conclude that muskrats may have contributed to the decline of emergent vegetation in MERP but they were not the principal ecological cause as the term "eatout" suggests. From 1985 to 1989 ≈35 to 45% of the emergent biomass in the medium and high MERP cells was eliminated, mostly due to death of *Scolochloa festucacea* and *Scirpus lacutris* (see Chapter 7). Neither of these species is a highly preferred food or lodge-building material of muskrats. During the same period belowground biomass tripled, increasing fairly steadily during the period primarily due to expansion of rhizome systems of *Typha glauca*. The latter species is a primary lodge-building and winter food of muskrats (Clark 1994). By August 1987, the emergent vegetation in the cells had already begun to decrease and the percentage of open water in medium and high treatments exceeded that in normal cells by 30% (van der Valk et al. 1994; see Chapter 7). The highest densities of muskrats occurred in October 1987 and winter survival declined dramatically (Clark and Kroeker 1993) after the reduction in emergent vegetation had been detected. Consumption and wastage by muskrats probably interacted with the flooding effects (Squires and van der Valk 1992; van der Valk 1994) because growth and survival of flooded emergents is known to decline after cutting (Weller 1975; Lieffers and Shay 1981; Sale and Wetzel 1983; Ball 1990). The most obvious consumption by muskrats is of senescent aboveground emergent plant parts, although consumption of belowground parts during winter must reduce stored reserves of emergents. Consumption in 1986 could have hastened the rate of increase in open water relative to emergent cover that was observed in 1987 by van der Valk et al. (1994). The vegetation data from the deep-flooding period of MERP and the low impact of muskrats during the reflooding years showed that prolonged high water was sufficient to eliminate emergents in the absence of muskrats.

Conversely, the decline in abundance of food and cover and the patchy habitat quality that resulted from the flooding effect on emergent vegetation ultimately caused the observed changes in muskrat mortality. The interaction of factors is similar to the pattern of changes that explain the dynamics of other microtines (Batzli 1992) and snowshoe hares (*Lepus americanus*) (Keith 1983; Krebs et al. 1995), although the role of predation on muskrats has not been studied recently. Flooding, perhaps enhanced by consumption, reduces the abundance of emergents that provide food and cover for muskrats. Winter survival is sharply reduced among both adults and juveniles, as the relative proportion of good habitat is decreased and predation increases. Density-dependent decline in growth and reproduction, combined with the decreased survival, accentuates the population decline. For snowshoe hares, deep snow is functionally similar to deep water, restricting access to adequate food (Pease et al. 1979; Keith 1983) and influencing the value of habitat as cover from predators (Wolff 1980).

Similar density-dependent habitat shifts occur among muskrats during the ice-free season in response to changing water conditions, and there are associated density-dependent differences in survival (Virgl and Messier 1996). Food is not limiting during the ice-free period per se (Virgl and Messier 1997) but rather habitat selection is more substantially influenced by predation risk. Spacing behavior plays a key role in population dynamics of muskrats during both the breeding and winter periods, as it does in other microtines (Krebs 1996). During summer individuals forced into marginal habitats presumably suffer higher losses to predators such as mink (Virgl and Messier 1996), and during winter mortality in marginal habitats is probably influenced by the combination of poor access to food and shelter, and vulnerability to predation. The observed timing of changes in vegetation and muskrat demography in MERP suggests that muskrat populations responded to the declining emergent vegetation rather than were its principal cause.

Decomposition and vegetation diversity.   Muskrat feeding and lodge-construction activities may potentially have a great effect on litter decomposition. The major processes causing breakdown of standing litter of emergents are leaching, fragmentation, and toppling (Davis and van der Valk 1978; see Chapter 7). Activities of muskrats short-circuit the standing litter stage by directly transferring material into the fallen litter stage. The freshly cut plant has substantially different nutrient qualities than standing dead material (Davis and van der Valk 1978) and therefore are colonized at different rates by decomposers. Furthermore, the material is not distributed uniformly as it would be if it dropped in place because it is concentrated either in lodges or feeding platforms or it drifts into piles after floating away. Decaying lodges have been shown to last >5 months (Nicholson and Davis 1957). Concentrated litter from decaying lodges or wind-drifted vegetation may significantly retard the regrowth of emergents from the seedbank (van der Valk 1986), thus increasing spatial variation in emergent stands. Active and decaying muskrat lodges provide substrates for a variety of wetland plants (Kangas and Hannan 1985), further increasing vegetation diversity.

MANAGEMENT

Our increasing understanding of muskrat population dynamics and habitat relationships should influence practical management. Biologists have considerably improved methods of estimating vital statistics, thereby providing more quantitative data on density dependence, compensatory interactions, and the effects of harvest on populations.

Estimating Vital Statistics

Capture and marking.   Muskrats can be trapped easily using wire mesh, single- or double-door, collapsible $15 \times 15 \times 45$ cm traps set on houses, feeding platforms, or in runways. Traps can be baited with apples or carrots, but unbaited traps set in runways or suspended at water level are very effective (Perry 1982). Covering traps with vegetation or burlap increases capture success and reduces mortality in the trap. Clay and Clark (unpublished data) modified Snead's (1950) original design of a multiple-capture trap, the bottom of which is submerged in runways (Figure 11.5). With this design it is possible to capture as many as a dozen animals representing multiple litters from the same den. Muskrats in livetraps can be carefully removed by the tail and placed in a handling cone (Figure 11.6; McCabe 1982). Muskrats have been marked with #1 monel metal tags in the ears, with tags inserted through the skin on the back (Errington and Errington 1937), and with bands inserted between the Achilles tendon surrounding the tibia (Takos 1943; Clay and Clark 1985). Clark (1987) estimated that the rate of individual ear- and leg-tag losses was 0.137, suggesting that multiple marking is necessary to ensure recovery.

Sexing and aging techniques.   Sexing and aging of live animals can be done by examining the appearance of external genitalia (Boutin and Birkenholz 1987). Dead animals can be sexed by necropsy or by examining the dried pelt. Aging has been done by examining the primeness of the dried pelt, by examining the fluting of the molar teeth (Boutin and Birkenholz 1987), or by measuring the molar-crown-to-total-length ratio (Doude Van Troostwijk 1976). Moses and Boutin (1986) concluded that the pelt primeness method classified adults more accurately than molar fluting. However, the crown-to-total length technique has been most reliable for estimating age ratios in harvested samples (Doude Van Troostwijk 1976; Erb 1993).

Density.   Estimating population trends from counts of lodges is commonly used for management purposes (Perry 1982; Proulx and Gilbert 1984; Boutin and Birkenholz 1987), but the relationship of these counts to actual density is unknown. Houses can be counted accurately and precisely directly from aircraft, aerial imagery, or on the ground by using line transect methods (Burnham et al. 1980; Clark and Andrews 1982). However, the reported ratio of muskrats:house varies from less than one to five (Boutin and Birkenholz 1987) and many factors, especially water levels, can influence the number of active houses in autumn (Proulx and Gilbert 1984). Aerial surveys of habitat variables along riverine wetlands associated with muskrat presence have been effectively used over large areas (Nadeau et al. 1995).

Although mark–recapture methods such as those used in MERP are labor-intensive, they are the most reliable way to estimate population size and density. A trapping effort of 5 or 6 days with ≈ 10 traps per hectare gives sufficient samples and recaptures for precise estimation. By using the methods of Otis et al. (1978), we have estimated population size with coefficients of variation <20% under a variety of conditions (Clay and Clark 1985; Clark and Kroeker 1993). The use of mark–recapture methods could be extended practically to management situations by using multistage sampling, combining estimates from intensively sampled representative habitats with estimates of habitat area or even lodge counts (Skalski and Robson 1992).

Reproduction.   Reproduction is most often estimated by counting placental scars in samples of harvested females (Perry 1982; Boutin and Birkenholz 1987). Although the method certainly has problems associated with interpretation of scars (Payne 1982), it is the most direct way to obtain large samples. Ratios of juveniles to adults in the harvest also are used to measure recruitment, although such ratios confound natality with postna-

Figure 11.5   Using a family trap (modified from Snead 1950) set in submerged runways, as many as a dozen muskrats have been captured from the same burrow or lodge (Photo by R. Clay).

Figure 11.6   Muskrats can be sexed by external examination of genitalia and marked without chemical restraint in a handling cone (McCabe 1982) made with wire rods that are easily separated (Photo by D. Dyck).

tal mortality, are highly variable (Perry 1982; Boutin and Birkenholz 1987), and are known to be affected by harvest and density (Boutin and Birkenholz 1987; Clark 1987).

When more intensive study is possible, counting the number of young in lodges is an accurate estimate of natality (Dorney and Rusch 1953; Errington 1963; Caley 1987). Recruitment also can be very effectively estimated from mark–recapture data by using Jolly–Seber statistics (Pollock et al. 1990; Clark and Kroeker 1993) and it may be possible to separate natality from postnatal mortality if age-specific models are adequate (Pollock et al. 1990).

Mortality.   Mortality rates have been inferred from comparison of placental scar counts with age ratios in the fall harvest (Perry 1982; Boutin and Birkenholz 1987), but are subject to the same biases outlined for each of the component parts. More accurate and precise estimates of mortality have come from mark–recapture samples. Mortality estimates have been made by comparing estimated population change between two points in time (Clay and Clark 1985). Clark (1987) demonstrated that banding methods (Brownie et al. 1985) could be used to estimate age-specific survival over large areas by combining marking methods with recovery during subsequent harvest seasons. Recapture of young marked in lodges has been effectively used to estimate postnatal mortality (Boutin et al. 1988). Muskrat populations are well-suited to analyses of recaptures based on Jolly–Seber statistics (Pollock et al. 1990) and extensions (Lebreton et al. 1992), and this approach has been applied to estimation of mortality of muskrats (Clay and Clark 1985; Clark and Kroeker 1993).

Effects of Harvest on Populations

Biologists have always been interested in the effects of harvest on populations because of the importance of muskrats as a furbearing species and because managers viewed harvest as a means of influencing muskrat effects on wetlands (Errington 1963). Muskrat populations are often harvested heavily and reported rates range from 50 to 90% (Perry 1982). Researchers have attempted to determine the sustainability of such rates. Parker and Maxwell (1984) manipulated harvest and concluded that removing 60% of the population in a marsh in New Brunswick in autumn or spring did not reduce populations. Based on survival and reproductive data, Clark (1987) estimated maximum sustained harvest rate of 60–68% for populations along the upper Mississippi River. Smith et al. (1981) projected maximum sustained harvest rate of 74% of estuarine populations in Connecticut based on a production model. Sustained yields of 7–9 muskrats per hectare were reported for wetlands in South Dakota (Aldous 1947) and estimated to be two muskrats per hectare in poor habitat and eight per hectare in good habitat on the upper Mississippi River (Clark 1987).

Observing changes in population levels, without estimating how changes in harvest rates influence vital statistics such as reproduction and survival is a misleading approach to quantifying harvest effects (Clark 1990). Reproductive rates clearly respond to reduction in numbers in a density-dependent fashion (Errington 1963; Clark and Kroeker 1993). Because of low annual survival, there is substantial ability for changes in nonharvest sur-

vival to compensate for increased harvest rate (Clark 1990; Clark and Fritzell 1992).
There is not a distinct difference in vulnerability to harvest of juveniles and adults among
muskrats (Clark 1987) as there is among some other longer-lived furbearing species
(Clark and Fritzell 1992). Based on mark–recovery methods, Clark (1987) concluded that
although harvest rates varied, annual survival of juveniles and adults remained constant.
Furthermore, Clay and Clark (1985) and Clark (1987) provided data that suggested that
the compensatory mechanism was increased survival of juveniles from birth to autumn.
This density-dependent increase in survival after substantial harvest partially accounts for
the direct relationship observed between autumn age ratios and harvest rate (Mathiak
1966; Smith et al. 1981; Perry 1982; Parker and Maxwell 1984; Clark 1987).

Regulating harvest.   The best guideline for management is the sustainable harvest rate of
≈67% of the autumn population, but this target may be difficult to achieve in practice.
Management agencies attempt to regulate harvest by the extent and timing of trapping seasons (Sather 1958; Perry 1982). Because of great biotic potential and substantial compensatory mortality, it is difficult to overharvest a population, although overharvest is
occasionally reported on a local scale (Errington 1940; Mathiak 1966; Smith et al. 1981;
Clark 1987). Restricting harvest seldom increases the rate at which populations recover.
Underharvest is often of greater concern because of the perceived influence of high densities of muskrats on wetland vegetation (Mathiak 1966; Bishop et al. 1979). Clark (1986)
estimated that >90% of harvest occurred within the first 3 weeks of the season, so length-
ening the season is not likely to increase total harvest. Harvest rates are affected more by
the initiation of the season relative to environmental conditions, such as freezing temperatures or high water (Clark 1986), which affect trapper effort. Harvesting muskrats in both
spring and autumn causes population declines (Parker and Maxwell 1984) and may be the
most effective way to control populations.

RESEARCH NEEDS

All of the components of prairie wetland ecosystems, including muskrats, deserve continued scientific investigation. The MERP provided a major improvement over previous studies of muskrats because it was a designed, direct manipulation of water level that affected muskrat habitats and population dynamics. The replicated mark–recapture approach
enabled me to accurately and precisely estimate vital statistics of population size, survival,
and recruitment, strengthening interpretations about population mechanisms. Although
macrohabitat conditions were directly manipulated, I do not know how microhabitat
selection within cells specifically affected survival and reproduction of individuals. We
still need to estimate how spatial variation in reproduction and survival contributes to variation in observed density-dependent effects (Clark 1994; Morris 1987; Pulliam and
Danielson 1991; Messier et al. 1990). Questions also remain about age-specific density-
dependent effects. Telemetry studies of individual habitat selection, reproductive performance, and survival would be a logical next step.

Recent estimates of sustainable harvest rates are based on solid, theoretical concepts and quantitative data. Perhaps we need research on how the behavior of trappers influences whether sustainable use of the resource is actually achieved. Wildlife managers also want to know if manipulating harvest really has any impact on outbreaks of diseases that could cause declines in muskrats. Researchers have only described the epizootiology and know essentially nothing about relative importance of disease in muskrat population dynamics.

Most importantly, we need to design controlled exclosure studies to determine the effects of muskrat herbivory on vegetation dynamics and detrital pathways in prairie wetlands. Most observations of eatouts are largely circumstantial and we need repeatable estimates of consumption by muskrats under natural conditions. We also must understand whether herbivory decreases regrowth and increases shoot mortality and whether muskrat activity enhances flotation and wave action before we can understand their role in the dynamics of emergent plants. Effective exclosures must be large and substantial enough to avoid destruction by the muskrats themselves. The data on the vegetation–animal interactions would not only be valuable in determining the relative importance of flooding and herbivory to changes in emergent vegetation in wetland ecosystems but also would elucidate the physiological basis for effects of deteriorating habitat on the populations of muskrat.

ACKNOWLEDGMENTS

I especially thank graduate students D.W. Kroeker and R.T. Clay, whose thesis work stimulated many of the ideas herein and provided data to support them. R.A. MacArthur willingly contributed input on estimates of consumption. J.J. Majure conducted the GIS analyses, and L.M. Fischer and K.J. Neilson created other figures. I appreciate the helpful comments on a draft of this chapter by K. Danell, E.K. Fritzell, T.R. McCabe, and D.L. Watson. Besides the primary support of the Marsh Ecology Research Program (MERP), a project of Ducks Unlimited Canada and the Delta Waterfowl and Wetlands Research Station, support was also provided by the Department of Animal Ecology, the Iowa Cooperative Fish and Wildlife Research Unit, Iowa State University, and the Institute for Wetland and Waterfowl Research (IWWR). This is MERP paper No. 104 and Journal Paper J-15445 of the Iowa Agriculture and Home Economics Experiment Station, Projects 2401 and 3299.

LITERATURE CITED

Aldous, S.E. 1947. Muskrat trapping at Sand Lake National Wildlife Refuge, South Dakota. Journal of Wildlife Management 11:77-90.

Allen, D.L. 1962. *Our Wildlife Legacy.* New York: Funk and Wagnalls.

Ball, J.P. 1990. Influence of subsequent flooding depth on cattail control by burning and mowing. Journal of Aquatic Plant Management 28:32-36.

Batzli, G.O. 1992. Dynamics of small mammal populations: a review. In *Wildlife 2001: Populations.* (Eds.) D.R. McCullough and R.H. Barrett, pp.831-850. London, UK: Elsevier Science.

Bellrose, F.C., and L.G. Brown. 1941. The effect of fluctuating water levels on the muskrat population of the Illinois River Valley. Journal of Wildlife Management 5:206-212.

Bellrose, F.C., and J.B. Low. 1943. The influence of flood and low water levels on the survival of muskrats. Journal of Mammalogy 24:173-188.

Bishop, R.A., R.D. Andrews, and R.J. Bridges. 1979. Marsh management and its relationship to vegetation, waterfowl, and muskrats. Proceedings of the Iowa Academy of Science 86:50-56.

Boutin, S., and D.E. Birkenholz. 1987. Muskrat and round-tailed muskrat. In *Wild Furbearer Management and Conservation in North America*. (Eds.) M. Novak, J. A. Baker, M.E. Obbard, and B. Malloch, pp.314-325. North Bay: Ontario Trappers Association.

Boutin, S., R.A. Moses, and M.J. Caley. 1988. The relationship between juvenile survival and litter size in wild muskrats (*Ondatra zibethicus*). Journal of Animal Ecology 57:455-462.

Boyce, M.S. 1978. Climatic variability and body size variation in the muskrats (*Ondatra zibethicus*) of North America. Oecologia (Berl.) 36:1-19.

Brownie, C., D.R. Anderson, K.P. Burnham, and D.S. Robson. 1985. Statistical inference from band recovery data—a handbook, 2nd edition. United States Fish and Wildlife Service Resource Publication Number 156.

Burnham, K.P., D.R. Anderson, and J.L. Laake. 1980. Estimation of density from line transect sampling of biological populations. Wildlife Monographs 72.

Butler, L. 1962. Periodicities in the annual muskrat population figures for the Province of Saskatchewan. Canadian Journal of Zoology 40:1277-1286.

Caley, M.J. 1987. Dispersal and inbreeding avoidance in muskrats. Animal Behavior 35:1225-1233.

Campbell, K.L., and R.A. MacArthur. 1994. Digestibility and assimilation of natural forages by muskrat. Journal of Wildlife Management 58:633-641.

Campbell, K.L., and R.A. MacArthur. 1996. Seasonal changes in gut mass, forage digestibility, and nutrient selection of wild muskrats (*Ondatra zibethicus*). Physiological Zoology 69:1215-1231.

Campbell, K.L., G.L. Weseen, and R.A. MacArthur. 1998. Seasonal changes in water flux, forage intake and assimilated energy of free-ranging muskrats. Journal of Wildlife Management 62:292-299.

Chubbs, T.E., and F.R. Phillips. 1993. Unusually high number of embryos in a muskrat, *Ondatra zibethicus*, from Central Labrador. Canadian Field-Naturalist 107:363.

Clark, W.R. 1986. Influence of open season and weather on the harvest of muskrats. Wildlife Society Bulletin 14:376-380.

Clark, W.R. 1987. Effects of harvest on annual survival of muskrats. Journal of Wildlife Management 51:265-272.

Clark, W.R. 1990. Compensation in furbearer populations: current data compared with a review of concepts. Transactions of the North American Wildlife and Natural Resources Conference 55:491-500.

Clark, W. R. 1994. Habitat selection by muskrats in experimental marshes undergoing succession. Canadian Journal of Zoology 72:675-680.

Clark, W.R., and R.D. Andrews. 1982. Review of population indices applied in furbearer management. In *Midwest Furbearer Management*. (Ed.) G.C. Sanderson, pp.11-22. Wichita: Kansas Chapter of the Wildlife Society.

Clark, W.R., and E.K. Fritzell. 1992. A review of population dynamics of furbearers. In *Wildlife 2001: Populations*. (Eds.) D.R. McCullough and R.H. Barrett, pp.899-910. London, UK: Elsevier Science.

Clark, W.R., and D.W. Kroeker. 1993. Population dynamics of muskrats in experimental marshes at Delta, Manitoba. Canadian Journal of Zoology 71:1620-1628.

Clay, R.T., and W.R. Clark. 1985. Demography of muskrats on the Upper Mississippi River. Journal of Wildlife Management 49:883-890.

Danell, K. 1978a. Intra- and interannual changes in habitat selection by the muskrat. Journal of Wildlife Management 42:540-549.

Danell, K. 1978b. Population dynamics of the muskrat in a shallow Swedish lake. Journal of Animal Ecology 47:697-709.

Danell, K. 1985. Population fluctuations of the muskrat in coastal northern Sweden. Acta Theriologica 30:219-227.

Davis, C.B., and A.G. van der Valk. 1978. Litter decomposition in prairie glacial marshes. In *Freshwater Wetlands: Ecological Processes and Management Potential*. (Eds.) R.E. Good, D.F. Whigham, and R.L. Simpson, pp.99-113. New York: Academic Press.

Dorney, R.S., and A.J. Rusch. 1953. Muskrat growth and litter production. Wisconsin Conservation Department Technical Bulletin 8.

Doude Van Troostwijk, W.J. 1976. Age determination in muskrats, *Ondatra zibethicus* (L.), in the Netherlands. Lutra 18:33-43.

Earhart, C.M. 1969. The influence of soil texture on the structure, durability, and occupancy of muskrat burrows in farm ponds. California Fish and Game Journal 55:179-196.

Erb, J.D. 1993. Age-specific reproduction of riverine muskrats. M.S. thesis, University of Missouri, Columbia, Missouri.

Errington, P.L. 1937. Drowning as a cause of mortality in muskrats. Journal of Mammalogy 18:497-500.

Errington, P.L. 1938. Reaction of muskrat populations to drought. Ecology 20:168-186.

Errington, P.L. 1939. Observations on young muskrats in Iowa. Journal of Mammalogy 20:465-478.

Errington, P.L. 1940. Natural restocking of muskrat-vacant habitats. Journal of Wildlife Management 4:173-185.

Errington, P.L. 1943. An analysis of mink predation upon muskrats in north-central United States. Iowa State College of Agriculture Experiment Station Research Bulletin 320:798-924.

Errington, P.L. 1946. Predation and vertebrate populations. Quarterly Review of Biology 21:144-177,221-245.

Errington, P.L. 1954a. On the hazards of overemphasizing numerical fluctuations in studies of "cyclic" phenomena in muskrat populations. Journal of Wildlife Management 18:66-90.

Errington, P.L. 1954b. The special responsiveness of minks to epizootics in muskrat populations. Ecological Monographs 24:377-393.

Errington, P.L. 1963. *Muskrat Populations*. Ames: Iowa State University Press.

Errington, P.L., and C.S. Errington. 1937. Experimental tagging of young muskrats for purposes of study. Journal of Wildlife Management 1:49-61.

Errington, P.L., R.J. Siglin, and R.C. Clark. 1963. The decline of a muskrat population. Journal of Wildlife Management 27:1-8.

Finerty, J.P. 1980. *The Population Ecology of Cycles in Small Mammals*. New Haven: Yale University Press.

Friend, M., G.E. Cummings, and J.S. Morse. 1964. Effect of changes in winter water levels on muskrat weights and harvest at the Montezuma National Wildlife Refuge. New York Fish Game Journal 11:125-131.

Fritzell, E.K. 1989. Mammals in prairie wetlands. In *Northern Prairie Wetlands*. (Ed.) A.G. van der Valk, pp.268-301. Ames: Iowa State University Press.

Grodzinski, W., and B.A. Wunder. 1975. Ecological energetics of small mammals. In *Small Mammals: Their Productivity and Population Dynamics*. (Eds.) F.B. Golley, K. Petrusewicz, and L. Ryszkowski, pp.173-204. Cambridge, UK: Cambridge University Press.

Jelinski, D.E. 1989. Seasonal differences in habitat use and fat reserves in an arctic muskrat population. Canadian Journal of Zoology 67:305-313.

Kangas, P.C., and G.L. Hannan. 1985. Vegetation of muskrat mounds in a Michigan marsh. American Midland Naturalist 113:393-396.

Keith, L.B. 1983. Role of food in hare population cycles. Oikos 40:385-395.

Krebs, C.J. 1996. Population cycles revisited. Journal of Mammology 77:8-24.

Krebs, C.J., S. Boutin, R. Boonstra, A.R.E. Sinclair, J.N.M. Smith, M.R.T. Dale, K. Martin, and R. Turkington. 1995. Impact of food and predation on the snowshoe hare cycle. Science 269:1112-1115.

Kroll, R.W., and R.L. Meeks. 1985. Muskrat population recovery following habitat re-establishment near southwestern Lake Erie. Wildlife Society Bulletin 13:483-486.

LeBoulenge, E., and P.Y. LeBoulenge-Nguyen. 1981. Ecological study of a muskrat population. Acta Theriologica 26:47-82.

Lebreton, J.D., K.P. Burnham, J. Clobert, and D.R. Anderson. 1992. Modeling survival and testing biological hypotheses using marked animals: a unified approach with case studies. Ecological Monographs 62:67-118.

Lieffers, V.J., and J.M. Shay. 1981. The effects of water level on growth and reproduction of *Scirpus maritimus* var. *paludosus* on the Canadian prairies. Canadian Journal of Botany 59:118-121.

Lynch, J.J., T.O'Neil, and D.W. Lay. 1947. Management significance of damage by geese and muskrats to Gulf Coast marshes. Journal of Wildlife Management 11:50-76.

MacArthur, R.A. 1978. Winter movements and home ranges of the muskrat. Canadian Field-Naturalist 92:345-349.

MacArthur, R.A. 1986. Metabolic and behavioral responses of muskrats *(Ondatra zibethicus)* to elevated CO_2 in a simulated winter microhabitat. Canadian Journal of Zoology 64:738-743.

MacArthur, R.A.. and M. Aleksiuk. 1979. Seasonal microenvironments of the muskrat (*Ondatra zibethicus*) in a northern marsh. Journal of Mammalogy 60:146-154.

Marinelli, L., and F. Messier. 1993. Space use and the social system of muskrats. Canadian Journal of Zoology 71:869-875.

Marinelli, L., and F. Messier. 1995. Parental-care strategies among muskrats in a female-biased population. Canadian Journal of Zoology 73:1503-1510.

Marinelli, L., F. Messier, and Y. Plante. 1992. Use of DNA fingerprinting to determine parentage in muskrats (*Ondatra zibethicus*). Journal of Heredity 83:356-360.

Marinelli, L., F. Messier, and Y. Plante. 1997. Consequences of following a mixed reproductive strategy in muskrats: the importance of parental care, location, and litter order. Journal of Mammalogy 78:163-172.

Mathiak, H.A. 1966. Muskrat population studies at Horicon marsh. Wisconsin Conservation Department Technical Bulletin 36.

McCabe, T.R. 1982. Muskrat population levels and vegetation utilization: a basis for an index. Ph.D. dissertation, Utah State University, Logan, Utah.

McKwan, E.H., N. Aitchison, and P.E. Whitehead. 1974. Energy metabolism of oiled muskrats. Canadian Journal of Zoology 52:1057-1062.

Messier, F., and J.A. Virgl. 1992. Differential use of bank burrows and lodges by muskrats, *Ondatra zibethicus*, in a northern marsh environment. Canadian Journal of Zoology 70:1180-1184.

Messier, F., J.A. Virgl, and L. Marinelli. 1990. Density-dependent habitat selection in muskrats: a test of the ideal free distribution model. Oecologia (Berl.) 84:380-385.

Mitsch, W.J., and J.G. Gosselink. 1986. *Wetlands.* New York: Van Nostrand Reinhold.

Morris, D.W. 1987. Tests of density-dependent habitat selection in a patchy environment. Ecological Monographs 57:269-281.

Moses, R.A., and S. Boutin. 1986. Molar fluting and pelt primeness techniques for distinguishing age classes of muskrats: a reevaluation. Wildlife Society Bulletin 14:403-406.

Murkin, H.R. 1989. The basis for food chains in prairie wetlands. In *Northern Prairie Wetlands.* (Ed.) A.G. van der Valk, pp.316-338. Ames: Iowa State University Press.

Nadeau, S., R. Decarie, D. Lambert, and M. St-Georges. 1995. Nonlinear modeling on muskrat use of habitat. Journal of Wildlife Management 59:110-117.

Nicholson, W.R., and D.E. Davis. 1957. The duration of life of muskrat houses. Ecology 38:161-163.

Odum, E.P. 1971. *Fundamentals of Ecology, 3rd edition.* Philadelphia: W.B. Saunders.

Olsen, P.F. 1959. Muskrat breeding biology at Delta, Manitoba. Journal of Wildlife Management 23:40-53.

O'Neil, T. 1949. The Muskrat in the Louisiana Coastal Marshes. New Orleans: Louisiana Department of Wild Life and Fisheries.

Otis, D.L., K.P. Burnham, G.C. White, and D.R. Anderson. 1978. Statistical inference for capture data from closed populations. Wildlife Monographs 62.

Parker, G.R., and J.W. Maxwell. 1980. Characteristics of a population of muskrats (*Ondatra zibethicus*) in New Brunswick. Canadian Field-Naturalist 94:1-8.

Parker, G.R., and J.W. Maxwell. 1984. An evaluation of spring and autumn trapping seasons for muskrats, *Ondatra zibethicus*, in Eastern Canada. Canadian Field-Naturalist 98: 293-304.

Payne, N.F. 1982. Assessing productivity of furbearers. In *Midwest Furbearer Management.* (Ed.) G.C. Sanderson, pp.39-50. Wichita: Kansas Chapter of the Wildlife Society.

Pease, J.L., R.H. Vowles, and L.B. Keith. 1979. Interaction of snowshoe hares and woody vegetation. Journal of Wildlife Management 43:43-60.

Pelikán, J., J. Svoboda, and J. Kvet. 1970. On some relations between the production of *Typha latifolia* and a muskrat population. Zoologika Listy 19:303-320.

Perry, H.R., Jr. 1982. Muskrats (*Ondatra zibethicus* and *Neofiber alleni*). In *Wild Mammals of North America.* (Eds.) J.A. Chapman and G.A. Feldhammer, pp.282-325. Baltimore: Johns Hopkins University Press.

Pollock, K.H., J.D. Nichols, C. Brownie, and J.E. Hines. 1990. Statistical inference for capture-recapture experiments. Wildlife Monographs 107.

Proulx, G., and B.M.L. Buckland. 1986. Productivity and mortality rates of southern Ontario pond- and stream-dwelling muskrat, *Ondatra zibethicus*, populations. Canadian Field-Naturalist 100:378-380.

Proulx, G., and F.F. Gilbert. 1983. The ecology of the muskrat, *Ondatra zibethicus*, at Luther Marsh, Ontario. Canadian Field-Naturalist 97:377-390.

Proulx, G., and F.F. Gilbert. 1984. Estimating muskrat population trends by house counts. Journal of Wildlife Management 48:917-922.

Proulx, G., J.A. McDonnell, and F.F. Gilbert. 1987. The effect of water level fluctuations on muskrat, *Ondatra zibethicus*, predation by mink, *Mustela vison*. Canadian Field-Naturalist 101:89-92.

Pulliam, H.R., and B.J. Danielson. 1991. Sources, sinks, and habitat selection: a landscape perspective on population dynamics. American Naturalist 137:S50-S66.

Sale, P.J.M., and R.G. Wetzel. 1983. Growth and metabolism of *Typha* species in relation to cutting treatments. Aquatic Botany 15:321-334.

Sather, J.H. 1958. Biology of the Great Plains muskrat in Nebraska. Wildlife Monographs 2.

Seabloom, R.W., and J.R. Beer. 1964. Observations of a muskrat population decline in North

Dakota. Proceedings of the North Dakota Academy of Science 17:66-70.

Simpson, M.R., and S. Boutin. 1993. Muskrat life history: a comparison of a northern and southern population. Ecography 16:5-10.

Sipple, W. 1979. A review of the biology, ecology, and management of *Scirpus olneyi*. Vol. II: a synthesis of selected references. Maryland Department of Natural Resources Wetlands Publication 4.

Skalski, J.R., and D.S. Robson. 1992. Techniques for Wildlife Investigations. San Diego: Academic Press.

Smith, H.R., R.J. Sloan, and G.S. Walton. 1981. Some management implications between harvest rate and population resiliency of the muskrat (*Ondatra zibethicus*). In *Proceedings of the Worldwide Furbearer Conference*. (Eds.) J.A. Chapman and D. Pursley, pp.425-442. Annapolis: Maryland Wildlife Administration.

Smith, L.M., and J.A. Kadlec. 1985. Fire and herbivory in a Great Salt Lake marsh. Ecology 66:259-265.

Snead, I.E. 1950. A family type live trap, handling cage, and associated techniques for muskrats. Journal of Wildlife Management 14:67-79.

Sprugel, G., Jr. 1951. Spring dispersal and settling activities of central Iowa muskrats. Iowa State College Journal of Science 26:71-84.

Squires, L., and A.G. van der Valk. 1992. Water-depth tolerances of emergent macrophytes at the Delta Marsh, Manitoba. Canadian Journal of Botany 70:1860-1867.

Stewart, R.W., and J.R. Bider. 1974. Reproduction and survival of ditch-dwelling muskrats in southern Quebec. Canadian Field-Naturalist 88:429-436.

Takos, M.J. 1943. Trapping and banding muskrats. Journal of Wildlife Management 7:400-407.

Taylor, R.J. 1984. *Predation.* New York: Chapman & Hall.

van der Valk, A.G. 1986. The impact of litter and annual plants on recruitment from the seed bank of a lacustrine wetland. Aquatic Botany 24:13-26.

van der Valk, A.G. 1994. Effects of prolonged flooding on the distribution and biomass of emergent species along a freshwater wetland coenocline. Vegetatio 110:185-186.

van der Valk, A.G., L. Squires, and C.H. Welling. 1994. Assessing the impacts of an increase in water level on wetland vegetation. Ecological Applications 4:525-534.

Virgl, J.A., and F. Messier. 1992a. Seasonal variation in body composition and morphology of adult muskrats in central Saskatchewan, Canada. Journal of Zoology (London) 228:461-477.

Virgl, J.A., and F. Messier. 1992b. The ontogeny of body composition and gut morphology in free-ranging muskrats. Canadian Journal of Zoology 70:1381-1388.

Virgl, J.A., and F. Messier. 1993. Evaluation of body size and body condition indices in muskrats. Journal of Wildlife Management 57:854-860.

Virgl, J.A., and F. Messier. 1995. Postnatal growth and development in semi-captive muskrats, *Ondatra zibethicus*. Growth, Development and Aging 59:169-179.

Virgl, J.A., and F. Messier. 1996. Population structure, distribution, and demography of muskrats during the ice-free period under contrasting water fluctuations. Ecoscience 3:54-62

Virgl, J.A., and F. Messier. 1997. Habitat suitability in muskrats: a test of the food limitation hypothesis. Journal of Zoology (London) 243:237-253.

Welch, C.E. 1980. Relationship between the availability, abundance, and nutrient quality of *Typha latifolia* and *Scirpus acutus* to summer foraging and use of space by muskrats (*Ondatra zibethica*) in south-central Alberta. M.S. thesis, University of Alberta, Edmonton, Alberta.

Weller, M.W. 1975. Studies of cattail in relation to management for marsh wildlife. Iowa State Journal of Science 49:383-412.

Weller, M.W. 1981. *Freshwater Marshes.* Minneapolis: University of Minnesota Press.

Weller, M.W., and C.E. Spatcher. 1965. Role of habitat in the distribution and abundance of marsh birds. Iowa Agriculture and Home Economics Experiment Station Special Report 43.

Willner, G.R., J.A. Chapman, and J.R. Goldsberry. 1975. A Study and Review of Muskrat Food Habits with Special Reference to Maryland. Annapolis: Maryland Wildlife Administration.

Willner, G.R., G.A. Feldhamer, E.E. Zucker, and J.A. Chapman. 1980. *Ondatra zibethicus*. Mammalian Species 141:1-8.

Wilson, K.A. 1953. Raccoon predation on muskrats near Currituck, North Carolina. Journal of Wildlife Management 17:113-119.

Wilson, K.A. 1968. Fur production on southeastern coastal marshes. In *Proceedings of the Marsh and Estuary Management Symposium*. (Ed.) J.D. Newsom, pp.149-162. Baton Rouge: Louisiana State University Division of Continuing Education.

Wobeser, G. 1981. Tyzzer's disease. In *Infectious Diseases of Wild Mammals, 2nd edition*. (Eds.) J.W. Davis, L.H. Karstad, and D.O. Trainer, pp.347-351. Ames: Iowa State University Press.

Wobeser, G., D.B. Hunter, and P.Y. Daoust. 1978. Tyzzer's disease in muskrats: occurrence in free-living animals. Journal of Wildlife Diseases 14:325-328.

Wolff, J.O. 1980. The role of habitat patchiness in the population dynamics of snowshoe hares. Ecological Monographs 50:111-130.

Part IV

Summary

12

Marsh Ecology Research Program: Management Implications for Prairie Wetlands

Henry R. Murkin
Arnold G. van der Valk
William R. Clark
L. Gordon Goldsborough
Dale A. Wrubleski
John A. Kadlec

ABSTRACT

The Marsh Ecology Research Program (MERP) was developed and designed to provide new experimental data to improve the management of prairie wetlands. This chapter reviews the implications of the MERP results for the management of these dynamic systems. Although MERP was not a hydrological study, a great deal of effort was spent on evaluating ways to determine important components of the water budgets, particularly evapotranspiration and seepage. The MERP nutrient budgets indicated that although nutrients may be lost from wetlands during drawdown management, these losses are easily replaced from the large nutrient stores in the sediments. By managing the diversity and productivity of macrophytes within a wetland, the manager is, in effect, managing the nutrients in the system. MERP provided new information on a variety of aspects of macrophyte management. There was no difference in the vegetation response to the 1- and 2-year drawdowns, although timing of drawdown within a year did affect species composition of the resulting vegetation community. The design of seed-bank studies is important to provide useful information for developing drawdown studies. With reliable seed-bank information, managers are better equipped to manage problem seed banks and manipulate recruitment from seed banks. In addition, an understanding of the life-history requirements of wetland vegetation is necessary for the development of reliable management plans for wetland vegetation. The abundance and diversity of vegetation also determines the resources available for algae and invertebrates within the system. Although muskrat management is often considered an important tool in vegetation management during MERP the muskrats did not cause the decline in vegetation observed in the experimental cells. The hemi-marsh does provide optimal conditions for a wide range of species within the wetland, however like all stages of the wet–dry cycle, it is transitory. By maintaining the entire wet–dry cycle, managers are providing for the greatest productivity and diversity of species over the long term.

INTRODUCTION

Advances in prairie wetland management have largely been accomplished by trial-and-error and not based on experimental studies. The need for long-term experimental research to advance the theory and practice of prairie wetland management for wildlife has been recognized for many years (Weller 1978). Ducks Unlimited Canada and the Delta Waterfowl and Wetlands Research Station implemented the Marsh Ecology Research Program (MERP) as a direct response to this need for experimental study that would improve management of prairie wetlands.

Although prairie wetlands have a wide range of functions and values (Murkin 1998), management of these habitats has been by public and private agencies primarily interested in waterfowl use and production. Most management still focuses on developing and maintaining the hemi-marsh mixture of vegetative cover and open water as originally outlined by Weller and Spatcher (1965) and Weller and Fredrickson (1974). Their ideas drew attention to the influence of natural and artificial water-level fluctuations on the productivity of these habitats for wildlife. Following the early work by Weller and his colleagues, the classic paper by van der Valk and Davis (1978) related the changes in vegetation in prairie wetlands to the various stages of the wet–dry cycle characteristic of the prairie environment. They emphasized the importance of these fluctuations between wet and dry conditions for maintaining the long-term productivity of these systems. More recently, Kadlec and Smith (1992) reviewed habitat management for breeding waterfowl and reemphasized the need for experimental data on habitat management practices, particularly water-level control.

MERP was a water-level manipulation experiment designed to simulate the effects of the wet–dry cycle in prairie wetlands. The objective of this chapter is to review the implications of the results of MERP for the management of wildlife and the various other system components contributing to the overall productivity of these wetland systems. The chapter is organized into the implications for each of the major topics addressed by MERP (hydrology, nutrient budgets, vegetation, algae, invertebrates, birds, and muskrats), as well as the general implications of the MERP results for prairie wetland management.

Wetland management might be characterized by four primary activities: preservation, creation, restoration, and enhancement of existing wetland systems. Preservation, as the name implies, simply consists of protecting wetland areas from being lost from the landscape or being disturbed, and may or may not include further management activities. Creation involves the development of wetlands in areas where wetlands did not exist previously. Restoration refers to efforts to recreate wetlands where they had formerly occurred but were lost through drainage or other disturbances. Enhancement normally entails the use of management intervention to improve the productivity of existing wetlands and to restore functions that have been degraded over time.

A variety of techniques has been used to create, restore, and enhance wetlands (Kadlec and Smith 1992; Galatowitsch and van der Valk 1994). These include both artificial and natural methods that have resulted in varying degrees of success (Weller 1978). Artificial methods include mechanical manipulation of vegetation (e.g., mowing) and substrates (e.g., level ditching, blasting), whereas natural methods include burning, water-level manipulation, and in some cases, management of herbivores (e.g. muskrats, cattle grazing). Artificial methods are usually labor-intensive, expensive, and often provide only temporary results. Of the natural methods, water-level manipulation is by far the most commonly used technique to restore or enhance productivity within wetland basins. With wetland creation and restoration, the management usually involves plugging drainage outlets and/or installation of dikes (either fixed crest or with water-level controls) to hold water within the basin (Galatowitsch and van der Valk 1994). Enhancement also frequently involves diking and the installation of control structures to allow maintenance and manipulation of water levels. The MERP studies were intended to provide information on water-level changes in prairie wetlands and therefore have direct implications for both natural and artificial water-level management in these systems.

HYDROLOGY

Winter (1989) and LaBaugh et al. (1998) provided excellent overviews of the hydrology of northern prairie wetlands. The hydrology of the MERP experimental cells was manipulated to simulate water levels during a wet–dry cycle in prairie wetlands. Dikes were constructed and pumps installed to allow water levels to be raised and lowered as required by the experimental design. MERP was never intended to be a hydrological study, nevertheless, the hydrological results from MERP are relevant to management because much time and effort were spent on trying to improve and evaluate water budgets (Kadlec 1983, 1993). Wetland managers who control water levels either by pumping or by gravity flow often need to develop simple hydrologic budgets to determine the quantity of water needed for a wetland to reach full supply level or to offset losses due to seepage and evapotranspiration over time.

The major inputs of water into the MERP experimental cells were precipitation, pumping, and seepage. The major outputs were evapotranspiration, pumping, and seepage. Changes in storage were measured using a combination of staff gauges and water-level recorders and detailed tables of volume versus water level derived from contour maps (10-cm intervals) of each cell (see Appendix 1). Both pumping (all the pumps were metered) and precipitation (there were two weather stations at the site) were measured directly. Because they are difficult to measure, evapotranspiration and seepage were the two most uncertain components of the MERP hydrologic budgets (Kadlec 1993). Winter (1989) observed that measurement or estimation of evapotranspiration is one of the main problems in developing hydrologic budgets for wetlands as proved to be the case for the MERP experimental cells (Kadlec 1993). Although MERP did not develop any new approaches to measuring or estimating any of the parameters in determining water budgets, much of our effort with respect to hydrology was spent evaluating ways to estimate seepage through the dikes and to measure evapotranspiration within the diked experimental cells.

Three methods were used to estimate evapotranspiration from the MERP cells: lake evaporation calculations with data provided by Environment Canada from their Delta Marsh monitoring site at the University of Manitoba Field Station at the west end of the marsh, adjusted Class A pan evaporation data from the MERP site, and a modified Penman equation using data from the MERP weather stations (Kadlec 1993). There was a high correlation among the estimates from all three methods and therefore no compelling evidence to suggest that one method gave more accurate estimates than the others. For the MERP study, the most direct estimate and the one most closely linked to site conditions was evapotranspiration = 0.7 x pan evaporation at the site. This was similar to the comparison of evapotranspiration estimation techniques by Kadlec and Knight (1996) who also found a close relationship between the rule-of-thumb calculation of 0.7 to 0.8 times pan evaporation and other more detailed methods. Note, however, that this technique provides estimates for the entire growing season and that the area of the wetland in question affects the relationship between pan evaporation and true evapotranspiration. Evapotranspiration in very small wetlands can be up to 2 × pan evaporation (Bavor et al. 1988). However, in most management situations, calculation from local evaporation pan data should apply well. If more refined budgets are necessary, modifications of the pan calculation method proposed by Christiansen (1968) have been found to work well in wetlands (Kadlec et al. 1987) or the more detailed Penman estimator can be used (see Kadlec and Knight [1996] for description of these procedures).

The need to determine seepage varies among wetland management projects. In the MERP study, the dikes were built from available materials from borrow pits within the cells (mostly sand and silt) with no attempt to incorporate an impervious core ("key") as is the case with most wetland management dike construction (Kadlec and Knight 1996). Thus, the MERP dikes were very "leaky" and seepage was a dominant factor in the water budget (Kadlec 1983, 1993). In management scenarios where the entire project is ringed by impervious dike construction (e.g., clay "key" or sheet-metal barrier incorporated into the dike itself), seepage estimates may not be a major concern. Vertical seepage was not a major factor in the MERP annual water budgets (Kadlec 1983, 1986). For prairie wetlands in general, vertical flow into regional groundwater is a very small fraction of the annual water budget (van der Kamp and Hayashi 1998). In many management situations, however, natural topographical highs (e.g., ridge, hillsides) are incorporated into the wetland design to hold water, and seepage into these areas may need to be estimated to obtain water budgets for the planning process.

As in most hydrological studies (Winter 1981), seepage was estimated by difference in the MERP water budgets (i.e., it is the one unknown variable in the water budget equation [Kadlec 1983]). Although this procedure results in incorporating measurement error into the seepage term, subsequent checks on seepage rates among the MERP replicated treatments found the overall error to be small (Kadlec 1983). The most accurate measures of seepage during MERP were made by simply recording changes in water levels within the cells during periods when evapotranspiration was minimal (late fall and early spring). With no evaporation losses, volume of water gained or loss through seepage could then be determined from these changes in water levels.

In summary, MERP results suggest that both evapotranspiration and seepage can be estimated simply and cheaply for diked wetland projects where rough water budgets are sufficient for planning needs. The former can be estimated from Class A pan evaporation data in the immediate area of the project (evapotranspiration = 0.7 × pan evaporation), and the latter can be estimated by measuring water losses (water level changes) in early spring or late fall when evapotranspiration losses are minimal. Horizontal seepage may be a significant factor in the hydrology of managed wetlands especially in situations like MERP, where there was no opportunity to "key" the dikes.

NUTRIENTS AND NUTRIENT BUDGETS

An important consideration regarding managing the overall productivity of prairie wetlands is that, unlike wetland systems in other areas (Murkin et al. 1991), prairie wetlands are generally not nutrient limited to any great degree (see Chapter 8). Although nutrient limitation may exist for some nutrients at some times of the year (Hooper-Reid and Robinson 1978; Robarts and Waiser 1998) and nutrient availability may be influenced by flooding depth in some parts of the wetland (Neill 1990a, b), it is unlikely that prairie wetlands experience significant nutrient limitation for extended periods given the normally eutrophic conditions and the general fertility of the prairie landscape. It is more probable that prairie wetlands may be affected by an overabundance of nutrients due to anthropogenic inputs from the surrounding landscape (Crumpton and Goldsborough 1998).

Inputs–Outputs

Considering input and output of nutrients, there is some concern that pumping water to and from wetland impoundments may affect the overall nutrient budgets. During the MERP studies the inputs–outputs through pumping were minor compared with the sizes of the pools and fluxes in the system (particularly those associated with macrophytes), even during the deep-flooding years when considerable volumes of water were added to the cells (see Chapter 4). Although the water added to the experimental cells was marsh water pumped directly from the Delta Marsh, inputs of some nutrients, especially inorganic forms, were highest in precipitation even during years of extensive pumping (see Tables 4.2 and 5.2 in Chapters 4 and 5, respectively). It would appear, therefore that managers do not need to be concerned about the nutrients added to the impoundment during initial flooding or during maintenance of water levels, when the water source has nutrient levels representative of the natural wetlands in the area. If the water source has elevated nutrient levels due to inputs from agricultural, industrial, or urban sources, there should be some consideration of the nutrient loading and the ability of the wetland to handle that loading (Kadlec and Knight 1996).

An oft-repeated concern with drawdown management in wetlands is the loss of nutrients from the system during drawdown (see Chapter 4). Nutrients associated with the surface water are lost during drawdown and at certain times could represent a significant mass of nutrients. During the MERP deep-flooding experiments, more nitrogen (N) and phosphorus (P) were pumped out of the cells than was initially pumped in at the beginning of the experiment (representing a loss of 10% of the N and 25% of the P in the nonsediment pools). Although this was a significant portion of the nutrients not contained in the sediments, it should not be of concern for a couple of reasons. Firstly, this was an unusual flooding regime where water levels were raised 1 m above normal to kill the existing vegetation and then drawn down to expose the substrate. It is very unlikely this type of treatment would be repeated on an annual basis in a normal wetland management program. Secondly, the actual amount of loss is a small fraction of the nutrients in the sediments and would probably be replaced quickly by release from the sediment pool (see Chapter 6).

During the MERP drawdown and subsequent reflooding experiment (see Chapter 5), the water-level manipulation scenario was most similar to that of a regular wetland management scheme (i.e., drawdown followed by refilling to normal full supply level). During this part of the study, the experimental cells actually retained nutrients so that the losses due to subsequent drawdown were less than the amount of nutrients added to the cells during the initial filling and regular additions to maintain water levels.

In summary, using water from adjacent water bodies with "normal" background nutrient levels to raise water levels in prairie wetlands should not result in excessive addition of nutrients to the system over time. In addition, most drawdown management plans (periodic drawdown after several years of flooding) should not result in significant long-term loss of nutrients from the system.

Intrasystem Cycling

Prairie wetland sediments contain large pools of N, P, and carbon (C). Even during periods when all nonsediment pools are at their maximum sizes, they represent <10 to 25% of the nutrients within the sediments (see Chapter 6). Because of this large residual pool within the sediments, the issue for management related to overall productivity and nutrient budgets in prairie wetlands is the location of the nutrients (pools) within the system and the movements among these pools (fluxes) rather than the total amount of any particular nutrient in the wetland. The productivity of the overall system increases as the proportion of nutrients outside the sediments increases. For example, as a wetland moves to the lake marsh stage and the low productivity associated with this stage of the wet–dry cycle (see Chapter 6; van der Valk and Davis [1978]), a greater and greater proportion of the nutrients is found in the sediments as the macrophytes, algae, and other aboveground pools are reduced or eliminated. Only by reestablishing the nonsediment pools and fluxes can productivity be restored to the system.

Although the proportion of available nutrients outside the sediment and pore-water pools is important, the diversity of pools and fluxes also plays a major role in overall productivity, especially for wetland wildlife and the resources on which they depend. For example, the largest nutrient pools recorded during the MERP studies, other than the sediments, were the above- and belowground macrophytes in the undiked reference cells (see Chapters 4 and 5). During the MERP reflooding experiments (1985–1989), the macrophyte pools in the diked experimental cells never attained the levels found in the undiked reference cells in the main Delta Marsh (see Chapter 5). However, as mentioned in many of the previous chapters, overall productivity as well as avian use was higher in the diked experimental cells than in the undiked reference cells during the reflooding experiment. Despite the size of the macrophyte pools, the monotypic nature of the macrophyte stands in the reference cells (i.e., few other pools and fluxes) resulted in the reference cells generally being considered unproductive throughout the MERP studies.

Although the sediments and associated pore water represented the largest pool during all stages of the wet–dry cycle, the variety and size of other nutrient pools and fluxes changed during the various stages of the wet–dry cycle. The lake marsh stage is generally considered to be the period of lowest productivity during the cycle (van der Valk and Davis 1978) and it is during this period that the number and size of pools and fluxes outside the sediments is greatly reduced as nutrients become "tied up" in the sediments and pore-water pool. In contrast, during the transition between the regenerating and degenerating stages, there is a diversity in available pools and fluxes. At the same time, overall productivity is at or near the maximum found during the wet–dry cycle (see discussion of hemi-marsh phase in Chapter 6). Therefore, the goal of managers interested in overall wetland productivity should be to maintain a diversity of nutrient pools and fluxes within the system.

During the MERP simulated wet–dry cycle, the macrophyte pools and associated fluxes dominated the nutrient budgets. Therefore, by managing the macrophytes within the basin, the manager is by proxy managing the nutrients within the wetland. During the dry marsh stage, reestablishing the macrophytes on the drawdown surface is critical to returning productivity to the system. Plant uptake serves as the dominant flux to reestablish the aboveground nutrient pools that, in turn, support further primary (e.g., algae) and secondary production in the system. Although the aeration of the sediments allowing more efficient aerobic microbial decomposition is the oft-cited reason for the increased productivity following drawdown conditions, reestablishment of the plant uptake flux during the drawdown stage is critical to making nutrients available to other pools again.

During the MERP dry marsh stage, uptake of N and P exceeded the mass of dissolved inorganic nutrients in the pore water, indicating rapid inputs to the pore water, probably through increased mineralization of organic compounds in the sediments and pore water (see Chapter 5). Although this result supports the suggestion of increased aerobic decomposition during the drawdown stage, the plant uptake was actually the main path to move these nutrients above ground. Without the germination of macrophytes and annual plants on the drawdown surface, the impact of drawdown on productivity would be substantially reduced, and probably limited to a short-term response by algae. With drawdown management, the manager is attempting to reestablish the plant uptake flux within the system.

The type of vegetation establishing on the drawdown surface has important implications for the nutrient budgets. For example, annual plants incorporate nutrients into aboveground biomass that is not translocated below ground at the end of the growing season. Thus, nutrients in the annual plant litter pool are moved into other pools through leaching and other decomposition processes at the end of the growing season. Therefore, annual plants cycle nutrients relatively quickly to other pools in the system, whereas perennial emergent species move a portion of the aboveground nutrients below ground through translocation to support early growth the following spring. Thus, the proportion of nutrients made available to other aboveground pools may be somewhat smaller in an emergent-dominated drawdown response. Even within the perennial macrophytes there is a difference between short-lived and long-lived species (see Chapter 5). Short-lived species die out sooner moving their associated nutrients to other pools at a faster rate than long-lived species.

Once the wetland is reflooded at the end of the dry stage, leaching and decomposition of the vegetation move nutrients to the other potential pools in these systems. A large pool of nutrients is released very quickly into the surface water through leaching. During the MERP studies, up to 30% of the nutrients in the litter was released by leaching within 2 days of inundation. However, water sampling did not reveal any large increases in surface-water nutrient pools 1 month into the growing season. So, although there was undoubtedly a large release of nutrients into the surface water, it was rapidly taken up by algae and "fueled" the high production of algae observed early in flooding (see Chapter 8).

Following leaching, subsequent movement of nutrients from the litter is through decomposition. Decomposition is often considered a "major control point" in nutrient cycling because it controls the movement of nutrients from the large plant-litter pool to other pools in the system (see Chapter 6). Although there is little practical management that can be done to control this flux of nutrients, managers should be aware of a couple of important points. Firstly, the decomposition process is relatively slow in emergent species. For example, litter of *Phragmites* and *Typha* retained >50% of the original N and P after 2 to 3 years of inundation (Murkin et al. 1989). Secondly, there is a difference in decomposition rates among the various species of perennial emergents and annual plants with the larger, more "woody" species decomposing at a much slower rate than some of the smaller species. The sequence of species dying following reflooding has important implications for maintaining the diversity of pools and fluxes within the wetland. The annual species are killed immediately upon flooding, and through leaching, add nutrients to the other pools. The short-lived emergent species die within 1 or 2 years of flooding and add another pulse of nutrients at that time. After 3 years or so, the long-lived macrophytes dominate nutrient cycling within the system. This sequence of species allows for an extended period of high productivity following flooding.

With continued flooding, the wetland proceeds through the degenerating stage and into the lake marsh stage where once again the nutrients within the system become "locked" in the sediment and pore-water pools. Depending on the management goals, the manager may want to initiate a drawdown to reestablish macrophytes and the movement of nutrients into other pools in the system.

MACROPHYTES

By the statement "the use of natural forces and processes in management is most likely to stimulate natural events, conditions, and results", Weller (1981) provided the philosophical framework for most wetland management. For prairie wetlands, natural vegetation management primarily has involved manipulating water levels to avoid the two least desirable stages of the wet–dry cycle, the lake stage during which the emergent vegetation is reduced to a fringe around the periphery and an arrested regenerating marsh stage during which dense emergent vegetation covers the entire wetland.

van der Valk and Davis (1978) demonstrated that seed banks played a central role in understanding the vegetation changes during wet–dry cycles in wetlands. Seeds of nearly all of the species that dominate each of the stages of a wet–dry cycle are present in the seed bank (van der Valk and Davis 1978, 1980). They proposed that the composition of the vegetation at any stage in a wet–dry cycle can be predicted from a knowledge of the conditions under which species can become established and of the conditions under which they can be extirpated (i.e., the life-history characteristics of each species). The MERP provided an experimental test of this life-history model and of its underlying assumption that the best way to understand and predict vegetation change is from a reductionist perspective (van der Valk 1982, 1985). One of the most important features of this model is the central role of seed banks in prairie wetlands.

There are three major aspects of wetland management and restoration to which the MERP vegetation studies have made significant contributions. These are the effects of water-level manipulations on vegetation composition and distribution (emergent macrophytes and water-level manipulations, optimum length and timing of drawdowns), factors controlling recruitment from seed banks (design of a seed-bank study, absent or problem seed banks, manipulating recruitment from seed banks), and life-history-based management.

Water Level Manipulations and Vegetation Composition and Distribution

Emergent Macrophytes and Water-Level Manipulations

As noted previously, natural wetland management largely involves water-level manipulations. The most common effect of a prolonged (i.e., >1 year) increase in water level is the elimination first of annual species and eventually of emergent perennial species (van der Valk and Davis 1978; van der Valk 1994; van der Valk et al. 1994). A prolonged increase in water level also can result in the migration of emergent species up slope (van der Valk and Davis 1976). Thus, there are two seemingly conflicting ideas about how emergent species respond to a permanent increase in water level: extirpation or migration. Because water-level manipulations are such a powerful tool for wetland managers, it is worthwhile to examine the consequences of water-level changes in more detail, especially for perennial emergent species.

The extirpation model predicts that a significant increase in water level eventually kills emergent species. According to this model, there is little or no compensating migration up the elevation ranges within which species are found. This result is what happened in the MERP cells during the deep-flooding years and during the reflooding years in the medium and high treatments (van der Valk 1994; van der Valk et al. 1994). However, it may take several years for the emergents to die.

With changes in water levels, the migration model predicts that emergents, over time, move into areas where water depths are comparable to those they previously occupied. There should be no net loss of emergent species; they simply move upslope approximately the same amount as the mean water level increased. According to this model, only basin morphometry (i.e., the area within a given elevation range at a given water level) determines the area covered by different emergent species (de Swart et al. 1994).

Although it is still widely held that raising and lowering water levels in wetlands simply results in a shift in vegetation zones up or down slope, unless there are physical barriers preventing such a shift, the MERP results indicate that this is not always true. Most emergent species did not migrate upslope in the medium- and high-water-level treatments during the reflooding years (de Swart et al. 1994). Instead, they either died out after 3 years because they could not survive permanent flooding (*Scolochloa festucacea* and *Scirpus lacustris*) or stayed where they were (*Phragmites australis*). Only *Typha glauca* became more widespread because of expansion into areas dominated formerly by other species eliminated by flooding. Most emergents were unable to move upslope because when growing in suboptimal water depths, they did not appear to have the additional energy needed for clonal growth upslope (Squires and van der Valk 1992). Furthermore, they were not able to establish themselves upslope from seed because their seeds cannot germinate under water. Therefore, permanently raising water levels normally results in declines in many emergent plant populations, at least in the short term. The magnitude of increase in water levels determines the extent to which emergent zones are affected.

There appear to be no such impediments for emergents to move downslope if water levels are dropped for an extended period of time (see Chapter 7). Thus, raising and lowering water levels in a wetland does not produce the same results. Raising water levels tends to eliminate or reduce emergent zones, whereas lowering water levels can result in the migration of emergent zones downslope. Managers should always keep in mind that raising and lowering water levels may have very different impacts on wetland vegetation.

In summary, the extirpation and migration models are probably both valid, but under different circumstances for different species. The migration model may be applicable to situations where there is a decrease in water levels, whereas the extirpation model seems appropriate in situations where there are increases in water levels as shown during the MERP study.

Length and Timing of Drawdowns

One of the treatment comparisons built into the overall MERP water-manipulation experiment was a 1-year versus 2-year drawdown. There were two arguments in favor of longer drawdowns: longer drawdown would provide a longer period to recruit emergent species, and plants recruited during a longer drawdown would be more likely to survive reflooding. The main disadvantage of longer drawdown is that the wetland cannot be used as habitat by wetland wildlife for a longer period of time. The MERP results clearly showed that there is no practical difference in vegetation response between 1- and 2-year drawdowns (Welling et al. 1988a, b). Most of the recruitment occurs during the first couple of months of drawdown, and survival of emergent seedlings established for 1 and 2 years did not differ when the cells were reflooded.

During MERP, Merendino et al. (1990) and Merendino and Smith (1991) conducted an experimental study to determine the best time during the growing season to begin a drawdown. Specifically, they examined the effect of May, June, July, and August drawdowns on emergent recruitment and survival after reflooding. Recruitment, on the basis of total shoot density, was highest for the May drawdown. Survival of plants was also highest when they were reflooded in the May drawdown treatment. May drawdowns also reduced the recruitment of potential problem species, including *Typha* spp. and *Lythrum salicaria*. In short, the longer the growing season during a drawdown, the larger the plants will be when they are reflooded and the better their chances of survival following flooding. In the northern prairie region, drawdowns should begin in May and one complete growing season is sufficient duration.

Factors Affecting Recruitment from Seed Banks

Design of a Seed-Bank Study

One of the most extensive seed-bank studies of any wetland was carried out during MERP (Pederson 1981, 1983). Unlike most previous seed-bank work, this study was done in a shelter in the field rather than in an environmentally controlled greenhouse. Several subsidiary MERP seed-bank studies also were conducted both in greenhouses and in the field. The following comments and recommendations for doing seed-bank studies are based on experiences gained from conducting these studies.

A seed bank consists of all viable seeds present in the soil at a given time. Small vegetative propagules (e.g., turions, of flowering plants) and spores of mosses, liverworts, and ferns are functionally equivalent to seeds and usually considered to be part of the seed bank. Normally only the seeds present in the upper few centimeters of soil germinate (Galinato and van der Valk 1986). Consequently, most seed-bank samples are only taken to a depth of ≈5 cm with corers or other suitable sampling devices. Sometimes deeper samples have to be taken in poorly consolidated sediments to collect sufficient volume of sample (van der Valk and Rosburg 1997). Large vegetative propagules (bulbs, tubers, rhizomes, etc.) are usually not considered to be part of the seed bank and are manually removed from the samples.

Estimating the composition of a seed bank can be done in one of two ways: physical separation of seeds (seed assays) and seedling emergence (seedling assays). Seed assays involve passing the seed-bank sample through a series of increasingly finer sieves to trap the seeds. Seedling assays involve placing the soil sample under moisture and temperature conditions suitable for seed germination and counting the number of seedlings that emerge. Seed assays are quicker than seedling assays, provided that enough labor is available, but have two major drawbacks: they normally require additional work to determine if the seeds are viable, and seed identification is difficult because there are no seed identification manuals for wetland plant species.

Seedling assays require less actual labor, but have a lag time usually of 6 months or more, between when a sample is collected and when the data are available. They also require a greenhouse or similar facility in which to keep the samples. Another drawback is that seedling assays may introduce bias to the results. Because seed-bank samples can be exposed to only a few environmental conditions (often only two, moist soil and shallowly flooded), the germination of seeds of some species is favored over others. This, however, can be an advantage if conditions can be set up similar to field conditions. This is most easily done by conducting the seedling assay in the field. At a minimum, seedling assays should be conducted by placing subsamples of each seed-bank sample under two conditions, moist soil and shallowly flooded (3–5 cm) to provide conditions for the germination of most species in the seed bank of prairie wetlands.

The seedling assay technique is recommended for seed-bank studies used to develop management plans because it is easier, simpler, and possibly cheaper than a seed assay. Comparative studies also have shown that there is no meaningful difference between the results of seed and seedling assays of the same seed bank (Poiani and Johnson 1988; van der Valk and Rosburg 1997). In fact, seedling assays often are better for estimating the number of species in a seed bank than seed assays, which is advantageous for management purposes. As noted previously, a seedling assay does require either greenhouse space or a suitable shelter in which to keep the samples and the ability to water the samples on a daily basis. Most importantly, however, seedling assays require time. Consequently, seedling assays normally need to be done the year prior to the scheduled start of a management plan.

If no greenhouse space is available, a suitable shelter can easily be built (Pederson 1981) to prevent samples from being washed away during rainstorms and to prevent contamination of the sample by wind-blown seed. The latter can be corrected for by placing trays filled with sterilized soil at random within the seed-bank shelter or greenhouse. Occasionally, it is feasible to float trays in the wetland being investigated. Drilling holes in the floating trays can eliminate the need for a watering system. Such field studies need to be done in well-protected areas that are not subjected to wave action during storms. In all cases, seed-bank shelters need to be set up in areas free of shade. Shading from plant canopies inhibits germination of many species.

Seed-bank samples should be collected as early as possible in the spring and placed in shallow trays that maximize the surface area of the sample. Ideally there should be three or more replicate trays for each sample placed at random among the treatments in question. If feasible, the entire tray should be filled with a seed-bank sample, or if not, a thin layer (1 to 3 cm) of seed-bank sample can be placed over a layer of sterilized potting soil or sand.

Seeds in the soil typically have a very clumped distribution, therefore, it is usually best to take a large number of small samples or to subdivide a site and take many small subsamples and combine them into one sample for each area. The latter is more practical because it reduces the number of samples to be processed. It is important, however, to ensure that all parts of a wetland are sampled, especially areas free of emergent vegetation (see next section). For wetland management projects, a precise estimate of seed densities is usually not required. A list of species present and their relative abundance is usually sufficient. Both seed and seedling assay techniques are described in more detail in Roberts (1981), Leck (1989), and van der Valk and Rosburg (1997).

Absent or Problem Seed Banks

One of the implicit assumptions of natural wetland management in the prairie region is that a suitable seed bank is present. This is true for both extant wetlands and for wetlands being restored. However, sometimes seed banks are absent (Pederson and van der Valk 1985; van der Valk et al. 1989; Weinhold and van der Valk 1989). Likewise, there are situations in which the composition of the seed bank can cause other problems (e.g., presence of noxious weeds) for managers. To avoid such situations, seed-bank studies should be done prior to the initiation of major projects.

The extensive seed-bank study done by Pederson (1981, 1983) showed that deep bays in the Delta Marsh that were free of emergent vegetation had a very sparse seed bank or no seed bank at all. Consequently, a drawdown to reestablish emergent vegetation in these bays would most likely fail. In other situations, an entire wetland might lack an effective seed bank because of excessive sedimentation (van der Valk and Pederson 1989). In such situations, conventional drawdowns do not work, however, it may still be possible to restore the vegetation with modified drawdown methods. One of the simplest is to use partial drawdowns that drop the water level in the wetland to just below the outermost belt of emergent vegetation. Once vegetation has become established in this exposed zone, water levels are dropped again and so on, until vegetation is reestablished throughout the basin or as required by the management plan. Depending on the size of the wetland, it could take several years to revegetate the entire basin. Managers should note, however, that it is not possible to establish vegetation below contour levels at which seedlings cannot survive after the wetland has been reflooded.

An alternative to partial drawdowns over a number of years is to flush seeds from areas around the periphery into areas without a seed bank during a full drawdown. To expedite this, it may be desirable to dig or scrape shallow trenches from the existing vegetation downslope (Pederson 1981); however, this approach is likely to be very expensive compared with conventional drawdowns and to date has not been tested in a rigorous manner (R. Pederson, personal communication). One specialized case where the managed dispersal of seeds into a wetland without a seed bank is likely to be effective is in situations where water and associated seeds can be taken from a nearby, extant marsh and pumped or gravity fed into the wetland without a seed bank. This should be done in the fall after seed set.

Another alternative is to use donor seed banks from an extant marsh to inoculate the wetland without a suitable seed bank. This approach has been used with some success in wetland restoration projects (van der Valk and Pederson 1989). Care has to be taken with the selection of the donor seed bank. It must contain seeds of the desired species and not seeds of unwanted species. It needs to be emphasized that the composition of the seed bank often bears little resemblance to the composition of the vegetation from where the seed-bank sample was collected. Consequently, the use of donor seed banks is only recommended when previous analyses have established the composition of the seed bank in the donor wetland.

Unfortunately seed banks also can contain undesirable species. Not very much can be done to eliminate this problem because wetland species often have very long-lived seeds. However, sometimes it is possible to manipulate recruitment from the seed bank to reduce the establishment of unwanted species (see the next section). For wetland restoration projects, it may be better to remove the contaminated seed bank or bury it under soil that is free of seeds of undesirable species.

Manipulating Recruitment from Seed Banks

Seed banks often contain the seeds of many species with very different germination requirements. Consequently, environmental conditions during a drawdown (soil temperature, soil moisture, salinity, etc.) can have a major effect on the composition of draw down and postdrawdown vegetation. This could be seen in the MERP cells in which different annual species dominated the vegetation in cells drawn down in 1983 and 1984. Similar results were obtained by Merendino et al. (1990) in their study of timing of drawdowns. Comparative experimental studies of annual and emergent seed germination (Galinato and van der Valk 1986; Naim 1987) as well as experimental studies of recruitment from seed banks (van der Valk and Pederson 1989) have confirmed that it is theoretically possible to manipulate the composition of vegetation recruited from seed banks.

There are two ways that wetland managers traditionally have manipulated environmental conditions to increase recruitment from the seed bank or to promote the establishment of desirable species. Soil-surface salinity in brackish and saline wetlands is often too high during drawdowns to allow seed in the seed bank to germinate. Consequently, salinity during a drawdown has to be reduced. This can be accomplished by having a partial drawdown to prevent the buildup of salts on the soil surface due to evaporation, or by flushing the soil surface with fresh water to reduce surface salinity. The timing of drawdowns also has been used to manipulate recruitment patterns in wetlands (Fredrickson and Taylor 1982). In fact, this is a key feature of most moist-soil management plans. Table 12.1 illustrates the effects of drawdowns begun at different times of the year on the composition of the vegetation. The vegetation in the May drawdown treatment was dominated by *Scirpus maritimus* (44% of all seedlings), whereas vegetation in the June drawdown treatment was dominated by *Typha* spp. (36%). Recruitment of other species (e.g., *Phragmites australis*), however, was not as greatly influenced by the timing of drawdown.

The results of MERP field and laboratory studies indicate that fine control of water levels during drawdowns can be used to control recruitment during drawdowns. Soil moisture is one of the most important factors regulating seed germination of wetland species (Naim 1987; van der Valk and Pederson 1989). If feasible, controlling soil temperature through the timing of the drawdown and soil moisture by raising and lowering water levels as needed would give managers almost complete control of recruitment patterns in freshwater wetlands. In reality, there are limitations to this type of controlled management. Primarily, not enough is known about the seed germination requirements of most wetland species and the environmental conditions that inhibit or promote seed germination for a given species. Secondly, most wetlands do not have adequate water-control structures to finely adjust water levels or adequate sources of water during all times of the year. Most water-level manipulations in the past have consisted of either pulling all the

Table 12.1. Seedling density (per square meter) at the end of August for drawdown treatments initiated in mid-May, June, July, and August

Species	May	June	July	August
Emergents				
Phragmites australis	16	21	11	0
Scirpus lacustris	300	220	130	0
Scirpus maritimus	620	110	70	0
Scholochloa festucacea	10	39	58	0
Typha spp.	140	500	310	18
Lythrum salicaria	71	180	120	22
Annuals				
Aster brachyactis	29	27	13	0
Chenopodium rubrum	94	120	140	27
Rumex crispus	9	17	20	8
Others	27	48	15	0
Total	1,400	1,380	887	92

Source: Adapted from Merendino et al. (1990).

logs out of a stop-log structure or replacing them all at once. The possibility for finely manipulating vegetation establishment during drawdowns is very real, and much more can and should be done by managers to develop drawdown protocols to take advantage of this opportunity.

Life History-Based Management

The management of wetland vegetation ultimately involves manipulating the sizes and distribution of the various plant populations in the wetland. Therefore, it is necessary to understand what controls the establishment and spread of the species as well as what controls mortality rates. During MERP, a series of studies was conducted on the life history and management of whitetop, *Scolochloa festucacea* (Neckles 1984; Neckles et al. 1985; Neckles and Wetzel 1989; Neill 1990a, b; 1992; 1993a, b; 1994). These studies were designed to understand what controlled its growth to provide management recommendations for its harvest as wild hay that were compatible with maintaining *Scolochloa*-dominated areas as breeding waterfowl habitat. Cultivation of seasonally flooded wetlands or wetland zones where *Scolochloa* is the dominant species often results in the soils becoming highly saline and incapable of growing crops. Thus, a management scheme that allows harvest of *Scolochloa* for hay, while preventing disruption of breeding waterfowl would create a beneficial situation for both farmers and wetland managers.

To determine if such a management plan is feasible, a series of field and greenhouse studies was conducted during MERP. These studies showed that *Scolochloa* is a cool-season grass that grows best when shallowly (<40 cm) flooded in the spring with water levels dropping to the surface of the marsh by the beginning of July (Neill 1994). This maximizes aboveground and belowground seed production. Spring flooding is important not only to provide an ideal moisture regime but also to lower salinity levels (Neill 1993b). The other major factor controlling *Scolochloa* growth is litter buildup. The greater the litter accumulation, the lower the production. Burning or mowing to remove accumulated litter would be needed to sustain high annual production (Neckles et al. 1985).

These studies indicated that management to optimize *Scolochloa*-dominated wetlands for use by breeding waterfowl and for wild hay production is feasible. *Scolochloa* stands should be shallowly flooded in spring and then drawn down in June. Although it takes 3 to 5 weeks of dry conditions before harvesting equipment can be brought into these wetlands, this schedule allows haying to be done in mid-July in most years. July harvesting allows enough new shoot growth to occur before fall so that snow can be trapped by the standing litter during the winter. This is needed to provide an adequate supply of fresh water the following spring.

During droughts when *Scolochloa* stands are not flooded in the spring, they should not be hayed because this reduces belowground reserves which, in turn, reduces the following year's production and allows the establishment of other species. A model of the effects of harvesting *Scolochloa* stands was developed by Neckles and Wetzel (1989). This simulation model, which is based on known causal relationships and rate coefficients based on field measurements, provides a powerful tool for exploring how various management options could effect *Scolochloa* growth and survival.

The MERP studies of the effects of flooding and harvesting on *Scolochloa* were simple field studies of the growth of this species at different water depths at different times over the growing season. They can be easily replicated for other species. This experimental approach combined with simple models, such as the one developed by Neckles and Wetzel (1989), shows great promise for generating meaningful and realistic management plans for wetland vegetation.

ALGAE

From the standpoint of providing energy to wetland herbivores, epiphyton (algae growing on submersed plant surfaces) is probably the most desirable algal assemblage because it consists of small, easily assimilable cells (more so than large filamentous green algae comprising metaphyton), is relatively abundant compared with epipelon and plocon (algae associated with the substrate surface), and is less prone to erratic fluctuations in biomass characteristic of phytoplankton-dominated shallow systems. Also present in a wetland dominated by epiphyton are abundant submersed and emergent macrophytes. Hence, maintenance of the hemi-marsh or open state (i.e., open water interspersed with emergent vegetation) is the normal goal of wetland management. This can be achieved through water-level management as described in the previous macrophyte section of this chapter, although prevention of metaphyton (floating mats of filamentous algae) dominance such as occurred in MERP cells (Robinson et al. 1997) and other experimentally flooded wetlands (Wu and Mitsch 1998) may be an issue in some situations.

Metaphyton was abundant during the deep-flooding experiment from 1981 to 1982, when water levels in the experimental cells were raised 1 m above normal, and again during the reflooding experiments from 1985 to 1989. Metaphyton proliferation in wetlands appears to be promoted by inorganic nutrient enrichment (Gabor et al. 1994; Murkin et al. 1994; McDougal et al. 1997). Although such enrichment was not detected in the post-flooding phases of MERP, this may be the result of rapid algal assimilation of nutrients leached from the inundated plant litter. Profuse metaphyton is a problem for two reasons. Firstly, the thick, cohesive metaphyton mats cause decreased growth of other algal groups and submersed plants through shading and marked water-column deoxygenation during their decomposition. Secondly, limited data suggest that high metaphyton production does not lead to high secondary production because metaphytic algae is not consumed, to a large extent, by herbivores.

Metaphyton may provide habitat for aquatic invertebrates, including construction material for tube-building caddisflies (Ross and Murkin 1993). However, stable C and N isotope studies in the MERP cells and adjacent areas found that invertebrates did not appear to feed on metaphyton (Neill and Cornwell 1992). Unfortunately, options for controlling metaphyton are limited. The manager may have no choice but to let the algal community proceed through the steps indicated in the model presented in Chapter 8, for eventually the system will move to other, more ecologically desirable algal groups. This is, of course, unless high external nutrient loading permits the metaphyton to persist. Where feasible, nutrient inputs should be minimized.

INVERTEBRATES

One of the primary objectives of many wetland management projects is to provide adequate food resources for birds and other wildlife over the annual cycle. Aquatic invertebrates are significant components in the trophic structure of prairie wetlands, providing food resources for a variety of birds and other wetland wildlife (see Chapter 10). The habitat and food resources available to invertebrates fluctuate greatly over the wet–dry cycle in prairie wetlands (see Chapter 9) and water-level manipulation through management can be expected to have a major influence on invertebrates in these systems.

During the dry marsh stage, terrestrial annuals as well as emergent macrophytes dominate the substrate surface (see Chapter 7). Upon reflooding, the litter produced by these plants provides abundant habitat for aquatic invertebrates and the algal resources on which they feed (Murkin and Ross 1999; Wrubleski 1999; see Chapter 9). Mudflat annuals, which make up a significant part of the plant community during drawdown (van der Valk and Davis 1976, 1978; Davis and van der Valk 1978; Welling et al. 1988a, b), provide excellent habitat for invertebrate production when prairie marshes are reflooded. In addition, the developing emergent macrophytes provide habitat for many epiphytic invertebrates. With continued flooding, the death of the short-lived emergent species and subsequently the longer-lived species provides a continuous supply of litter as food and habitat. Algae, an important food source for many invertebrate groups, flourishes during the early stages of flooding. Eventually the open areas that develop with the death of the emergent vegetation support larger benthic-dwelling invertebrate species (Murkin and Kadlec 1986;

Wrubleski 1991). This differential use of the various plant species following reflooding ensures an abundant and diverse invertebrate community during the initial years of flooding.

Shallowly flooded wetlands with extensive stands of dense emergent vegetation are not productive invertebrate habitats (Murkin and Ross 1999), and are also not preferred habitat for most vertebrate species (see bird section following). Wetlands with an interspersion of emergent and submersed macrophytes are usually productive invertebrate habitats (Murkin et al. 1992; see Chapter 9). By managing for a diverse plant community, wetland managers are ensuring the development of a diverse invertebrate community and abundant algal food resources during the same period of time (see Chapter 9). Food requirements by wetland birds and other wildlife vary among species and over the annual cycle; therefore, an abundant and diverse invertebrate community would meet the needs of a diverse and ever-changing group of consumers.

Flooding regimes also modify the composition of invertebrate communities because of different life-history requirements. Many wetland taxa (e.g., some species of Cladocera, Ostracods, and Culicidae) are adapted to seasonal flooding regimes. Permanent flooding greatly reduces the abundance of these taxa (Neckles et al. 1990). Alternatively, seasonal flooding regimes restrict invertebrate communities to species with adaptations to drought conditions. Invertebrates, such as amphipods, require permanent water and are not present in seasonally flooded habitats (Kantrud et al. 1989).

Whitman (1974) suggested that optimum conditions for invertebrate production in newly flooded wetlands occurred during the first 1.5 to 4 years and recommended drawdown between 5 to 7 years to improve food and cover for waterfowl. In the MERP study, invertebrate production and diversity were high from the first application of water and continued to be high through the 5 years following reflooding, first within the flooded mudflat annual habitats and then within the drowned emergent macrophyte habitats (Murkin and Ross 1999; Wrubleski 1999; see Chapter 9). It was anticipated that this production and diversity would decline as the emergent macrophytes continued to disappear and decomposition proceeded to reduce the remaining litter. If submersed macrophytes develop within these areas, continued high invertebrate production and diversity may be expected (Wrubleski and Rosenberg 1990; Murkin and Ross 1999). However, with no submersed vegetation development, invertebrate production in these open areas would be expected to be very low (Wrubleski and Rosenberg 1990). At such a point, water-level manipulations would be required to restart the cycle.

BIRDS

Kadlec and Smith (1992) provided a general review of wetland management for breeding waterfowl and discussed a variety of techniques available to wetland managers to enhance waterfowl use and production on managed wetland areas. They highlighted water regime as the dominant factor controlling wetland structure and function and argued that the most productive waterfowl habitats are those with fluctuating water levels. The importance of water-level fluctuations has been known for some time and water-level management has been a common practice on managed waterfowl production areas for many years (Kadlec 1962; Whitman 1974). The general principle has been that once an impounded area becomes unproductive (i.e., in the lake marsh stage), it should be drawn down for a period of time to reestablish the macrophytes, and then reflooded.

The MERP study provided the opportunity to monitor the avian response to drawdown management on the Delta Marsh. In Chapter 2, Batt described how long-term stable water levels have led to much reduced productivity and bird use of the marsh. The drawdown and reflooding experiments in MERP allowed comparison of bird use of these managed sites with unmanaged sites in the main Delta Marsh. During the baseline year in 1980, the waterfowl use (all duck species combined) of the diked experimental cells (cells 1–10) was the same as the undiked reference cells in the main marsh (cells 11 and 12) at 1.12 + 0.28 (mean ± SE) birds observed per hectare. During all years (and seasons) following reflooding after the drawdown, the waterfowl densities in the diked cells were higher than the reference cells in the main marsh (Table 12.2). Interestingly, waterfowl use increased following flooding and was still at elevated levels after 5 years. Moving the diked areas of the Delta Marsh from the lake marsh stage, characteristic of the open-water areas of the main marsh (see Chapter 2), to the earlier stages of the wet–dry cycle increased the bird use of these areas by a factor of four or more.

Several factors are responsible for this increased bird use during the early flooded stages of the wet–dry cycle. Chapter 10 discusses the resources required and used by birds in prairie marshes and how available resources change during the wet–dry cycle. Cover provided by the expanding stands of emergent vegetation during the early reflooding stages provides visual isolation, nesting sites, and cover from predators and weather for a variety of birds species. As noted previously in this chapter, invertebrate food resources are also higher during these early stages than the lake marsh stage. Murkin and Kadlec (1986) found that waterfowl use of the MERP experimental cells was related to invertebrate levels during the spring breeding period.

The goal of water-level management for waterfowl and other wetland birds is often the hemi-marsh stage, the point in the marsh cycle when open water and cover are present in approximately equal proportions in an interspersed pattern. This goal follows from the work of Weller and Spatcher (1965) and Weller and Fredrickson (1974) who found the highest avian use of a wetland in Iowa was during the hemi-marsh stage. This phenomena has been confirmed experimentally in the Delta Marsh by Kaminski and Prince (1981) for dabbling ducks and by Murkin et al. (1982, 1989) for dabbling ducks and blackbirds. Kadlec and Smith (1992) reported that the attractiveness of hemi-marsh conditions is related to the abundant cover and food resources provided during this stage, and the proximity of the food and cover provided by the interspersion within the basin. Weller (1978) reviewed the role of edge in the hemi-marsh concept and reported that birds are attracted to the abundant interface between cover and water provided by this interspersion. During MERP, Murkin et al (1989) reported that for red-winged blackbirds there appeared to be a preferred distance to the edge for nest sites so the abundant edges provided by the hemi-marsh stage would attract nesting blackbirds and other species with similar habitat requirements. Murkin et al. (1982) reported that the mixture of cover and water during the

Table 12.2. Mean (± SE) waterfowl densities (all species combined) (number per hectare) following reflooding of the MERP experimental cells (1985–1989), Delta Marsh, Manitoba

Year	Year of flooding[a]	Spring[b]		Summer[c]		Fall[d]	
		Managed[e]	Unmanaged[f]	Managed	Unmanaged	Managed	Unmanaged
1985	1	3.77 (0.33)	0.98 (0.22)	2.74 (0.34)	0.52 (0.18)	3.44 (0.50)	1.20 (0.48)
1986	2	3.57 (0.39)	0.81 (0.16)	4.02 (0.33)	1.07 (0.41)	5.47 (0.53)	1.25 (0.41)
1987	3	4.35 (0.35)	0.61 (0.18)	4.01 (0.47)	0.62 (0.20)	6.93 (0.55)	1.10 (0.33)
1988	4	6.51 (0.42)	1.18 (0.31)	5.04 (0.33)	1.37 (0.26)	6.31 (0.42)	1.04 (0.09)
1989	5	5.87 (0.34)	1.28 (0.23)	5.34 (0.42)	0.91 (0.32)	6.75 (0.52)	1.06 (0.63)

[a]Diked experimental cells were reflooded in the spring of 1985 following 1 or 2 years of drawdown.
[b]May and June.
[c]July and August.
[d]September and October.
[e]Diked MERP experimental cells 1–10.
[f]Undiked MERP reference cells 11–12.

hemi-marsh stage provided the pair space (open water) and visual isolation (vegetative cover) to allow the maximum number of breeding pairs to use the available habitat. This would be particularly important during the spring breeding season when many species are territorial.

It also has been suggested that the cover–water edges or interface in wetlands are the areas of highest invertebrate abundance and, therefore, the abundant edges during the hemi-marsh stage result in abundant food resources being available to feeding birds. During the MERP study, however, Murkin et al. (1992) found that peaks in invertebrate numbers did not occur at the open water–emergent cover interface at various sampling sites within the Delta Marsh. They suggested that the high invertebrate abundance and diversity in interspersed wetland habitats was related more to the mixture of habitat types rather than to the actual amount of interface present. Different invertebrate groups prefer different habitat types so interspersed habitats such as the hemi-marsh have a diverse invertebrate community. These diverse invertebrate food resources probably contribute to avian use of these habitats. Different bird species feed on different invertebrate groups (see Chapter 10); therefore, diverse and abundant invertebrate faunae provided by habitats such as the hemi-marsh support a greater variety and abundance of marsh birds.

Kadlec and Smith (1992) suggest that the hemi-marsh concept is applicable primarily during the breeding season. The MERP avian data supported this observation. Although the avian monitoring program on the experimental cells indicated that hemi-marsh conditions resulted in maximum use for a wide range of species during the spring breeding season (Murkin et al. 1997), they did not meet the requirement of all species in all seasons (see Figures 10.4-10.7 in Chapter 10). In fact, no species were found in equal abundance during all stages of the wet–dry cycle. All stages of the cycle have unique combinations of resources available to birds; thus, different stages of the cycle may be more suitable to some bird species than others. This fact coupled with the importance of the wet–dry cycle for maintaining the long-term productivity of these systems has important implications for managers who are interested in maintaining an abundant and diverse bird community. A regular wet–dry cycle ensures the highest avian use, diversity, and productivity over time. Once the managed wetland reached the lake marsh stage, use is limited primarily to birds such as diving ducks and other species preferring open-water sites. Depending on the management objectives, the manager will want to impose a drawdown to restore vegetation to the central parts of the basin, and with reflooding begin a new cycle. If several managed wetlands are in close juxtaposition within a wetland complex, keeping the wet–dry cycles in the individual basins out of phase (i.e., have the drawdowns in different years) ensures the maximum avian use and diversity in the complex at any point in time.

MUSKRATS

Muskrats are an integral component of prairie wetland ecosystems and are the most important resident consumer of emergent vegetation in these systems (see Chapter 11). Through their foraging and lodge-building activities, they directly affect vegetation structure, distribution, and primary productivity, and indirectly influence decomposition, invertebrate communities, and habitat use by birds and other wetland wildlife. Managers frequently visualize muskrats as an indicator of conditions for wildlife and overall productivity within prairie wetlands, and incorporate direct manipulation of their populations into management planning. In Chapter 11 Clark summarizes the information needs to manage muskrats population effectively and the techniques to collect and interpret that information.

Most attention toward muskrats and wetland management has been focused on the destruction of wetland vegetation during foraging and lodge building. Observations that under stable water conditions (with sufficient depth for overwinter survival), muskrat populations expand rapidly leading to eatout situations that are widespread in the management literature (see Chapter 11). Although on cursory examination, this appeared to happen in the MERP experimental cells during the reflooding years (1985–1989) (see Figure 5.1 in Chapter 5), Clark concluded that although muskrats may have contributed to the decline of the emergent vegetation in the experimental cells, they were not the principle cause as the term eatout implies. He calculated that muskrats consumed a small fraction (≈10%) of the annual belowground macrophyte production even at the highest population levels. Collectively, the MERP data support the conclusion that the decline in vegetation in the MERP cells was due to the intolerance of the various plant species to the depth of the long-term flooding treatments and would have occurred in the absence of muskrat feeding and lodge building.

The timing of the observed changes in vegetation standing crops and muskrat demography during the MERP study suggested that muskrat populations were responding to the declining vegetation rather than causing the decline. The greatest densities of muskrats occurred in the experimental cells in 1987 and then declined dramatically, after the reduction in emergent vegetation had been detected. The MERP results suggest that managers should recognize the interaction between water levels and muskrat activity. Managers must carefully evaluate the timing of changes in both vegetation and muskrats before assuming that controlling muskrat populations will result in arresting the degeneration of the marsh at the hemi-marsh stage of vegetation.

The MERP data also showed that muskrat habitat use changed as the marsh underwent successional changes. Data in Chapter 11 and the references cited therein show that declines in food and cover provided by the emergent vegetation caused the observed changes in muskrat condition and survival and ultimately affected population levels. Food is not normally limiting during the ice-free growing season so habitat selection during this period is most strongly related to predation risk (see Chapter 11). During the winter, however, availability of food and shelter, water depth, and vulnerability to predation all play major roles in survival. The MERP results indicated that *Typha* and *Scirpus* were the best overall vegetation types for lodge building and overall winter survival (Clark 1994). As these preferred habitats become fully occupied during population expansion, some individuals are forced into less suitable habitats such as *Scholochloa* because of density-dependent competition for areas with the highest value for reproduction and survival. Overwinter survival in these less suitable habitats is very low due to the combined effects of shallow water, freezing, limited access to food, and increased predation. Managers should be aware that lodge-building activities in these less suitable habitats indicates that the preferred habitats are filling to capacity and survival of muskrats in these less suitable areas will be very low. At this point, managers may want to assess their muskrat harvest program to take advantage of the surplus animals that are unlikely to survive and reproduce.

WETLAND MANAGEMENT ON PRAIRIES

The importance of the wet–dry cycle in maintaining the productivity of wetlands is the primary consideration in any discussion of the management of prairie wetlands. The fluctuation between wet and dry conditions is critical to the long-term productivity of these systems. Managers have long understood this phenomena and used drawdown management to enhance the productivity of prairie wetlands. In addition, maintaining the hemi-marsh has been the goal of management for waterfowl and other marsh birds. However, each stage, including the hemi-marsh stage of the cycle, is transitory and regular fluctuations in water levels ensure the highest productivity over the long term. In addition, certain stages of the cycle are more suitable for some species than others and by maintaining the progression among stages, managers also are creating conditions for the highest diversity of species.

An important lesson from the MERP study is that the macrophytes appear to "drive" the overall productivity and nutrient budgets of these systems. By managing the macrophytes, the manager is essentially managing the nutrient budgets of the system and the other components linked to macrophytes. Plant uptake and subsequent decomposition are the dominant control points for prairie wetland nutrient budgets. Macrophytes are important in determining habitat quality and food availability for aquatic invertebrates. Macrophytes influence the submersed surface areas and nutrient availability for algal growth. The cover resources provided by macrophytes are important in determining avian use and productivity in prairie wetlands. Macrophyte standing crops influence muskrat survival and reproduction. Managing macrophytes by maintaining the wet–dry cycle provides the conditions for the greatest long-term productivity of these systems.

Generally, artificial water manipulation is not required in prairie wetlands. Natural climatic fluctuations force wetland basins and impoundments through the wet–dry cycle. The only management required in most situations is to ensure that the basin will hold water during wet periods. This is normally accomplished by blocking natural or artificial drains. Once the basin can hold water, the natural cycle of wet periods and drought results in the regulation of water levels. Only under circumstances where water levels have been artificially stabilized (e.g. the Delta Marsh [see Chapter 2]) will active water-level management be required to maintain long-term productivity by creating an artificial wet–dry cycle.

To provide the greatest diversity of habitat and overall productivity at any point in time, conservation programs should target wetland complexes with a variety of wetland classes (Stewart and Kantrud 1971). The deeper classes hold water during most years, whereas the shallower ponds proceed through the wet–dry cycle at different rates. An ideal goal is to maintain several managed wetlands that are in close juxtaposition so that drawdowns (either natural or artificial) in different years keep the wet–dry cycle out of phase within the complex. Such management ultimately provides the maximum spatial and temporal diversity of habitat conditions.

ACKNOWLEDGMENTS

This chapter is dedicated to the Ducks Unlimited biologists and field staff who have patiently waited for this information to appear in print. Bruce Batt, Pat Caldwell, Brian Gray, and Lisette Ross reviewed an earlier draft of the manuscript. This is Paper No. 105 of the Marsh Ecology Research Program, a joint project of Ducks Unlimited Canada and the Delta Waterfowl and Wetlands Research Station.

LITERATURE CITED

Bavor, H.J, D.J. Roser, S.A. McKersie, and P. Breen. 1988. Treatment of secondary effluent. Report to Sydney Water Board, Sydney.

Christiansen, J.E. 1968. Pan evaporation and evapotranspiration from climatic data. Transactions of the International Committee on Irrigation Drainage 3:23.569-23.596.

Clark, W.R. 1994. Habitat selection by muskrats in experimental marshes undergoing succession. Canadian Journal of Zoology 72:675-680.

Crumpton, W.G., and L.G. Goldsborough. 1998. Nitrogen transformation and fate in prairie wetlands. Great Plains Research 8: 57-72.

Davis, C.B., and A.G.van der Valk. 1978. Litter decomposition in prairie glacial marshes. In *Freshwater Wetlands: Ecological Processes and Management Potential.* (Eds.) R.E. Good, R.L. Simpson, and D.F. Whigham, pp.99-113. New York: Academic Press.

de Swart, E.O. A.M., A.G. van der Valk, K.J. Koehler, and A. Barendregt. 1994. Experimental evaluation of realized niche models for predicting responses of plant species to a change in environmental conditions. Journal of Vegetation Science 5:541-552.

Fredrickson, L.H., and T.S. Taylor. 1982. Management of seasonally flooded impoundments for wildlife. U.S. Department of Interior, Fish and Wildlife Service, Resources Publication 148.

Gabor, T.S., H.R. Murkin, M.P. Stainton, J.A. Boughen, and R.D. Titman. 1994. Nutrient additions to wetlands in the interlake region of Manitoba, Canada: effects of a single pulse addition in spring. Hydrobiologia 280:497-510.

Galatowitsch, S.M., and A.G. van der Valk. 1994. *Restoring Prairie Wetlands.* Ames: Iowa State University Press.

Galinato, M.I., and A.G. van der Valk. 1986. Seed germination traits of annuals and emergents recruited during drawdowns in the Delta Marsh, Manitoba, Canada. Aquatic Botany 26:89-102.

Hooper-Reid, N.M., and G.G.C. Robinson. 1978. Seasonal dynamics of epiphytic algal growth in a marsh pond: composition, metabolism, and nutrient availability. Canadian Journal of Botany 56:2441-2448.

Kadlec, J.A. 1962. The effect of a drawdown on the ecology of a waterfowl impoundment. Ecology 43:267-281.

Kadlec, J.A. 1983. Water budgets for small diked marshes. Water Resources Bulletin 19:223-229.

Kadlec, J.A. 1986. Effects of flooding on dissolved and suspended nutrients in small diked marshes. Canadian Journal of Fisheries and Aquatic Science 43:1999-2008.

Kadlec, J.A. 1993. Effect of depth of flooding on summer water budgets for small diked marshes. Wetlands 13:1-9.

Kadlec, J.A., and L.M. Smith. 1992. Habitat management for breeding areas. In *Ecology and Management of Breeding Waterfowl*. (Eds.) B.D.J. Batt, A.D. Afton, M.G. Anderson, C.D. Ankney, D.H. Johnson, J.A. Kadlec, and G.L. Krapu, pp.568-589. Minneapolis: University of Minnesota Press.

Kadlec, R.H., and R.L. Knight. 1996. *Treatment Wetlands*. New York: Lewis Publishers.

Kadlec, R.H., R.B. Williams, and R.D. Scheffe. 1987. Wetland evapotranspiration in temperate and arid climates. In *The Ecology and Management of Wetlands*. (Eds.) D.D. Hook, W.H. McKee, H.K. Smith, J. Gregory, V.G. Burrell, M.R. DeVoe, R.E. Sojka, S. Gilbert, R. Banks, L.H. Stolzy, C. Brooks, T.D. Matthews, and T.H. Shear, pp.146-160. Portland: Timber Press.

Kaminski, R.M., and H.H. Prince. 1981. Dabbling duck and aquatic macroinvertebrate responses to manipulated wetland habitat. Journal of Wildlife Management 45:1-15.

Kantrud, H.A., G.L. Krapu, and G.A. Swanson. 1989. Prairie basin wetlands of the Dakotas: a community profile. United States Fish and Wildlife Service Biological Report 85.

LaBaugh, J.W., T.C. Winter, and D.O. Rosenberry. 1998. Hydrologic functions of prairie wetlands. Great Plains Research 8:3-15.

Leck, M.A. 1989. Wetland seed banks. In *Ecology of Soil Seed Banks*. (Eds.) M.A. Leck, V.T Parker, and R.L. Simpson, pp.283-305. San Diego: Academic Press.

McDougal, R.L., L.G. Goldsborough, and B.J. Hann. 1997. Responses of a prairie wetland to inorganic nitrogen and phosphorus: production by planktonic and benthic algae. Archiv für Hydrobiologie 140:145-167.

Merendino, M.T., and L.M. Smith. 1991. Influence of drawdown date and reflood depth on wetland vegetation establishment. Wildlife Society Bulletin 19:143-150.

Merendino, M.T., L.M. Smith, H.R. Murkin, and R.L. Pederson. 1990. The response of prairie wetland vegetation to seasonality of drawdown. Wildlife Society Bulletin 18:245-251.

Murkin, H.R. 1998. Freshwater functions and values of prairie wetlands. Great Plains Research 8:3-15.

Murkin, H.R., and J.A. Kadlec. 1986. Responses by benthic macroinvertebrates to prolonged flooding of marsh habitat. Canadian Journal of Zoology 64:65-72.

Murkin, H.R., and L.C.M. Ross. 1999. Northern prairie marshes (Delta Marsh, Manitoba): macroinvertebrate responses to a simulated wet-dry cycle. In *Invertebrates in Freshwater Wetlands of North America: Ecology and Management*. (Eds.) D. Batzer, R.D. Rader, and S.A. Wissinger, pp. 543-569. New York: Wiley.

Murkin, H.R., R.M. Kaminski, and R.D. Titman. 1982. Responses by dabbling ducks and aquatic invertebrates to an experimentally manipulated cattail marsh. Canadian Journal of Zoology 60:2324-2332.

Murkin, HR., R.M. Kaminski, and R.D. Titman. 1989. Responses by nesting red-winged blackbirds to manipulated cattail habitat. In *Freshwater Wetlands and Wildlife*. (Eds.) R.R. Sharitz and J.W. Gibbons, pp. 673-680. DOE Symposium Series No. 61, USDOE Office of Scientific and Technical Information, Oak Ridge, TN.

Murkin, H.R., E.J. Murkin, and J.P. Ball. 1997. Avian habitat selection and prairie wetland dynamics. Ecological Applications 7:1144-1159.

Murkin, E.J., H.R. Murkin, and R.D. Titman. 1992. Nektonic invertebrate abundance and distribution at the emergent vegetation - open water interface in the Delta Marsh, Manitoba, Canada. Wetlands 10:45-52.

Murkin, H.R., J.B. Pollard, M.P. Stainton, J.A. Boughen, and R.D. Titman. 1994. Nutrient additions to wetlands in the interlake region of Manitoba, Canada: effects of periodic additions throughout the growing season. Hydrobiologia 280:483-495.

Murkin, H.R., M.P. Stainton, J.A. Boughen, J.B. Pollard, and R.D. Titman. 1991. Nutrient status of wetlands in the interlake region of Manitoba, Canada. Wetlands 11:105-122.

Murkin, H.R., A.G. van der Valk, and C.B. Davis. 1989. Decomposition of four dominant macrophytes in the Delta Marsh, Manitoba. Wildlife Society Bulletin 17:215-221.

Naim, P.A., 1987. Wetland seed banks: implications in vegetation management. Ph.D. dissertation, Iowa State University, Ames, Iowa.

Neckles, H.A. 1984. Plant and macroinvertebrate response to water regime in a whitetop marsh. M.S. thesis, University of Minnesota, St. Paul, Minnesota.

Neckles, H.A., and R.L. Wetzel. 1989. Effects of forage harvest in seasonally flooded prairie marshes: simulation model experiments. In *Freshwater Wetlands and Wildlife.* (Eds.) R.R. Sharitz and J.W. Gibbons. USDOE Symposium Series 61. Oak Ridge: USDOE Office of Scientific and Technical Information.

Neckles, H.A., H.R. Murkin, and J.A. Cooper. 1990. Influences of seasonal flooding on macroinvertebrate abundance in wetland habitats. Freshwater Biology 23:311-322.

Neckles, H.A., J.W. Nelson, and R.L. Pederson. 1985. Management of whitetop (*Scolochloa festucacea*) marshes for livestock forage and wildlife. Technical Bulletin No. 1. Delta Waterfowl and Wetlands Research Station, Portage la Prairie, Manitoba.

Neill, C. 1990a. Effects of nutrients and water levels on species composition in prairie whitetop *Scolochloa festucacea* marshes. Canadian Journal of Botany 68:1015-1020.

Neill, C. 1990b. Effects of nutrients and water levels on emergent macrophyte biomass in a prairie marsh. Canadian Journal of Botany 68:1007-1014.

Neill, C. 1992. Life history and population dynamics of whitetop (*Scolochloa festucacea*) shoots under different levels of flooding and nitrogen supply. Aquatic Botany 42:241-252.

Neill, C. 1993a. Growth and resource allocation of whitetop (*Scolochloa festucacea*) along a water depth gradient. Aquatic Botany 46:235-246.

Neill, C. 1993b. Seasonal flooding, soil salinity and primary production in northern prairie marshes. Oecologia (Berl.) 95:499-505.

Neill, C. 1994. Primary production and management of seasonally flooded prairie marshes harvested for wild hay. Canadian Journal of Botany 72:801-807

Neill, C., and J.C. Cornwell. 1992. Stable carbon, nitrogen, and sulfur isotopes in a prairie marsh food web. Wetlands 12:217-224.

Pederson, R.L. 1981. Seedbank characteristics of the Delta Marsh, Manitoba: Applications for wetland management. In *Selected Proceedings of the Midwest Conference on Wetland Values and Management.* (Ed.) B. Richardson, pp.61-69. Navarre: The Freshwater Society.

Pederson, R.L. 1983. Abundance, distribution, and diversity of buried viable seed populations in the Delta Marsh, Manitoba. Ph.D. dissertation, Iowa State University, Ames, Iowa.

Pederson, R.L., and A.G. van der Valk. 1985. Vegetation change and seed banks in marshes: ecological and management implications. Transactions of the North American Wildlife and Natural Resource Conference 49:271-280.

Poiani, K.A., and W.C. Johnson. 1988. Evaluation of the emergence method in estimating seed bank composition of prairie wetlands. Aquatic Botany 32:91-97.

Robarts, R.D., and M.J. Waiser. 1998. Effects of atmospheric change and agriculture on the biogeochemistry and microbiol ecology of prairie wetlands. Great Plains Research 8:113-136.

Roberts, H.A. 1981. Seed banks in soils. Advances in Applied Biology 6:1-55.

Robinson, G.G.C., S.E. Gurney, and L.G. Goldsborough. 1997. Response of benthic and planktonic algal biomass to experimental water level manipulation in a prairie lakeshore wetland. Wetlands 17:167-181.

Ross, L.C.M., and Murkin, H.R. 1993. The effect of above-normal flooding of a northern prairie marsh on *Agraylea multipunctata* Curtis (Trichoptera: Hydroptilidae). Journal of Freshwater Ecology 8:27-35.

Squires, L., and A.G. van der Valk. 1992. Water-depth tolerances of the dominant emergent macrophytes of the Delta Marsh, Manitoba. Canadian Journal of Botany 70:1860-1867.

Stewart, R.E., and H.A. Kantrud. 1971. Classification of natural ponds and lakes in the glaciated prairie region. Washington, U.S. Fish and Wildlife Service. Resource Publication 92.

van der Kamp, G., and M. Hayashi. 1998. The groundwater recharge function of small wetlands in the semi-arid northern prairies. Great Plains Research 8:39-56.

van der Valk, A.G. 1982. Succession in temperate North American wetlands. In *Wetlands: Ecology and Management*. (Eds.) B. Gopal, R.E. Turner, R.G. Wetzel, and D.F. Whigham, pp.169-179. Jaipur, India: International Scientific Publications.

van der Valk, A.G. 1985. Vegetation dynamics of prairie glacial marshes. In *Population Structure of Vegetation*. (Ed.) J. White, pp.293-312. . The Hague, Netherlands: Junk.

van der Valk, A.G. 1994. Effects of prolonged flooding on the distribution and biomass of emergent species along a freshwater wetland coenocline. Vegetatio 110:185-196.

van der Valk, A.G., and C.B. Davis. 1976. Changes in the composition, structure, and production of plant communities along a perturbed wetland coenocline. Vegetatio 32:87-96.

van der Valk, A.G., and C.B. Davis. 1978. The role of the seed banks in the vegetation dynamics of prairie glacial marshes. Ecology 59:322-335.

van der Valk, A.G., and C.B. Davis. 1980. The impact of a natural drawdown on the growth of four emergent species in a prairie glacial marsh. Aquatic Botany. 9:301-322.

van der Valk, A.G., and R.L. Pederson. 1989. Seed banks and the management and restoration of natural vegetation. In *Ecology of Soil Seed Banks*. (Eds.) M.A. Leck, V.T. Parker, and R.L. Simpson, pp.329-346. New York: Academic Press.

van der Valk, A.G., and T.R. Rosburg. 1997. Seed bank composition along a phosphorus gradient in the northern Florida Everglades. Wetlands 17:228-236.

van der Valk, A.G., L. Squires, and C.H. Welling. 1994. An evaluation of three approaches for assessing the impacts of an increase in water level on wetland vegetation. Ecological Applications 4:525-534.

van der Valk, A.G., C.H. Welling, and R.L. Pederson. 1989. Vegetation change in a freshwater wetland: a test of a priori predictions. In *Freshwater Wetlands and Wildlife*. (Eds.) R.R. Sharitz and J.W. Gibbons, pp.207-217. USDOE Symposium Series 61. Oak Ridge: USDOE Office of Scientific and Technical Information.

Weinhold, C.E., and A.G. van der Valk. 1989. The impact of duration of drainage on the seed banks of northern prairie wetlands. Canadian Journal of Botany 67:1878-1884.

Weller, M.W. 1978. Management of freshwater marshes for wildlife. In *Freshwater Wetlands: Ecological Processes and Management Potential*. (Eds.) R.E. Good, D.F. Whigham, and R.L. Simpson, pp.267-284. New York: Academic Press.

Weller, M.W. 1981. *Freshwater Marshes*. Minneapolis: University of Minnesota Press.

Weller, M.W., and L. H. Fredrickson. 1974. Avian ecology of a managed glacial marsh. Living Bird 12:269-291.

Weller, M.W., and C.S. Spatcher. 1965. Role of habitat in the distribution and abundance of marsh birds. Iowa Agriculture and Home Economics Experiment Station, Special Report 43, Ames, Iowa.

Welling, C.H., R.L. Pederson, and A.G van der Valk. 1988a. Recruitment from the seed bank and the development of zonation of emergent vegetation during a drawdown in a prairie wetland. Journal of Ecology 76:483-496.

Welling, C.H., R.L. Pederson, and A.G. van der Valk. 1988b. Temporal patterns in recruitment from the seed bank during drawdowns in a prairie wetland. Journal of Applied Ecology 25:999-1007.

Whitman, W.R. 1974. The response of macro-invertebrates to experimental marsh management. Ph.D. dissertation, University of Maine, Orono, Maine.

Winter, T.C. 1981. Uncertainties in estimating the water budgets of lakes. Water Resources Bulletin 17:82-115.

Winter, T.C. 1989. Hydrologic studies of wetlands in the northern prairie. In *Northern Prairie Wetlands*. (Ed.) A.G. van der Valk, pp.16-54. Ames: Iowa State University Press.

Wrubleski, D.A. 1991. Chironomidae (Diptera) community development following experimental manipulation of water levels and aquatic vegetation. Ph.D. dissertation. University of Alberta, Edmonton, Alberta.

Wrubleski, D.A. 1999. Chironomidae of the Delta Marsh. In *Invertebrates in Freshwater Wetlands of North America: Ecology and Management*. (Eds.) D. Batzer, R.D. Rader, and S.A. Wissinger, pp. 571-601. New York: Wiley.

Wrubleski, D.A., and D.M. Rosenberg. 1990. The Chironomidae (Diptera) of Bone Pile Pond, Delta Marsh, Manitoba, Canada. Wetlands 11:1-16.

Wu, X., and W.J. Mitsch. 1998. Spatial and temporal patterns of algae in newly constructed freshwater wetlands. Wetlands 18:9-20.

13

Summary and Recommendations

Henry R. Murkin
Arnold G. van der Valk
William R. Clark

ABSTRACT

The Marsh Ecology Research Program (MERP) was a long-term experimental study to assess the response of prairie wetlands to water fluctuations during a simulated wet–dry cycle. The duration of flooding and water levels applied to the experimental cells were within the ranges observed in natural prairie wetlands. The objectives of this chapter are to summarize the key results from MERP and identify remaining research needs. The MERP nutrient-budget analyses clearly documented the importance of macrophytes to the nutrient budgets in prairie wetlands. Besides the large above- and below-ground nutrient pools associated with macrophytes, plant uptake, litter leaching, and decomposition were dominant fluxes controlling the nutrient movement in these systems. Although the macrophyte pools and associated fluxes were eliminated during the lake marsh stage, the pools associated with sediments and pore water increased. Plant uptake was reestablished during the dry marsh stage, and at the same time the aerobic conditions during the dry period enhanced mineralization to support the increased plant uptake and growth. Following reflooding, the macrophyte response in the three water-level treatments was similar during the first 3 years. After that time, as the short-lived emergents died back, they were replaced by long-lived species in the low treatment but not in the deeper treatments. Seed-germination patterns influenced the initial distribution of plants along the elevation gradient in the cells; however, a variety of factors influenced the final distribution of plants following flooding. Although algal standing crops were low at any point in time, high productivity and turnover resulted in annual production that equaled or exceeded that of macrophytes. Invertebrate diversity and abundance fluctuated greatly over the course of the wet–dry cycle in response to the variations in food and habitat. The importance of the hemi-marsh to avian use of prairie marshes was confirmed for several species during the spring breeding season; however, other stages of the cycle were important at other times of the year and for a variety of other species. Although muskrat eatouts are commonly cited for prairie marshes, muskrats had a minor effect on vegetation within the experimental cells. The muskrat populations tracked the macrophyte abundance rather than caused the macrophyte decline. Several research needs have been identified, including the role of microbes and algae in nutrient cycling in prairie wetlands; information on macrophyte seed production, dispersal, and longevity; habitat use by birds other than waterfowl and blackbirds; vegetation consumption rates by muskrats; and the interaction of factors such as salinity with the wet–dry cycle in prairie wetlands.

Introduction

The Marsh Ecology Research Program (MERP) was a long-term experimental study of the wet–dry cycle in prairie wetlands. This cycle was first described in general terms by Weller and Spatcher (1965) and Weller and Fredrickson (1974) and later in more detail by van der Valk and Davis (1978). The general objective of MERP was to develop a better understanding of the response of prairie wetlands to fluctuations in water levels characteristic of the prairie environment (see Chapter 1). Thus, a series of experimental impoundments or cells ($\approx$5 ha each) was subjected to a simulated wet–dry cycle to determine the effects of water-level changes on the major ecosystem components. The objectives of this chapter are to summarize the resulting advances in our understanding of prairie wetland ecology and to highlight the remaining priority research needs.

An important consideration in reviewing the MERP information is the applicability of results from the experimental cells to prairie wetlands in general. Hence, we begin this chapter with an evaluation of the MERP experimental cells as model prairie wetlands. We address this issue by comparing the flooding durations and water depths of the simulated wet–dry cycle in the MERP study to those observed during natural wetland cycles.

The MERP Experimental Cells as Prairie Wetlands

Construction of the MERP experimental complex in the Delta Marsh in 1979 created a series of 10 contiguous experimental cells separated from each other by dikes. The Delta Marsh is a large lacustrine marsh, separated from Lake Manitoba by a sand ridge barrier with only a few connecting channels (see Chapter 2). Until the 1960s, water levels in the Delta Marsh fluctuated $\approx$1.5 m from periods of low water (drought) to periods of high water (prolonged above-normal precipitation) (see Chapter 2). Since the early 1960s when a water-control structure was installed on Lake Manitoba, there has been little fluctuation of water levels within the marsh (see Figure 2.5 in Chapter 2).

Comparison of the water-level treatments during MERP (see Chapter 1) with water levels during wet–dry cycles in natural prairie wetlands is difficult. There is relatively little published information about water-level changes during natural cycles, although existing information suggests that natural cycles are highly variable. For example, the length of an entire wet–dry cycle often varies from cycle to cycle within the same wetland as illustrated in Figure 5.4 in Kantrud et al. (1989), which summarized spring and fall water levels in a prairie wetland in Saskatchewan for >20 years. During this period, there were three wet–dry cycles that lasted $\approx$10, 8, and 5 years. Only during the first cycle of 10 years was the flooding of sufficient duration to cause reduction in emergent vegetation over most of the basin. Thus, the degenerating and lake stages as described by van der Valk and Davis

(1978) were reached only once in more than 20 years. During the postdrawdown flooding experiments in MERP (1985–1989), the cells in the low treatment were still in the regenerating stage after 5 years of flooding, whereas the medium and deep treatments had entered the degenerating stage and appeared headed for the lake marsh stage (see Chapter 5). Hence, the length of the simulated cycle in the MERP cells seems to be comparable to that found in Canadian prairie wetlands. Similar wet–dry cycle lengths and ranges in water depths can be seen in long-term hydrographs for other prairie wetlands in North Dakota (Winter 1989; Poiani and Johnson 1993).

The maximum water depth in prairie wetlands is determined by topographic contours of the basin; however, annual precipitation patterns can cause tremendous variation in depth among years (Kantrud et al. 1989). During MERP, water depths varied ≈1.5 m (from the deep-flooding period to drawdown), well within the range found in natural prairie wetlands. For example, Kantrud et al. (1989) reported the range of water depths during the three wet–dry cycles in the Saskatchewan wetland as 2.1, 1.4, and 1.4 m. The rapid rates at which water levels changed between years in the MERP cells also were comparable to prairie wetlands. Water-level changes can occur very rapidly in prairie wetlands because the basins are generally small and there are often immediate responses to major storm events and prolonged precipitation (Kantrud et al. 1989).

One feature of water levels in the MERP cells that differed from natural cycles was the lack of seasonal water-level fluctuations. When prairie wetlands are not dry, seasonal water-level fluctuations typically range from ≈30 to 80 cm (Kantrud et al. 1989). Water levels are typically highest in spring and decline throughout the summer, although variations in this pattern do occur in some years. The most extreme within-year water-level fluctuations are found when the wetland is reflooding at the end of a drought or when water levels are dropping at the onset of a drought. In the MERP cells, water levels did fluctuate slightly (±10 cm) due to precipitation events and pumping lag times, but these fluctuations were much smaller than natural seasonal fluctuations in prairie wetlands.

In summary, the MERP water-level treatments were based on maximum and minimum water depths observed during previous cycles in the Delta Marsh (see Figure 2.5 in Chapter 2), and this range of depths also fell within the ranges recorded elsewhere for prairie wetlands. In addition, the length of the simulated cycle was well within the range of natural cycle lengths that have been recorded for prairie potholes. Only the absence of seasonal water-level fluctuations makes the MERP simulated cycle different from natural cycles, but we suggest that these small fluctuations would not have a significant effect on the overall changes that would occur during a complete wet–dry cycle. Several short-term MERP studies outside the experimental cells addressed the importance of seasonal water-level fluctuations on the ecology of prairie wetlands (Neckles et al. 1990; Neill 1992, 1993).

Summary of Key Results

The previous chapters in this volume described in detail the results of the MERP study for each of the major ecosystem components monitored. Herein we briefly summarize the most important contributions made by MERP to our understanding of wetland ecology.

Nutrient Budgets (Chapters 4–6)

The pools and fluxes associated with the macrophytes clearly dominated the nutrient budgets within the experimental cells during much of the simulated wet–dry cycle. Plant uptake, litter leaching, and decomposition are critical fluxes in moving nutrients from the sediments and pore water to aboveground pools. Plant uptake results in the development of aboveground litter that is important as submersed substrate for other primary producers (algae) and as habitat for the wide range of consumers found in these systems. The subsequent leaching and decomposition of this litter moves nutrients to the other pools and fluxes, thereby affecting the productivity of the entire wetland system. During the lake marsh stage as the macrophytes are eliminated, nutrients become "locked" in the pore-water and sediment pools. With the sequestering of nutrients in these two pools, the overall productivity of the wetland declines during this stage. Productivity remains low, until the dry stage allows germination of annual and emergent plants and thus the reestablishment of plant uptake, leaching and decomposition as dominant nutrient fluxes in the system. An important feature of the drawdown stage is the aerobic conditions that result in enhanced mineralization (microbial conversion of organic matter to inorganic matter) within the substrate. This mineralization "fuels" the increased plant uptake and growth. With reflooding, the macrophytes expand through the regenerating stage thereby supporting the other pools and fluxes and related productivity in the system.

Macrophytes (Chapter 7)

Results of the MERP vegetation studies largely confirmed the structure of the model presented by van der Valk and Davis (1978) on vegetation changes during the wet–dry cycle in prairie wetlands. The MERP deep-flooding experiments (1981 and 1982) were intended to move the experimental cells to the lake marsh stage; however, although the macrophytes were greatly reduced, they were not completely eliminated even after 2 years of flooding at 1 m above normal. It became apparent that belowground roots and rhizomes can survive deep flooding for up to 2 years, even though aboveground growth was eliminated in the first year. The delayed mortality of belowground tissues has important implications for the nutrient budgets of prairie wetlands. In addition, management practices to eliminate emergent macrophytes need to address survival of belowground roots and rhizomes.

Recruitment from the seed bank resulted in the revegetation of the MERP cells during the dry marsh stage of the simulated wet–dry cycle. Unlike many other prairie wetlands where the seed bank is uniformly distributed over the basin, the seed bank in the MERP experimental cells was most dense at the drift line associated with the long-term stabilized water levels of the Delta Marsh. However, this uneven seed distribution had only minor implications for the development of vegetation zones later in the cycle (see Chapter 7).

Reflooding the MERP cells at three different levels following the drawdown (1985–1989) was designed to assess the effects of water depth on macrophyte survival and growth. Cells at all three depths lost significant aboveground biomass and production after 3 years of flooding due to the elimination of short-lived emergent species (e.g., *Scirpus lacustris*) at that time. These species can tolerate flooding for only 2 or 3 years at the range of depths used in this study. After the third year, there was a difference in the vegetation response among the three flooding treatments. In the low treatment, areas formerly occupied by short-lived emergent species were invaded by the longer-lived species (e.g., *Typha* spp.). Thus, the regenerating stage continued in the low treatment after the third year of flooding. In the medium and deep treatments, the onset of the degenerating stage occurred soon after the short-lived emergents were eliminated because the greater depths prevented invasion of the areas formerly occupied by these species (see Chapter 7). In addition, the long-lived emergents began to die back in areas where the water depth was beyond their tolerance. In summary, the low treatment was still in the regenerating stage after 5 years of flooding, whereas both the medium and deep treatments had proceeded to the degenerating stage over the same time period. However, the amount of belowground biomass surviving after 5 years in these deeper treatments indicated that they had not reached the lake marsh stage.

The MERP studies made important contributions to our understanding of the development of vegetation zones in prairie wetlands. Competition, herbivory, seed dispersal, and environmental conditions all interact to determine the distribution of plant species in wetlands (see Chapter 7). In the MERP cells, seed dispersal had a minimal effect on the development of vegetation zones because most species started out with similar distributions of seeds along the elevation gradient. In contrast, seed-germination patterns did influence the distribution of plants along the elevation gradients. Some species (e.g., *Typha glauca*, *Phragmites australis*) had their highest seedling densities at elevations lower than their maximum seed densities, whereas for other species (e.g., *Scholochloa festucacea*) maximum seedling densities occurred at elevations higher than the maximum seed densities. These differences were due to varying germination requirements related to soil moisture, temperature, salinity, and in some situations, litter accumulation on the substrate surface.

Following reflooding, differential survival due to water-depth effects resulted in only minor changes in the distribution of emergent species. It appears that combinations of factors influence the distribution of individual species along the environmental gradient, and no one factor is responsible for the final distribution of all species. For example, in the MERP experimental cells during the reflooding years, differential seed germination was primarily responsible for the distribution of *Phragmites*, whereas several factors, including seed dispersal, differential germination, and seedling and adult mortality determined the final distribution of *Scholochloa*.

Vegetation models are essential for understanding the complexity of factors affecting vegetation distribution in prairie wetlands. Several models are presented in Chapter 7, however, the spatially explicit models developed by Poiani and Johnson (1993) and Seabloom (1997) appear to provide the most useful approach for examining the factors that control vegetation composition and distribution in these systems. Regardless, future efforts in modeling must incorporate a hydrological model with a vegetation model like that of Seabloom (1997) to address the entire suite of environmental factors controlling the distribution and abundance of vegetation in prairie wetlands.

Algae (Chapter 8)

One of our most important advances through MERP has been the improved understanding of the role of algae in the overall productivity and function of prairie wetlands. Although algal standing crops at any point in time may not reach the levels found for macrophytes, the high productivity and rapid turnover of algal cells may equal or exceed macrophyte production on an annual basis (Robinson et al. 1997a, b). In addition, although it has been assumed that wetlands are detritus-based systems with respect to secondary productivity, the MERP experiments have shown that algae probably play a very important role in supporting secondary productivity in these systems (Campeau et al. 1994). Algae also provided important structural habitat for aquatic invertebrates (Ross and Murkin 1993).

The van der Valk and Davis (1978) model of prairie wetland cycles, although appropriate for macrophytes, has shortcomings when considering algal communities. Goldsborough and Robinson (1996) developed a revised model to include the role of algae in the structure and function of prairie wetlands. They recognized four wetland states, each dominated by a different algal community. Data from MERP supported this new model and strengthened our understanding of the factors affecting the distribution and abundance of algae in these systems. This new model does not assume that shifts between states are one directional. Rather, algae can respond quickly to changes in nutrient levels, grazing pressure, and water-column stability and therefore may fluctuate among states depending on environmental conditions. The end points of the algae model, the dry state (dominated by epipelon) and the lake state (dominated by phytoplankton), are similar to the van der Valk and Davis (1978) model. However, the other two states, the open state (dominated by epiphyton) and sheltered state (dominated by metaphyton), are less clearly related to the macrophyte model because both have features of the regenerating and degenerating stages (see Chapter 8). This algal model provides a functional framework to assess the effects of environmental changes on algal abundance and distribution.

Invertebrates (Chapter 9)

A wide range of factors can affect the abundance and diversity of aquatic invertebrates in prairie wetlands. The MERP results indicated that these factors varied among the various stages of the wet–dry cycle, thereby resulting in different invertebrate communities over time. For example, during the drawdown stage, the terrestrial conditions required that aquatic invertebrates either move to other flooded areas or adapt to the dry conditions by entering a drought resistant stage or by burrowing into the bottom substrate. There are no data available on the longevity of these dormant forms, however, there was no difference in the invertebrate response to reflooding after the 1- and 2-year drawdowns in the experimental cells. The MERP study also showed that some species required periodic dry conditions and eliminating the dry period would eliminate these groups from the invertebrate community (Neckles et al. 1990).

When the experimental cells were reflooded during the regenerating stage, there was an abundance of submersed surfaces available for invertebrate colonization. The abundant plant litter with its complex structure provided important habitat and food resources for invertebrates at this point in the cycle. During the early reflooding years, the highest densities of chironomids (Chironomidae) occurred in flooded annual litter as opposed to developing emergent vegetation. Algae also responded to reflooding, thereby providing an important source of food for a wide range of invertebrate groups. After several years of flooding, invertebrate densities declined in the annual plant litter as it decomposed and increased in the stands of emergent vegetation as they became more structurally complex over time.

During the MERP studies, the point of maximum invertebrate abundance and diversity occurred during the transition from the regenerating to degenerating stage. At this time the submersed habitat, available litter, and algae were all abundant and diverse. Some taxa reached their maximum levels in stands of emergent vegetation, whereas others peaked in open-water sites with submersed vegetation. The proximity of these habitats during this transition between stages was important to the overall diversity of invertebrates in the wetland. This transition period includes the hemi-marsh phase described by Weller and Fredrickson (1974) and the diversity of invertebrates during this period probably plays a role in the high avian use observed in prairie wetlands during hemi-marsh conditions.

As the MERP cells proceeded through the degenerating stage, the emergent vegetation declined along with the invertebrates dependent on this vegetation for food and habitat. With the loss of the sheltering effects of the emergent vegetation, the submersed vegetation and its associated invertebrate community declined as well. As open, unvegetated areas began to dominate the flooded basin, the diversity of invertebrates was low and the community was dominated by large benthic taxa, primarily chironomids. Although the invertebrate community was relatively simple at this time, it reached high densities presumably due to the fine litter and algal food resources associated with the substrate surface.

Birds (Chapter 10)

The food and cover resources provided for wetland birds changed dramatically during the simulated wet–dry cycle in the MERP experimental cells. Although hemi-marsh conditions (cover and water available in approximately equal proportions in an interspersed pattern) are often considered the period of maximum avian use and production (Weller and Fredrickson 1974), MERP results showed that this is true for only a subset of species during the breeding season. Most dabbling ducks, some diving ducks, and blackbirds showed highest use of interspersed conditions during the spring breeding season. During the brood-rearing and moulting periods, dabbling ducks were found in areas with more cover, whereas diving ducks moved to more open-water areas. In the fall, most species moved to areas with open water. It is obvious that no one stage of the cycle meets the needs of all species for all seasons. Fluctuating water levels are important not only for maintaining wetland productivity but also for creating a diversity of habitat types that are required by wetland birds during various stages of their annual cycles. Although hemi-marshes provide important habitats for wetland birds, so do the other stages of the wet–dry cycle. Wetland managers who can create a mosaic of wetlands at different stages of the wet–dry cycle will undoubtedly attract the greatest diversity of wetland birds.

Although the changes in vegetative cover were obvious over the course of the wet–dry cycle, the impacts on avian food resources were less conspicuous. Our invertebrate studies did indicate that potential invertebrate foods changed over the wet–dry cycle as different invertebrate communities responded to changing habitat conditions. In addition, spring waterfowl use was related to invertebrate densities within the experimental cells. Later in the fall, however, when invertebrates are not an important component of the diet of most waterfowl species, this relationship between waterfowl and invertebrate densities disappeared. However, the ability of birds to exploit superabundant resources was demonstrated during the fall drawdown period when large numbers of waterfowl were observed feeding on the high densities of invertebrates in the newly exposed substrates within the cells. It is this adaptable nature of the bird species that inhabit prairie wetlands that allows them to respond to the very different habitat conditions available during the various stages of the wet–dry cycle.

Muskrats (Chapter 12)

As in most prairie wetlands, muskrats were the dominant herbivore in the MERP experimental cells. In prairie wetlands, densities of muskrats can vary with geographic location, wetland type, and stage of the wet–dry cycle. Densities also can fluctuate greatly within years (from low prebreeding levels in early spring to peak levels in early fall) and among years based on wetland conditions. Our studies indicated that density dependence had a large effect on population dynamics of muskrats. Rate of increase from spring to fall and overwinter survival were both related to population size. Although movements of muskrats among the experimental cells was fivefold greater in the fall than in the spring, movements could not be correlated with population density.

The MERP trapping data indicated that muskrats foraged in a variety of vegetation types, however, *Scholochloa*, *Typha*, and *Scirpus* were the most important foraging habitats. Muskrats also preferred *Typha* and *Scirpus* for lodge sites probably due to the combination of available food and water depth (for foraging under the ice in winter) in these habitats. Habitat selection by muskrats in the MERP experimental cells was consistent with the theory that relative density in preferred habitats increases more slowly than in less-suitable habitats as overall population density increases because of density-dependent competition within habitats with the highest value for reproduction and survival.

Although there has been much discussion on the impact of muskrats on marsh vegetation and the role of eatouts in prairie wetland succession, our data indicated that muskrats consumed only a small fraction of the total macrophyte production within the cells. Based on average muskrat consumption rates and the standing crops of emergent vegetation in the MERP cells, we estimated that a total of 1–11% of the annual standing crop of macrophytes was removed by muskrats on an annual basis. Based on this information and the relative timing of decline of the vegetation and muskrat populations, we concluded that although muskrats may have contributed to the decline of emergent vegetation in the MERP cells, they were not the principle cause of its elimination during the reflooding years. Conversely, the reduction in food and cover provided by the declining emergent vegetation resulted in increased muskrat mortality, which led to the subsequent decline in muskrat populations with continued flooding. Hence, muskrats may have been a victim, rather than a cause, of declining emergent vegetation.

Implications for General Wetland Ecology and Management

MERP was designed to provide experimental data to improve our understanding of the ecology and management of prairie wetlands. Ever since the early work by Weller and Spatcher (1965) and Weller and Fredrickson (1974) and the more detailed description of the wet–dry cycle by van der Valk and Davis (1978), the need for data from replicated wetland units during a complete wet–dry cycle had been obvious. The information presented in this book indicates that MERP made many important advances in this regard, confirming several earlier assumptions and adding new data on the factors influencing the overall productivity of prairie wetlands.

Our basic conclusion from MERP is that the macrophyte community plays a central role in the productivity of prairie wetland systems. Besides being the largest nutrient pool (aside from the sediments), they provide the main control points (plant uptake and decomposition) in prairie wetland nutrient budgets. Macrophytes have a profound influence on algal communities by providing submersed surfaces for epiphyton and sheltered areas required for metaphyton. However, macrophytes limit light penetration and thereby restrict algal growth within dense stands. Macrophytes also provide food and habitat for aquatic invertebrates. In addition, macrophytes provide the wide range of cover required by birds and other wildlife in prairie wetland systems. Macrophyte abundance is strongly correlated to muskrat productivity and survival over the course of the wet–dry cycle. Although the hemi-marsh stage is important to many wetland birds during the breeding season, it does not meet the needs of all species in all seasons. Thus, the changes in the macrophyte community during the wet–dry cycle have important implications for all components of the wetland ecosystem.

Another important contribution from MERP was an improved understanding of the role of algae in the overall productivity of prairie wetlands. The high annual productivity of the various algal communities probably plays a major role in the abundance and diversity of invertebrates in these systems. The axiom that "freshwater marshes are detritus-based systems" has certainly been challenged by our results.

Future Research

The previous chapters in this volume contain specific recommendations for further research for each of the major system components. Herein we highlight what we recognize as the priority research needs for better understanding the wet–dry cycle and its influence the ecology of prairie wetlands.

Nutrient Budgets

An important shortcoming of MERP was the lack of information on microbes and the role that microbial processes play in the nutrient cycles in prairie wetlands. Important among these processes are carbon and nitrogen exchanges with the atmosphere. Nitrogen fixation and denitrification probably play a significant role in nitrogen cycling, especially with the internal environment of prairie wetlands alternating between aerobic and anaerobic conditions, both over time and location within the wetland. There is a need for detailed studies of both fixation and denitrification in prairie wetlands so that the magnitude of these processes can be compared with that of the other pools and fluxes within the system.

Besides nitrogen fluxes related to microbial activity, the role of microbes in phosphorus and carbon cycling in prairie wetlands also requires attention. It has been suggested that bacteria control the phosphorus supply in large lake systems (Davelaar 1993; Eckerrot and Pettersson 1993), and it is likely that bacteria play a major role in these systems as well (Robarts and Waiser 1998). The importance of microbes as nutrient pools and organic sources for higher trophic levels also requires investigation. Another area that requires attention is the role of microbes in the cycling of labile matter. Pools such as dissolved organic carbon and associated fluxes in both surface and pore water have received little attention, yet they are undoubtedly important for understanding the fluxes of methane and other greenhouse gases from prairie wetlands (Robarts and Waiser 1998).

One unique characteristic of the MERP experimental cells is that there was no attempt to incorporate a waterproof core in the dikes to reduce water movement through them. This resulted in considerable seepage during periods when different water levels resulted in a substantial "head" across the dikes. Future work in this regard should include attempts to "waterproof" the dikes to reduce seepage to levels more consistent with groundwater flows in natural wetlands.

Macrophytes

Before the development of more sophisticated models of prairie wetland vegetation dynamics can be developed, better information is needed on the factors affecting macrophyte recruitment and growth. Data on seed production, dispersal, and longevity are required for the dominant emergent species in prairie wetlands. An important gap in our knowledge is information on clonal growth rates under different water depths, especially for long-lived species such as *Phragmites* and *Typha*. Long-term (>4 years) flooding tolerance of emergent species under varying water depths also requires investigation. Some of these long-term studies should span multiple wet–dry cycles in the same location. These studies do not require large experiments like MERP. Monitoring small permanent quadrats installed in natural wetlands over a number of years will provide valuable data to address these important issues.

With this new information, more sophisticated models should be developed for wetland vegetation dynamics. Although the Poiani and Johnson (1993) and Seabloom (1997) models are recent advances in this regard, next-generation models must include the effects of temperature during the growing season, seedling growth, and adult mortality and hydrological submodels.

Algae

The role of all algal communities in the nutrient cycling of prairie wetlands requires further research, especially for overall system primary production and nitrogen fixation. The importance of algae in the trophic dynamics of prairie wetlands was only briefly addressed in MERP and requires much more research, especially the use of the various algal communities by invertebrate and vertebrate grazers. Further taxonomic work on algae of prairie wetlands also is required.

Invertebrates

The species of invertebrates inhabiting prairie wetlands have not been described in sufficient detail for many of the important taxonomic groups; hence, taxonomic studies are still a priority for prairie wetlands. Before the necessary studies on invertebrate productivity and trophic dynamics can proceed effectively, basic life-history information is required for the many species inhabiting prairie wetlands. In addition, determinations of the functional roles of wetland invertebrate species and life stages within species are required before the trophic dynamics of these systems can be examined in detail. Further, the impact of water chemistry, long-term fluctuations in water levels, habitat structure, and changes in food resources on the dominant invertebrate species in these systems requires investigation.

Birds

Most studies of wetland birds have focused on waterfowl and blackbirds, but more information is required on the habitat, food, and cover requirements of the many other bird species in prairie wetlands. This understanding is needed to effectively evaluate the importance of wetlands to the many bird species spending all or part of their annual cycle in the prairie landscape. This information also will allow better prediction of how wetland management affects the entire avian community.

Muskrats

The use of controlled exclosure studies to determine the effects of muskrat herbivory on vegetation dynamics and detrital pathways is an important next step in the evaluation of the role of muskrats in the ecology of prairie wetland systems. There is also a need for information on the consumption rates of the various plant species by muskrats and the response of plants to muskrat herbivory (e.g., decreased regrowth, increased shoot production).

General Wetland Ecology

The effects of seasonal water-level fluctuations on all components of the prairie wetland ecosystem require further investigation. The role of seasonal changes in the overall wet–dry cycle also requires examination.

Conclusions

MERP confirmed the dominant role that macrophytes play in the nutrient budgets and overall ecology of prairie wetlands. The importance of algae in the overall primary productivity and trophic dynamics of prairie wetlands has been identified. Our understanding of the changes in invertebrate communities during the various stages of the wet–dry cycle and their importance in determining bird use of these systems has been expanded. Regarding avian use of prairie wetlands, MERP also has helped put the concept of the hemi-marsh into perspective compared with the other stages in the wet–dry cycle. Although hemi-marsh conditions are important, other stages of the wet–dry cycle also provide important habitat for waterbirds throughout the annual cycle. Finally, MERP demonstrated the relative importance of flooding effects and muskrat herbivory on the vegetation dynamics of prairie wetlands, and thus should stimulate further discussion on this issue.

In spite of these advances, MERP was not the definitive study on prairie wetlands. Like any good research project, MERP generated more questions than it answered. The roles of microbes in the nutrient cycling and trophic dynamics of prairie wetlands require extensive work. Highlighting the importance of macrophytes in all aspects of prairie wetland ecology has emphasized the need for additional life-history information on the important macrophyte species in these systems. MERP focused on waterfowl, blackbirds, and muskrats; however, the habitat requirements of many other species of animals using prairie wetlands, including other wetland birds, amphibians, and fish need additional examination. Finally, MERP was primarily a study of the effect of water-level changes on the ecology of prairie marshes. Future studies need to incorporate other factors such as water quality, including salinity and nutrient loading, and the impacts of surrounding land uses to advance our understanding of the structure and function of these dynamic ecosystems in an ever-changing prairie landscape.

Acknowledgments

This chapter is dedicated to R. Howard Webster and H.R. MacMillan who provided fellowship funding for many of the MERP field assistants. Mike Anderson, Todd Arnold, and Lisette Ross reviewed earlier drafts of this chapter. This is Paper No. 106 of the Marsh Ecology Research Program, a joint project of Ducks Unlimited Canada and the Delta Waterfowl and Wetlands Research Station.

LITERATURE CITED

Campeau, S., H.R. Murkin, and R.D. Titman. 1994. The relative importance of plant detritus and algae to invertebrates in a northern prairie marsh. Canadian Journal of Fisheries and Aquatic Sciences 51:681-692.

Davelaar, D. 1993. Ecological significance of bacterial polyphosphate metabolism in sediments. Hydrobiologia 253:179-192.

Eckerrot, A., and K. Pettersson. 1993. Pore water phosphorus and iron concentrations in a shallow, eutrophic lake - indications of bacterial regulation. Hydrobiologia 253:165-177.

Goldsborough, L.G., and G.G.C. Robinson. 1996. Pattern in Wetlands. In *Algal Ecology in Freshwater Benthic Ecosystems*. (Eds.) R.J. Stevenson, M.L. Bothwell and R.L. Lowe, pp. 77-117. New York: Academic Press.

Kantrud, H.A., G.L. Krapu, and G.A. Swanson. 1989. Prairie basin wetlands of the Dakotas: a community profile. United States Fish and Wildlife Service Biological Report 85.

Neckles, H.A., H.R. Murkin, and J.A. Cooper. 1990. Influences of seasonal flooding on macroinvertebrate abundance in wetland habitats. Freshwater Biology 23:311-322.

Neill, C. 1992. Life history and population dynamics of whitetop (*Scholochloa festucacea*) shoots under different levels of flooding and nitrogen supply. Aquatic Botany 42:241-252.

Neill, C. 1993. Seasonal flooding, soil salinity, and primary production in northern prairie wetlands. Oecologia (Berl.) 95:499-505.

Poiani, K.A., and W.C. Johnson. 1993. A spatial simulation model of hydrology and vegetation dynamics in semi-permanent prairie wetlands. Ecological Applications 3: 279-293

Robarts, R.D., and M.J. Waiser. 1998. Effects of atmosphere change and agriculture on the biogeochemistry and microbial ecology of prairie wetlands. Great Plains Research 8:113-136.

Robinson, G.G.C., S.E. Gurney, and L.G. Goldsborough. 1997a. Response of benthic and planktonic algal biomass to experimental water level manipulation in a prairie lakeshore wetland. Wetlands 17:167-181.

Robinson, G.G.C., S.E. Gurney, and L.G. Goldsborough. 1997b. The primary productivity of benthic and planktonic algae in a prairie wetland under controlled water-level regimes. Wetlands 17:182-194.

Ross, L.C.M., and H.R. Murkin. 1993. The affect of above-normal flooding of a northern prairie marsh on *Agraylea multipunctata* (Trichoptera: Hydroptilidae). Journal of Freshwater Ecology 8:27-35.

Seabloom, E.W. 1997. Vegetation dynamics in prairie wetlands. Ph.D. dissertation, Iowa State University, Ames, Iowa.

van der Valk, A.G., and Davis, C.B. 1978. The role of the seed banks in the vegetation dynamics of prairie glacial marshes. Ecology 59:322-335.

Weller, M.W., and L.H. Fredrickson. 1974. Avian ecology of a managed glacial marsh. Living Bird 12:269-291.

Weller, M.W., and C.S. Spatcher. 1965. Role of habitat in the distribution and abundance of marsh birds. Iowa Agriculture and Home Economics Experiment Station, Special Report 43, Ames, Iowa.

Winter, T.C. 1989. Hydrologic studies of wetlands in the northern prairie. In *Northern Prairie Wetlands*. (Ed.) A.G. van der Valk, pp.16-54. Ames: Iowa State University Press.

Appendix 1

Summary of MERP Techniques and Sampling Program

Henry R. Murkin
William R. Clark
Sharon E. Gurney
John A. Kadlec
Gordon G.C. Robinson
Lisette C.M. Ross
Arnold G. van der Valk

INTRODUCTION

The Marsh Ecology Research Program (MERP) long-term monitoring program involved a variety of techniques that were used to provide comparable data sets across the experimental cells and over the 10 years of fieldwork. Some of the techniques used were new, but most were adopted from other researchers. All have been modified to suit the needs for efficiency, economy, and practicality for long-term large-scale ecosystem research. A brief summary of the MERP techniques is provided in this appendix. More detail can be found in Murkin and Murkin (1989) and the individual publications emanating from this work (see Appendix 3).

STUDY AREA

Experimental Cells

The diked experimental cells were numbered 1 to 10 from west to east (see Figure 1.1). The approximate dimensions of each diked cell was 150m x 300m. Two undiked cells of similar dimensions were established in the main Delta Marsh. Cell 11 was located immediately west of cell 1 and cell 12 was east of cell 10 (Figure 1.1).

Main Transects

Each year 10 transects were randomly established within each experimental cell (Figure A1.1). These transects or "work lanes" were reassigned each fall following the last day of sampling. All sampling within a year took place as near as possible to the work lane. To minimize disturbance, all foot traffic was restricted to work lanes. When a sampling point

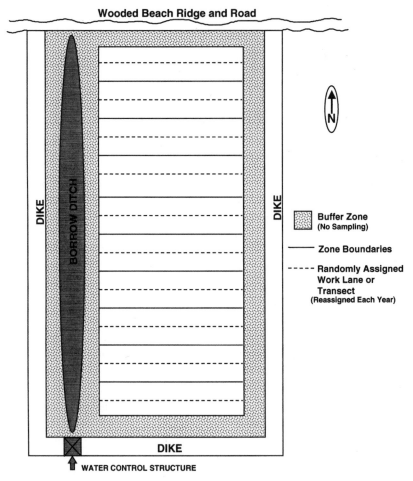

Figure A1.1   Schematic drawing of an experimental cell showing sampling zones, sampling transects or work lanes, buffer zones, and borrow ditch. Work lanes were reassigned each year.

was some distance from the work lane, a path perpendicular to the lane was established. No work lane or sampling station was located within a 10-m buffer zone around the inside edge of each experimental cell.

To assign transects and sampling stations, each experimental cell was divided into 10 equal zones (Figure A1.1). Zone boundaries were parallel to the southern dike. To determine the zones, 10 areas of equal width were established between a line 10 m north of the southern dike and a second line 10 m south of the road on the northern edge of the experimental cell. These zones were permanently marked so they could be located from year to year. Each season, a new transect was randomly selected within each of the zones. Both ends of the transect were marked with numbered stakes on the facing slopes of the dikes. The transects (and zones) were numbered 1 through 10 from north to south.

ANNUAL CLIMATIC FACTORS

The impacts of annual variations in local weather conditions on hydrology, physical–chemical conditions in the water and sediments, and many biological processes are well-known. Consequently, on-site measurements of standard meteorological characteristics were taken. The objectives of the measurements described in this section were to provide general background data on local weather and to provide data for use in computing water budgets.

Precipitation is reasonably well measured by standard rain gauges, which have the added advantage of being widely comparable to other data sets and studies. Evapotranspiration is less easily estimated. We chose several approaches: (1) standard class A evaporation pans, (2) standard class A evaporation pans installed in the marsh, and (3) estimation by a modified Penman equation (Eagleson 1970) by using data on temperature, wind, solar radiation, and relative humidity.

Weather Stations

Two weather stations were located just outside the experimental cells: one north of the complex and the second to the southwest of cell 1. Although most data were recorded daily, monthly averages were required for integration with the other data sets. Each of these stations included a rain gauge, a wet and dry-fall collection bottle, Stevenson screen with maximum-minimum thermometers, class A evaporation pan with maximum-minimum thermometers, and a totalizing anemometer. The stations were monitored daily from 1 May to 30 October each year. Two stations were used to better approximate average weather conditions over the entire study area. A third auxiliary weather station was established on the southern side near the center of the complex. This station consisted of a hygrothermograph (temperature and humidity) and a recording pyranograph (solar radiation).

Supplementary Weather Information

Two additional evaporation pans were placed in the experimental cells (exact locations varied from year to year to avoid disturbance and to sample different vegetation types). These pans were monitored daily in conjunction with the weather stations.

An elevated (15 m) totalizing anemometer was located in the center of the complex. It was monitored on a daily basis.

PHYSICAL ENVIRONMENT

Measures of the physical environment were required to provide data to help calculate or interpret processes important to biological productivity and nutrient budgets. These included soil and water temperatures, sediment bulk density and percentage of organic matter, import and export of suspended solids, and bottom contours. Data on bottom topography were necessary for calculating water level–volume relationships within the experimental cells. Similarly, amounts of sediment pumped with water into or out of the cells were measured to determine impacts on bottom-sediment composition and topography. Finally, sediment bulk density and organic matter determined the quantities and distributions of many chemical species in the sediments.

Soil and Water Temperatures

Soil and water temperatures were measured in conjunction with invertebrate monitoring. Each time an invertebrate station was sampled (weekly) the following temperatures were recorded: water column, surface of the sediments (top 1 cm), and 10 cm into the sediments. Twenty-four hour maximum and minimum water temperatures were recorded in representative stands of each major vegetation type on a weekly basis. Temperatures of interstitial water samples also were recorded at the groundwater well sites (see Water Chemistry section following).

Characterization of the Sediments

An initial soil survey in April 1979 provided soil particle size data. Ten samples were taken on a north-south transect on the eastern side of each of the experimental cells. Soil samples also were collected twice a year for determination of bulk density, percentage of water, and percentage of organic matter. Samples were taken with a 5-cm-ID corer at each of the groundwater well sites (see Water Chemistry section) during the last week of May and the last week of September. Analyses followed standard soil techniques.

Determination of Suspended Solids

The amount of silt and other suspended solids introduced to or removed from the experimental cells due to pumping or gravity flow through the control structures was determined by collecting weekly water samples from the pump discharge or flow through the control structure. The amount of suspended material per unit volume of water (dry weight) was determined by filtering procedures.

Bottom Contours

At the beginning of the study the bottom contours (10-cm intervals) of the experimental cells were determined using standard survey techniques. This information was digitized for each experimental cell. The cells were then resurveyed periodically (approximately every 2 years) to monitor changes in bottom contours in response to varying water levels. Staff gauges in each experimental cell also were surveyed when they were initially installed near the control structures and in the main marsh (see Hydrology section following). This allowed routine contour checks to be made by using water depths and staff-gauge readings. Water depths were recorded during all types of sampling, thereby connecting each sampling site with a specific contour. Staff gauges were resurveyed each year to correct for any changes caused by frost or settling.

HYDROLOGY

Accurate estimates of water budgets are essential prerequisites for studies of wetland nutrient budgets. The approach to hydrology used in this study was the concept of mass balance where inputs − outputs = change in volume. All three terms were estimated as part of the long-term monitoring program. Detailed description of methods used for determination of water budgets in the experimental cells can be found in Kadlec (1983, 1986a, 1993).

Changes in Volume

Changes in volume were calculated from daily records of water levels. Water levels in each experimental cell and the main marsh were recorded daily. To provide more detailed water-level data when required, two continuous water-level recorders were placed in stilling wells near the southern end of the borrow ditches in two experimental cells (one permanently located in experimental cell 6 and the other moved from marsh to marsh on a monthly basis). Water levels combined with the bottom contour data were used to calculate the volume of water in each experimental cell.

Water Inputs and Outputs

Input of water to the diked experimental cells was in three forms: precipitation, water pumped to maintain design levels, and seepage through dikes or the ridge forming the northern end of the experimental cells. Outputs of water were evaporation, evapotranspiration, pumping, and seepage. Precipitation inputs and evaporation were calculated from weather records and pumping (in and out) was metered. Seepage was estimated based on the other terms in the water-budget equation.

Supplementary Groundwater Sampling

To refine estimates of the inputs and outputs for the water budget of the experimental cells, several additional measurements were made to more accurately define the water storage, groundwater levels, and losses through seepage. Early in the 1982 field season, a series of groundwater wells was installed immediately adjacent to the northern and southern borders of each experimental cell. In addition, one well was established halfway along the western side of marsh 1 and another halfway along the eastern side of marsh 10. Wells to the south, east, and west of the experimental cells were placed at the outside base of the dike. Placement of the northern wells varied; however they were placed as close to the experimental cell as the terrain permitted.

These adjacent wells were made of PVC pipe (≈4 cm ID) and permanently capped at the bottom end. Installation consisted of using an auger to drill a hole to the appropriate depth. All wells had a sampling zone consisting of a ring of 1-cm holes (≈25 cm above the bottom of the well) covered by 1-mm mesh. The sampling zone was centered 50 cm below the groundwater table for wells to the south, east, and west and 30 cm below the groundwater table for wells to the north. The open tops of the wells extended above the ground surface and had removable caps. The elevations of the top of these wells were determined using standard survey techniques.

In 1985, additional groundwater wells were placed in the unflooded portions of experimental cells 3, 7, 9, and 10 to monitor water-table topography and groundwater chemistry. They were similar in design to the adjacent groundwater wells described previously, except that the sampling ports were ≈15 cm below the level of the groundwater table. As with the adjacent wells, the elevation of the top of each well was established each year by surveying. Sampling was done monthly for nutrients in conjunction with the regular groundwater sampling within the experimental cells (see following text). Each month the depth to the groundwater also was measured to determine actual elevation of the groundwater table.

The hydraulic conductivity was measured at the 22 adjacent water-table wells and the wells in the dry areas within the experimental cells twice during each field season (late May and early September). This was accomplished using a float apparatus (dowel [5-mm-diameter] calibrated in centimeters attached to a plastic bottle [3-mm-diameter] as a float). Five hundred milliliters of water was added to the well and the float apparatus inserted. The time required for the water level in the well to drop 10 cm was recorded.

WATER CHEMISTRY

In this section, we describe the measurements needed to estimate the major water-related components of the overall nutrient budgets. By multiplying nutrient concentration data by water volume, the mass of nutrients entering, leaving, or stored in water was calculated. Once again, we used the concept of mass balance: inputs − outputs = change in storage. Detailed methods can be found in Kadlec (1986a,b).

Nutrient Inputs and Outputs

For precipitation inputs, wet/dry-fall bottles were set out at each of the weather stations and near the evaporation pans within the experimental cells. After each rain event, samples were collected from these bottles and analyzed as described in the section following Analyses Performed. During periods of pumping, a 500-ml sample was taken weekly at the pump intake for each experimental cell. To monitor seepage inputs, on the last Monday of each month, when the water-table level in the adjacent wells was determined, a 500-ml water sample was collected from the well by using a hand pump and a piece of tubing long enough to reach the water table.

Nutrient Pools within the Wetland

Experimental cell 9 was selected for a detailed survey of surface and interstitial water. Within this experimental cell, one sampling station was randomly placed within the flooded area on each of the main sampling transects (a total of 10 stations). Beginning in 1986, sampling was reduced to five of these stations. At each sampling station, three groundwater wells (4-cm-ID PVC pipe) were installed with the sampling zone (ring of 1-cm holes) centered at one of three depths (15, 30, and 45 cm below the soil surface). In the remaining nine diked cells and undiked cells 11 and 12, one sampling station on five transects was randomly selected within the flooded area (a total of five sampling stations per cell). At each of these stations one groundwater well (15 cm in depth) was installed. The wells were sampled monthly from 1 May to 30 October by using a hand-held vacuum pump to withdraw the sample from the well. On the same schedule, a 500-ml surface-water sample was taken at each sampling station.

Analyses Performed

The following analyses were performed on the samples at the Analytical Chemistry Unit at the Freshwater Institute, Department of Fisheries and Oceans, in Winnipeg, Manitoba, following the techniques of Stainton et al. (1977). Temperature, conductivity, and pH measurements were made at the Delta station laboratory.

a. Surface water, pumped water (in and out), water flowing through the control structure (in and out), and rainwater: ammonium (NH_4), nitrate (NO_3), total dissolved nitrogen (TDN), soluble reactive phosphorus (SRP), total dissolved phosphorus (TDP), chloride (Cl), sulfate (SO_4), suspended phosphorus (SSP), suspended nitrogen (SSN), suspended carbon (SSC), dissolved inorganic carbon (DIC), and dissolved organic carbon (DOC) on all samples each month. The sodium (Na), potassium (K), calcium (Ca), and magnesium (Mg) concentrations were determined for all May and August samples.

b. Groundwater: NH_4, TDN, SRP, TDP, Cl, SO_4, DIC, and DOC on samples from all wells (22 adjacent wells and all wells in experimental cells) in May, August, and October. These same analyses were performed on all samples from wells within experimental cells only in June, July, and September. The Na, K, Ca, and Mg concentrations were determined for samples in May and August only.

ALGAL BIOMASS AND PRIMARY PRODUCTION

The objective of the algal monitoring program was to evaluate the influence of water depth on all algal communities within the MERP experimental cells.

Sampling Techniques

The following methods were used to determine the biomass of planktonic, epiphytic, epipelic, and metaphytic algal associations in all the experimental cells during the 5-year period following drawdown (1985–1989). Also described are the methods used to model the primary productivity of each algal association. Algal methods used prior to 1985 are described by Hosseini and van der Valk (1989 a, b).

Phytoplankton

For phytoplankton-biomass sampling, three permanent sites were selected within each experimental cell on a depth-stratified basis. At ≈ 3-week intervals throughout each ice-free period, integrated water-column samples were removed by the use of a 5.8-cm (ID) Plexiglas® tube. Three 500-ml subsamples were then filtered individually onto 4.25-cm GF/C filters and frozen for subsequent chlorophyll a determination. An additional 500-ml subsample was taken when conducting carbon assimilation experiments (see following text).

Epiphyton

For epiphyton sampling in 1985, from four to nine sites per cell were established to be representative of a relatively complete range of water depths and vegetation cover types. From 1986 to 1989 permanent sampling sites were established adjacent to the permanent macrophyte monitoring quadrats (see Macrophyte Sampling Methods following), which were situated in the flooded zones of each experimental cell. Epiphyton biomass was sampled by the use of extruded clear acrylic artificial substrata (Goldsborough et al. 1986), which had been positioned among the vegetation immediately following ice-out each year. At 3-week intervals, four substrata were removed from each site and sections of each (representing evenly distributed depths throughout the water column) were subsampled and frozen for chlorophyll a analyses. When samples were collected for productivity estimates, sections of the substrata were sampled and placed in culture tubes containing 25 ml of 0.45-mm filtered site water.

Epipelon

Epipelon was only sampled from areas of flooded open sediments. One sampling site was chosen at random from within each marsh on each 3-week sampling period. Triplicate sediment samples were taken at each site by aspirating the surface few centimeters of sediment enclosed within a 5.8-cm (ID) Plexiglas® coring cylinder. The method for sampling was essentially that of Eaton and Moss (1966), with the exception that to account for possible sequential migration of algal cells, lens tissue traps were removed and replaced three times during the cell collection. The lens tissue traps were individually frozen for subsequent chlorophyll a analysis. At times when productivity measurements were made (see following text), lens paper traps from each sampling period were placed together in 100 ml of filtered site water, shaken vigorously, then 25-ml allotments dispensed into culture tubes.

Metaphyton

Because of the temporal variability of metaphytic mat development, it was not possible to establish permanent sampling sites. Instead, on each 3-week sampling interval, sampling sites were chosen at random from all potential sites (those at which metaphyton occurred). Four 0.25-m^2 quadrats were positioned every 10 m along each work lane. If metaphyton occurred in any quadrat it was then considered as a potential site. At each chosen site, the percentage of cover of metaphyton was estimated within the quadrats. The water depth and emergent and submergent macrophyte densities also were recorded at each site. Following the metaphyton cover estimates, biomass and productivity sampling were conducted. The number of sites in each cell was usually three, unless there were fewer potential sites available. Sampling of metaphytic biomass is described by Gurney and Robinson (1988) and involved raising the metaphytic mat above the water surface on a Styrofoam® platform, after which three 1.54-cm² cores were removed and frozen individually for later chlorophyll determination. Metaphyton cores sampled for productivity estimates (see following text) were placed in culture tubes containing 25 ml of site water.

Biomass Estimation Methods

For all algal associations, biomass was expressed as chlorophyll *a* (corrected for phaeophytins) and determined by the method of Marker et al. (1980). Each extraction for phytoplankton, epiphyton, and epipelon pigment was conducted in 8 ml of 90% methanol, and in 20 ml of methanol for metaphyton. Extraction was conducted over a 24-h period in the dark at 5°C during which samples were shaken or vortexed three times. Subsamples of 4 ml were transferred to spectrophotometer cuvettes by using a syringe assembly fitted with a MSI glass G25 prefilter (1-m pore size).

Biomass Extrapolations

To estimate total algal biomass and productivity in each experimental cell, extrapolations of the measured biomass were required. For aerial estimates, the surface area of each 10-cm isobath was determined for each cell. Biomass estimates, as described previously, were assigned to their isobath of origin and values for intervening isobaths were derived by interpolation. Any isobath shallower than the shallowest measured value was ascribed that value, and any isobath deeper than the deepest measured value was ascribed that value. Mean biomass estimates (milligrams of chlorophyll *a* per square meter) were determined for the entire area of each isobath in each cell, and by summation for each cell as a whole. Detail for this computation as applied to each association can be found in Robinson et al. (1997a).

For phytoplankton biomass the extrapolation simply involved the multiplication of the mean chlorophyll *a* value of each isobath by the volume of water in each isobath. For epiphyton it was necessary to determine the surface area of macrophytes per square meter of wetland to permit the appropriate extrapolation of epiphyton biomass. This was done by first measuring and computing the submersed surface area of each standing macrophyte species associated with each epiphyton sampling site. Measurements from each site were used to account for possible water-depth and vegetation-density effects on the physical dimensions of macrophytes. The density of each species at each site then allowed for the derivation of the total surface area of standing natural submersed substratum per square meter of wetland. For epipelon it was necessary to estimate the area of unvegetated submersed sediment in each marsh. Calculation of metaphyton biomass required the use of multiple regressions (Gurney and Robinson 1988), which described the effect of vegetation cover and water depth on metaphyton cover.

Productivity Modelling

For the purposes of modelling the primary productivity of the four algal associations, Fee's (1973) approach, similar to the way it had been used to estimate epiphytic algal productivity by Jones (1984), was adopted. This approach incorporated available seasonal estimates of the biomass of each association expressed as area-based chlorophyll *a*, numerical relationships (P versus I) between biomass normalized carbon assimilation rates (specific productivity) and irradiance for each association, and available photosynthetically active radiation (PAR).

For the determination of the P versus I relationship for each association, experiments were conducted in which algal samples from all experimental cells were used. In each experiment samples of known biomass were incubated at ambient water temperatures in the presence of a known trace amount of ^{14}C sodium bicarbonate under known light conditions ranging from 0 to 2,000 μmole/m²/s for a 2-h period prior to individual filtration onto 0.45-mm membrane filters. Incubations were conducted in a water-filled incubator illuminated by a high-pressure sodium vapor lamp. The irradiance to which each sample was exposed in the incubator was determined with solar monitors. Following incubation and filtration, filters were fumed over concentrated HCl for 1 minute to remove residual inorganic ^{14}C, before being placed in scintillation vials containing a scintillation cocktail. Determination of the specific radioactivity (dpm) of each sample was conducted in a liquid scintillation counter, programmed for "H" number quench correction. Specific (chlorophyll normalized) productivity was determined according to the following equations:

$$\mu g \text{ C fixed/unit water volume(or unit area)/h/}\mu g \text{ Chl } a = \frac{\text{dpm(S)} \times C \times 1.05}{\text{dpm(T)} \times T \times \text{CHL}}$$

where: dpm(S) – specific radioactivity of each sample corrected for dark uptake;
dpm(T) – specific radioactivity of added ^{14}C bicarbonate;

C – dissolved inorganic carbon (DIC) of marsh water (μg) as determined from alkalinity (APHA 1980), pH and temperature;
CHL – phaeophytin-corrected chlorophyll *a* (μg);
1.05 – ^{14}C discrimination factor; and
T – incubation time (h)

The data base upon which the P versus I relationship for each association was determined consisted of 21 experiments yielding 1,141 specific productivity values with corresponding irradiance levels for phytoplankton; 34 experiments and 2,481 values for epiphyton; 17 experiments and 962 values for epipelon, and 22 experiments and 636 values for metaphyton.

The curve fitting equation of Platt et al. (1980) was used to describe the P versus I relationship for each association:

$$P_s = P_{max} \times (1 - e^{(-\alpha \times I)/P_{max}}) \times e^{(-\beta \times I)/P_{max}}$$

where: P_s – specific (chlorophyll normalized) productivity,
         P_{max} – maximum rate of P_s attained,
         α – slope of light-limited photosynthesis (photosynthetic efficiency),
         β – slope of photo-inhibition; and
         I – light intensity ($\mu mole/m^2/s$).

The parameters of each relationship were determined by nonlinear regression.

Based upon regular sampling of all four algal associations at sites within each cell (as described previously), area-based estimates of algal biomass (as chlorophyll a per square meter) were compiled for each 10-cm isobath in each cell over a period of 5 years (1985–1989 inclusive) for each sampling day, and by interpolation for all intervening days between sampling dates. These data formed a crucial component in the modelling of the primary productivity of each association in each cell.

To determine the productivity of each community in each isobath in each cell, hourly integrations of incident photosynthetically active radiation (PAR) were continuously recorded throughout the sampling periods by way of a solar monitor equipped with a quantum sensor positioned on an elevated platform within the MERP site. Depending upon the density of emergent vegetation, light reaching the water surface can be expected to be variably reduced by shading. In addition, emergent macrophyte density is probably water-depth dependent. Accordingly, to assess the degree of shading and any dependence on water depth, instantaneous values of PAR above the vegetation and at the water surface were accumulated on a 3-week cycle, in which these measurements and water depth were recorded at 10-m intervals along two randomly selected sampling lanes in each of the 10 cells, and a relationship between shading light loss and water depth was developed. A 10% reflectance loss (Hutchinson 1975) was universally applied. Extinction of light within the water column was assessed using the a 3-week cycle. At each 10-m interval along the sampling lanes light at 5-cm depth intervals was measured with an underwater quantum sensor and extinction coefficients derived (Wetzel and Likens 1991). It was not assumed that extinction coefficients might be independent of water depth, but rather a relationship between extinction coefficient and water depth was developed, which could then be universally applied to determine PAR availability for the purposes of computing productivity.

Throughout each of the 5 years of the study, for each day in each ice-free period, for each isobath that existed within each cell, a biomass (chlorophyll a per square meter) was either directly available or was derived by interpolation for each algal association. For each hour in which there existed measurable incoming PAR in each day a PAR value ($\mu mole/m^2/s$) was determined for each isobath. This value was determined according to the relationship between isobath depth and percentage of light reduction through shading, further reflective loss, and finally by the application of an appropriate extinction coefficient derived from the relationship between extinction coefficient and isobath depth. For each community, productivity values were then derived as according to Robinson et al. (1997b).

MACROPHYTE PRODUCTION

The major objective of the long-term monitoring of aquatic macrophytes was to estimate the above- and belowground net annual production. Standard harvest techniques were used because they are the most direct, simple, and reliable techniques available for estimating the standing crop of macrophytes per unit area.

Belowground Macrophyte Production

An estimate of net annual belowground macrophyte production was based on changes in belowground standing crop from late spring and early summer to fall. In late spring, when shoot initiation has used up most of the belowground store of carbohydrates, macrophytes have their minimum belowground standing crop. In the fall belowground biomass has reached its seasonal maximum. Spring sampling began when the cattails (*Typha* spp.) started to flower, the time when underground reserves are lowest for this species. Fall sampling began when the cattails were visibly senescing (i.e., turning yellow [near the end of September]). The difference between fall and spring standing crops provided an estimate of the net annual belowground production.

In both the spring and fall sampling periods two sample sites were randomly selected along each transect in each of the 12 experimental cells. All samples were collected from a 10-m transect running due north of the actual sampling site. Three coring sites were selected randomly along this transect. Cores were 15 cm in diameter and taken to a depth of at least 20 cm (deep enough to sample the entire root and rhizome zone) by using a coring device. The three cores collected at each site were placed together in a plastic bag. Average water depth at the coring sites was recorded.

All samples were washed and sorted into live and dead components by species. Samples were then dried to constant weight at 80°C and weighed. All samples were ground in a Wiley mill prior to analyses for nutrient content.

Aboveground Macrophyte Production

Maximum standing crop was used to estimate annual aboveground macrophyte production. Because most prairie emergents reach their maximum standing crop in late July or early August, sampling took place during this period. A single clip-plot method underestimates net primary production; however, this estimate can be corrected using the data from the turnover plots (see following text).

Four sampling sites were randomly located on each transect in each of the 12 experimental cells. Sites disturbed by spring belowground sampling were avoided. A 1 × 1 m (*Typha*, *Phragmites*, *Scirpus*) or a 0.25 × 0.25 m quadrat (all other species) was placed at each site. All standing vegetation (living and dead) within the quadrat was clipped. Water depth was recorded at flooded sites. Samples were sorted by species and living and dead components. Following sorting, all samples were dried to constant weight at 80°C and weighed. All samples were then ground in a Wiley mill prior to nutrient analysis.

Turnover Plots

Only the dominant species (*Typha*, *Phragmites*, *Scirpus*, *Scholochloa*) were monitored. The turnover plots were located in the 10-m buffer zones within the experimental cells and monitored every 2 weeks beginning the first week of June. During the first count of each year, all green stems in each plot were marked with orange flagging tape. The number of newly marked stems was recorded. Any dead stems from the previous year were retagged with yellow flagging tape and counted. During all subsequent counts, the fate of the marked stems was followed as well as the production and fate of new stems appearing throughout the year. Representative samples (50 stems) of each stem type (flowering or nonflowering) were collected from adjacent stands for nutrient analyses. A sample (500 g fresh weight) of living rhizomes for each species also was collected for subsequent nutrient analyses.

Photo Stations

Permanent photo stations were established on each experimental cell (two per cell) to provide a visual record of changes in macrophyte distribution and abundance. The observation tower overlooking each experimental cell (see Vertebrates section following) served as one station. The second station was located in the center of the southern boundary of each experimental cell. Color slide photographs (35-mm) were taken from each station on the first day of each month from 1 May to 1 November. The tower station included three photographs: northern half of the cell, center of the cell, and southern half of the cell. Fixed brackets in the tower were used to ensure the same scenes were photographed each time. The southern boundary photograph was taken from the southern dike and attempted to include as much of the experimental cell as possible.

General Vegetation Survey

Each August, vegetation was sampled in 40, 1 × 1 m quadrats in each experimental cell. Four quadrats were placed randomly along each of the 10 work lanes. In each quadrat, the cover of each species and its standing litter was estimated. The mean height of each species and the percentage of stems flowering also were recorded. From these data, the overall composition of the vegetation in the experimental cells was determined and comparisons of the vegetation among cells made. For example, the frequency of occurrence of each species in an experimental cell can be calculated as well as its total cover, and these data can be used to calculate the similarity of the vegetation in experimental cells in different treatments in any one year as well as the similarity of the vegetation from one year to another in a given cell.

Permanent Quadrats

Ten permanent quadrats (2 × 2 m) were established in each experimental cell early in the first year of drawdown. These quadrats were located by dividing each experimental cell in five equal zones from north to south (lengthwise) and two equal zones from east to west (widthwise). A plot was placed at random in each of these 10 zones. The corners and center of a plot were marked by iron fence posts, and its perimeter and diagonals strung with wire to delimit four triangular quadrats. The elevations of the plot at its four corners and center were obtained with a level transit, and averaged to estimate the mean elevation of a plot. Permanent quadrats are sampled three times each growing season (June, July, and August). The number of flowering and vegetative shoots of each species was counted in the northern and southern triangular quadrats (i.e., 2 m^2 were sampled). The densities of dead shoots of each dominant emergent species also were recorded. By using triangular quadrats, the vegetation in a permanent quadrat was never disturbed on two sides. During the drawdown years when seedling densities were very high, 1-m^2 subsamples were taken to estimate seedling density in a plot. Soil samples to determine soil moisture and salinity were taken adjacent to permanent quadrats. When they were flooded, water depth also was measured.

   Data from the permanent quadrats was used to monitor quantitative changes in the vegetation of an area over an elevational gradient (Welling et al. 1988a), during the growing season (Welling et al. 1988b), or from year to year (van der Valk and Welling 1988).

AERIAL PHOTOS AND COVER MAPS

To determine changes in the distribution and total area of the various vegetation types in the experimental cells, vegetation maps were made from low-level aerial photographs taken of each cell toward the end of each growing season. From these maps, changes in the vegetation present from year to year were established and the areas of each vegetation type in an experimental cell calculated. This information was used along with data from the vegetation studies to estimate total annual production and amounts of C, N, and P stored in the vegetation within the experimental cells. The vegetation maps were essential for extrapolating the many different types of MERP data to the entire experimental cells.

Aerial Photographs

Aerial infrared photographs were taken annually of each experimental cell with a 70-mm camera from a height of 610 m above ground level. The camera was equipped with a 50-mm lens corrected for aerial photography with a Wrattan B and W 16 orange filter. Prints of each experimental cell were made to a predetermined scale by using a standardized template. By following this procedure, a uniform scale was maintained among experimental cells and years.

Cover Mapping

Upon receipt of the aerial photograph prints, the experimental cells were cover mapped using mylar overlays on the prints. Mapping was done in the same field season as the photographs were taken. Each vegetation stand was traced and labelled on the mylar and identified by ground-truthing. During ground-truthing, five water depths were measured within each continuous dominant cover type. Each measurement site was located as precisely as possible on the cover map overlay by using visible land or vegetation features. The water level at the staff gauge within each experimental cell was recorded on the same day. Water-depth measurements were then converted to ground elevations by referencing them back to the staff-gauge readings.

The contour maps produced early in the study were overlaid on the vegetation maps to allow determination of vegetation changes according to contour levels and water depths throughout the study.

DECOMPOSITION

There are three components of the litter pool in prairie wetlands: standing litter, fallen litter, and dissolved organic compounds that leach from both standing and fallen litter (Davis and van der Valk 1978). The MERP decomposition studies were a series of short-term studies examining the decomposition rates of both above- and belowground components of the dominant emergent species and the transfer of litter from the standing to the fallen litter compartment within the experimental cells.

Litter-Bag Studies

Several decomposition studies were conducted over the course of the MERP experiments. Standard litter-bag techniques were used in all these studies. Examples of these studies follow.

Aboveground Shoot Decomposition Study (1986)

The objective of this study was to determine the decomposition rate of the five dominant emergent species from the 1985 growing season: *Phragmites australis*, *Scirpus lacustris*, *Scirpus maritimus*, *Scolochloa festucacea*, and *Typha glauca*. The above-water portion of shoots of each species was collected on 23 April 1986. Each species sample was a composite from several experimental cells. Samples were dried to a constant weight (for ≈48 h) at 20°C to determine fresh weight. To obtain "time zero" nutrient analyses, random control samples for each species were removed, dried to constant weight at 80°C, ground in a 40-mesh Wiley mill, and stored for later analysis. To estimate a weight ratio between fresh litter (20°C constant weight) and dry litter (80°C constant weight), 30 random preweighed fresh litter subsamples per species were dried to 80°C constant weight.

Litter bags were prepared by placing predetermined amounts of litter in sealed 20×40 cm polyethylene mesh bags (1-mm mesh size). One-hundred-fifty litter bags were prepared for each species. Five bags of each species were deployed at three different sites in each experimental cell on 26 May 1986. The collection schedule was one bag per site during each of the five collection periods: day zero, 48 h after deployment, July 1986, July 1987, and June 1988.

In the laboratory, the litter was removed from the bag and washed in distilled water to remove soil and extraneous plant and animal material. Samples were dried to constant weight at 80°C and weighed. Following weighing, each litter sample was ground in a 40-mesh Wiley mill and stored for later nutrient analyses (N, P, C). This nutrient information was combined with the dry weight data to determine mass of N, P, and C remaining for each species at each collection period.

Belowground Decomposition Study (1988)

The purpose of this study was to determine the decomposition rate of roots and rhizomes from four dominant emergent species in the Delta Marsh: *S. festucacea*, *S. lacustris*, *P. australis*, and *T. glauca*. Roots and rhizomes of the four designated species were collected in April 1988. A 5-cm segment of shoot was left attached to each sample. Each species sample was a composite from several experimental cells. To determine fresh weight, samples were dried to constant weight at 20°C. To obtain "time zero" nutrient analyses, random samples of each species were removed, dried to constant weight at 80°C, ground in a 40-mesh Wiley mill, and stored in labelled Whirlpak bags. The ratio between fresh litter (20°C) and dry litter (80°C) was determined as described previously.

Litter bags were prepared by placing 10 g of fresh weight in sealed 20×40 cm polyethylene mesh bags (1-mm mesh size). One-hundred-fifty litter bags were prepared for each species. Five bags of each species were deployed at three different sites in each experimental cell. Bags were inserted below the soil surface in slits created by inserting a spade and pulling it to the side. Bags were positioned so that the shoot protruded above the marsh substrate.

The collection schedule was one bag per site during each of the following collection periods: 48 h after deployment, July 1988, September 1988, June 1989, and August 1989. Laboratory procedures followed those described previously for the aboveground study.

Litter-Fall Study

The purpose of this study was to estimate changes in aerial litter standing crop (biomass and [N, P, C]) by monitoring seasonal litter-fall patterns of *T. glauca*, *S. lacustris*, *S. festucacea*, and *P. australis*. Aerial standing live shoots were marked and changes in the aerial standing litter component were measured over time. To initiate each experiment, four live shoots were marked with flagging tape at each of 70 sites. The criteria for site selection were the natural depth range for each species. New sites were selected in subsequent years when vegetation became sparse on previously selected sites. The marked shoots were followed on a monthly basis until they were completely toppled. Marked stems were used as a reference when collecting matching stems for weight and nutrient analyses. The first collection of matching stems was concurrent with the initial marking of permanent stems. A nearby stem, matching the marked stems as closely as possible in length, thickness, and number of leaves, was clipped, dried, weighed, and ground for nutrient analysis. Generally, each experiment had a time frame of 1 year from date of initiation.

Clip Quadrats

The purpose of this study was to estimate annual standing litter production by monitoring standing litter biomass and nutrient concentration (N, P, C) by using clip quadrats.

A total of 15 sites in each of the 10 diked experimental cells was selected. The criterion for site selection was to represent the natural depth range for each species. All vegetation within a 1 × 1m quadrat was clipped at all sites. The samples were separated into living and dead components by species. All samples were dried to constant weight (80°C) and weighed. All dried samples were ground in a 40-mesh Wiley mill and analyzed for total N, P, and C.

INVERTEBRATES

Sampling Design and Samplers

Invertebrate sampling sites within the experimental cells were stratified by cover (vegetation) type based on the dominant emergent plants and water depth. In the early years of the study (from 1981 to 1984) when all experimental cells were flooded at the same depth in any given year, sampling sites were stratified strictly by cover type. Six sampling sites were randomly selected for each cover type during these years. From 1985 to 1989, when water depths varied among cells, two water depths (30 cm [shallow] and 60 cm [deep]) strata were determined for each cover type. Six sampling sites were randomly chosen in each water depth for each cover type.

Density, biomass, and number of taxa were used as response variables. Because of taxonomic problems with the invertebrates of wetlands and the overlap of the trophic categories suggested by Cummins (1973), two broad trophic levels were distinguished: herbivore–detritivore and predator–parasite.

At each sampling site, there were two habitats sampled: the water column and substrate. Because no one sampler effectively samples the entire invertebrate community (Swanson and Meyer 1973), a variety of samplers was used to provide separate data on the nektonic, benthic, and epiphytic groups. Activity traps were used at sampling sites with standing water to provide an index to the abundance of free-swimming invertebrates (Murkin et al. 1983). Activity traps were set for a 24-h period biweekly at each site. An artificial substrate was placed midway in the water column at each sampling site with standing water to provide an index to gastropod and chironomid populations colonizing submersed surfaces. These were sampled on the same schedule as activity traps. Emergence traps were used to provide an index to aquatic insect population levels. The emergence trap used in this study was a modification of LeSage and Harrison's (1979) model "Week trap" with a base size of 0.5 m². Emergence traps were monitored on a weekly basis.

All macroinvertebrates in the samples were sorted by taxa and counted. Usually, identification was to family. Notable exceptions occurred in the Chironomidae, which were identified to subfamily. For many of the lower invertebrate groups, (e.g., Oligochaeta, Nematoda) identification was to order. Biomass was determined by drying subsamples of each taxon and extrapolating to the entire sample.

VERTEBRATES

Muskrats

The objectives of monitoring muskrat responses within each experimental cell were to document density within each water level treatment of the MERP experiment, and to determine which demographic parameters are responsible for the observed densities. Accurate estimates of population density require mark–recapture experiments (Clay and Clark 1985; Clark and Kroeker 1993), especially if there is a desire to measure other parameters such as survival and dispersal.

Muskrat density was estimated by sampling for 6 days by using capture–recapture techniques. Estimates of population size were calculated assuming a closed population (Program CAPTURE, Otis et al. [1978]). The goal was to determine two point estimates and associated variances for each experimental cell each year, one in May and another in September. Survival rates were estimated from the capture–recapture data by using a hybrid design described by Pollock et al. (1990). Captures from the 6-day periods were pooled and Jolly–Seber (Seber 1982) estimates of survival rate between the point estimates of population size were calculated using Programs JOLLY and JOLLYAGE (Pollock et al. 1990). Densities, survival rates, rates of increase, and changes in body condition were compared among treatments and seasons by using analysis of variance with an error term of variation among replicates within a treatment. Associations between density and the other variables were tested with simple, linear regression. Chi-square tests were used to identify differences in movements among experimental cells (Clark and Kroeker 1993).

Trapping periods began shortly after ice-out on all experimental cells and was completed in ≈6 weeks. Trapping resumed ≈1 September to allow ample time for completion before freeze in the fall. Each cell was trapped for six consecutive days and traps moved to a new cell on the seventh day. The cells were trapped in random order. Each experimental cell was set with 45 live traps arranged in a grid along the 10 work lanes. The goal was to get systematic coverage of the entire cell. New trap locations in the experimental cells were established each fall to accommodate the shifting work lanes. The same locations were used again the next spring. Traps were suspended above the water from two or more stakes. The door of the trap was at, or slightly below, the water line. The other end of the trap was above the water to allow the captured animals to remain dry. No bait was used, but the traps were covered with vegetation so they appeared as small muskrat shelters. Captured animals were transferred from the trap to a handling cone (McCabe 1983) for weighing. An animal was classified as a breeding adult if the weight exceeded 750 g (Clark and Kroeker 1993). In spring, all animals are classified as breeding adults regardless of weight. Animals were marked with identically numbered #1 monel fish tags in each ear. If the animal was previously marked, the tag number was recorded. Animals were sexed by external characteristics (Dimmick and Pelton 1994). Body length from nose to the base of the tail was recorded. The length of the hind food from heel to longest toe also was recorded.

Birds

Regular weekly counts were made on all experimental cells from 1 May to 30 June for blackbirds and from 1 May to 30 October for waterfowl and coots. Counts were completed in the morning by 0800 hours. Two observers were required to conduct a census. During a census, one observer walked or canoed through the experimental cell, while the second observer was stationed in an observation tower (5 m) located on the edge of each cell. The two observers communicated by walkie-talkie. The observer in the tower counted all blackbirds before the ground observer entered the cell. The ground observer moved through the entire cell noting all waterfowl and coots. The tower observer noted any birds that may have been recounted after being flushed by the ground observer.

Two nest counts were made each year: one during the third week of May and the second during the second week of June. Four randomly selected passes were made through each experimental cell with a 20-m nest drag. All active waterfowl and coot nests within the 20-m zone were recorded. The observers also scanned an area 5 m on each side of their path for active blackbird nests.

LITERATURE CITED

American Public Health Association (APHA). 1980. Standard methods for the examination of water and wastewater. 15th edition, Washington, D.C..

Clark, W.R., and D.W. Kroeker. 1993. Population dynamics of muskrats in experimental cells at Delta, Manitoba. Canadian Journal of Zoology 71:1620-1628.

Clay, R.T., and W.R. Clark. 1985. Demography of muskrats on the Upper Mississippi River. Journal of Wildlife Management 49:883-890.

Cummins, K.W. 1973. Trophic relations of aquatic insects. Annual Review of Entomology 18:183-206.

Davis, C.B., and A.G. van der Valk. 1978. Litter decomposition in prairie glacial marshes. In *Freshwater Wetlands: Ecological Processes and Management Potential*. (Eds.) R.E. Good, R.L. Simpson, and D.F. Whigham, pp.99-113. New York: Academic Press.

Dimmick, R.W., and M.R. Pelton. 1994. Criteria of sex and age. *In Research and Management Techniques for Wildlife and Habitats*. (Ed.) T.A. Bookhout, pp.169-214. 5th edition. Bethesda: The Wildlife Society.

Eagleson, P.S. 1970. *Dynamic Hydrology*. New York: McGraw Hill.

Eaton, J.W., and B. Moss. 1966. The estimation of numbers and pigment content in epipelic algal populations. Limnology and Oceanography 11:584-595.

Fee, E.J. 1973. Modelling primary production in water bodies: a numerical approach that allows vertical inhomogeneities. Journal of the Fisheries Research Board of Canada 30:1469-1473.

Goldsborough, L.G., G.G.C. Robinson, and S.E. Gurney. 1986. An exclosure/substratum system for
in situ ecological studies of periphyton. Archiv fur Hydrobiologie 106:373-393.

Gurney, S.E., and G.G.C. Robinson. 1988. VII. Small water bodies and wetlands. The influence of
water level manipulation on metaphyton production in a temperate freshwater marsh.
Verhandlungen Internationale Vereinigung fur Theoretische und Angewandte Limnologie
23:1032-1040.

Hosseini, S.M., and A.G. van der Valk. 1989a. Primary productivity and biomass of periphyton and
phytoplankton in flooded freshwater marshes. In *Freshwater Wetlands and Wildlife*. (Eds.) R.R.
Sharitz and J.W. Gibbons, pp.303-315. USDOE Symposium Series 61, Oak Ridge: USDOE
Office of Scientific and Technical Information.

Hosseini, S.M., and A.G. van der Valk. 1989b. The impact of above-normal flooding on metaphyton
in a freshwater marsh. In *Freshwater Wetlands and Wildlife*. (Eds.) R.R. Sharitz and J.W. Gibbons,
pp.317-324. USDOE Symposium Series 61, Oak Ridge: USDOE Office of Scientific and
Technical Information.

Hutchinson, G.E. 1975. Geography and Physics of Lakes. A treatise on Limnology. Volume 1. Part
1. New York: John Wiley & Sons.

Jones, R.C. 1984. Application of a primary production model to epiphytic algae in a shallow
eutrophic lake. Ecology 65:1895-1903.

Kadlec, J.A. 1983. Water budgets for small diked marshes. Water Resources Bulletin 19:223-229.

Kadlec, J.A. 1986a. Effects of flooding on dissolved and suspended nutrients in small diked marshes. Canadian Journal of Fisheries and Aquatic Sciences 43:1999-2008.

Kadlec, J.A. 1986b. Input-output nutrient budgets for small diked marshes. Canadian Journal of
Fisheries and Aquatic Sciences 43:2009- 2016.

Kadlec, J.A. 1993. Effect of depth of flooding on summer water budgets for small diked marshes.
Wetlands 13:1-9.

LeSage, L., and L.D. Harrison. 1979. Improved traps and techniques for the study of emerging
aquatic insects. Entomological News 90:65-78.

Marker, A.F.H., E.A. Nusch, H. Rai, and B. Reimann. 1980. The measurements of photosynthetic
pigments in freshwaters and standardization of methods: conclusions and recommendations.
Archiv fur Hydrobiologie Beihefte 14:91-106.

McCabe, T.R. 1983. Muskrat population levels and vegetation utilization: a basis for an index. Ph.D.
dissertation, Utah State University, Logan, Utah.

Murkin, E.J., and H.R. Murkin. 1989. Marsh Ecology Research Program long-term monitoring procedures manual. Delta Waterfowl and Wetlands Research Station Technical Bulletin 2, Portage la
Prairie, Manitoba.

Murkin, H.R., P.G. Abbott, and J.A. Kadlec. 1983. A comparison of activity traps and sweep nets
for sampling nektonic invertebrates in wetlands. Freshwater Invertebrate Biology 2:99- 106.

Otis, D.L., K.P. Burnham, G.C. White, and D.R. Anderson. 1978. Statistical inference from capture
data on closed animal populations. Wildlife Monographs 62.

Platt, T., C.L. Gallegos, and W.G. Harrison. 1980. Photoinhibition of photosynthesis in natural
assemblages of marine phytoplankton. Journal of Marine Research 38.

Pollock, K.H., J.D. Nichols, C. Browne, and J.E. Hines. 1990. Statistical inference for capture-
recapture experiments. Wildlife Monographs 107:1-97.

Robinson, G.G.C., S.E. Gurney, and L.G. Goldsborough. 1997a. Response of benthic and planktonic algal biomass to experimental water-level manipulation in a prairie lakeshore wetland.
Wetlands 17:167-181.

Robinson, G.G.C., S.E. Gurney, and L.G. Goldsborough. 1997b. The primary productivity of benthic and planktonic algae in a prairie wetland under controlled water regimes. Wetlands 17:182-194.

Seber, G.A.F. 1982. *The Estimation of Animal Abundance and Related Parameters, 2nd edition.* London, UK: Charles Griffin.

Stainton, M.P., M.J. Capel, and F.A.J. Armstrong. 1977. The Chemical Analysis of Freshwater. 2nd edition. Canadian Fisheries and Marine Services Special Publication 25.

Swanson, G.A., and M.I. Meyer. 1973. The role of invertebrates in the feeding ecology of Anatinae during the breeding season. In Waterfowl Habitat Management Symposium, pp.142-185. Moncton, New Brunswick.

van der Valk, A.G., and C.H. Welling. 1988. The development of zonation in freshwater wetlands: an experiment approach. Pages 145-158 *In Diversity and Pattern in Plant Communities.* (Eds.) H.J. During, M.J.A. Werger, and J.H. Williams. The Hague, Netherlands: SPB Academic Publishing.

Welling, C.H., R.L. Pederson, and A.G. van der Valk. 1988a. Recruitment from the seed bank and the development of zonation of emergent vegetation during a drawdown in a prairie wetland. Journal of Ecology 76:483-496.

Welling, C.H., R.L. Pederson, and A.G. van der Valk. 1988b. Temporal patterns in recruitment from the seed bank during drawdowns in a prairie wetland. Journal of Applied Ecology 25:999-1007.

Wetzel, R.G., and G.E. Likens. 1991. *Limnological Analysis.* New York: Springer–Verlag.

Appendix 2

Supplementary Nutrient Budget Information

Table A2.1. Mean (SE) annual surface and pore-water nutrient pools (kg/ha) in the MERP experimental cells by flooding treatment and year, 1980–1983

Treatment	Year	Flooded area (ha)	Surface water						Pore water			
			NH₄	DON	SRP	DOP	SSN	SSP	NH₄	DON	SRP	DOP
Baseline	1980 n = 10	3.02 (0.11)	0.67 (0.25)	10.6 (1.0)	0.23 (0.06)	0.14 (0.06)	4.02 (1.83)	0.86 (0.29)	7.80 (0.89)	18.6 (0.99)	0.94 (0.07)	0.44 (0.06)
	1981 n = 2	3.73 (0.37)	1.47 (0.98)	17.9 (3.6)	0.42 (0.38)	0.34 (0.07)	4.93 (0.29)	0.73 (0.09)	8.50 (1.80)	25.9 (2.10)	1.35 (0.14)	0.51 (0.16)
Flooded year one	1981 n = 8	6.16 (0.07)	1.65 (0.19)	22.7 (2.0)	0.42 (0.07)	0.75 (0.05)	2.24 (0.27)	0.36 (0.05)	14.6 (1.00)	28.8 (1.20)	3.20 (0.16)	0.94 (0.23)
	1982 n = 2	5.90 (0.12)	2.54 (0.83)	37.4 (1.68)	1.05 (0.21)	0.95 (0.08)	2.23 (0.91)	0.38 (0.13)	14.2 (1.38)	27.5 (0.90)	2.99 (0.29)	0.94 (0.09)
Flooded year two	1982 n = 8	6.03 (0.09)	2.42 (0.89)	30.3 (1.06)	1.43 (0.22)	0.68 (0.03)	2.10 (0.19)	0.38 (0.06)	23.3 (.86)	31.0 (1.02)	5.41 (0.29)	1.05 (0.24)
	1983 n = 2	5.68 (0.17)	2.15 (0.49)	31.9 (2.13)	1.31 (0.45)	0.63 (0.06)	2.66 (0.10)	0.33 (0.07)	23.1 (0.82)	30.5 (1.05)	4.68 (0.21)	0.91 (0.11)
Reference	1980 n = 2	3.02 (0.10)	0.56 (0.45)	5.25 (0.81)	0.35 (0.31)	0.56 (0.29)	1.12 (0.28)	0.33 (0.16)	7.60 (1.80)	15.1 (2.27)	1.30 (0.28)	0.49 (0.15)
	1981 n = 2	2.42 (0.11)	0.42 (0.13)	6.6 (1.2)	0.76 (0.56)	0.17 (0.01)	6.51 (2.79)	0.89 (0.33)	7.80 (3.10)	23.1 (2.40)	0.76 (0.16)	0.23 (0.13)
	1982 n = 2	2.15 (0.09)	1.30 (0.91)	10.2 (1.40)	1.13 (0.52)	0.23 (0.05)	6.85 (2.23)	0.96 (0.47)	12.0 (3.80)	25.8 (1.90)	1.95 (0.45)	0.57 (0.14)
	1983 n = 2	3.65 (0.27)	0.57 (0.27)	12.7 (2.70)	0.94 (0.43)	0.24 (0.04)	5.93 (2.44)	0.67 (0.33)	7.90 (1.71)	22.3 (2.95)	1.66 (0.26)	0.37 (0.07)

Note: Dissolved organic nitrogen (DON), soluble reactive phosphorus (SRP), dissolved organic phosphorus (DOP), suspended solids nitrogen (SSN), suspended solids phosphorus (SSP).

Table A2.2. Mean (SE) annual mass (kg/ha) of nitrogen (N) and phosphorus (P) in aboveground and belowground macrophyte and invertebrate pools in the MERP cells during the baseline and deep-flooding years and in undiked reference cells

Treatment	Year (sample size)	Aboveground Macrophytes		Belowground Macrophytes		Invertebrates	
		N	P	N	P	N	P
Unflooded	1980	58.7	18.5	83.1	11.8	0.56	0.06
	$n = 10$	(3.21)	(3.66)	(5.52)	(0.79)	(0.06)	(0.01)
	1981	69.9	20.6	83.9	8.81	0.72	0.07
	$n = 2$	(9.75)	(19.7)	(0.86)	(0.09)	(0.07)	(0.01)
Flooded	1981	15.0	3.91	77.4	7.56	0.94	0.09
1 yr.	$n = 8$	(4.40)	(0.94)	(8.63)	(0.83	(0.05)	(0.01)
	1982	0.50	0.45	80.4	8.76	1.14	0.11
	$n = 2$	(0.30)	(0.35)	(21.3)	(2.35)	(0.07)	(0.01)
Flooded	1982	17.8	3.24	71.5	7.79	3.20	0.32
2 yr.	$n = 8$	(3.55)	(0.35)	(6.53)	(0.71)	(0.20)	(0.02)
	1983	16.1	3.55	15.7	1.43	3.28	0.33
	$n = 2$	(2.25)	(1.55)	(2.80)	(0.31)	(0.04)	(0.05)
Reference	1980	78.3	19.9	83.8	11.9	0.49	0.04
	$n = 2$	(7.35)	(11.2)	(7.76)	(1.08)	(0.27)	(0.02)
	1981	86.5	20.2	84.9	8.80	0.69	0.06
	$n = 2$	(18.5)	(4.45)	(7.13)	(2.41)	(0.20)	(0.02)
	1982	89.2	22.0	93.6	13.7	0.69	0.06
	$n = 2$	(0.50)	(14.0)	(6.16)	(0.84)	(0.20)	(0.02)
	1983	106.0	28.1	94.9	9.41	0.54	0.05
	$n = 2$	(4.00)	(14.9)	(13.8)	(1.31)	(0.17)	(0.01)

Table A2.3. Mean (SE) annual (Oct. 1-Sept. 30) nitrogen (N) fluxes (kg/ha/yr) in the three flooding treatments in the diked experimental cells and the reference cells from 1985 to 1989

Flux	Low ($n = 4$)	Medium ($n = 3$)	High ($n = 3$)	Reference ($n = 2$)
Macrophytes				
Uptake				
1985	49.5(6.23)	47.5(7.21)	49.6(1.56)	72.4(14.3)
1986	47.1(11.9)	53.8(6.17)	67.7(0.35)	101.5(24.5)
1987	53.1(10.1)	63.8(16.1)	49.2(16.0)	150.1(12.8)
1988	72.9(13.7)	41.3(6.14)	65.5(13.9)	103.8(48.0)
1989	42.1(5.66)	33.1(17.7)	39.7(18.2)	60.9(25.9)

Continued

383

Table A2.3. (*Continued*)

Flux	Low (n = 4)	Medium (n = 3)	High (n = 3)	Reference (n = 2)
Translocation				
1985	27.9(0.88)	26.7(3.61)	30.6(1.15)	43.3(9.10)
1986	16.0(3.24)	24.6(3.47)	22.4(3.83)	35.7(6.02)
1987	25.5(2.43)	30.7(2.49)	30.5(7.85)	50.2(7.42)
1988	18.9(4.36)	22.3(5.20)	28.7(11.4)	48.0(14.5)
1989	26.9(5.16)	21.3(10.2)	28.7(12.6)	46.9(21.4)
Aboveground leaching				
1985	12.9(1.64)	12.3(1.66)	14.1(0.53)	20.0(4.20)
1986	7.79(1.47)	11.4(1.59)	10.6(1.63)	16.7(2.28)
1987	11.5(1.13)	13.9(0.94)	13.8(3.40)	22.7(3.34)
1988	8.94(1.93)	10.5(2.29)	13.3(5.16)	22.2(6.78)
1989	12.2(2.24)	9.89(4.51)	13.3(5.75)	21.7(9.65)
Belowground leaching				
1985	4.32(0.55)	4.15(0.72)	3.81(0.29)	5.82(1.04)
1986	2.29(0.98)	3.00(1.61)	1.44(0.72)	5.53(2.96)
1987	3.78(1.79)	4.70(1.57)	5.16(2.20)	7.40(1.33)
1988	4.78(1.35)	5.95(2.81)	4.10(2.09)	12.4(1.64)
1989	8.10(2.35)	7.30(3.25)	6.53(1.44)	9.60(0.63)
Aboveground litter				
1985	15.8(1.87)	13.4(1.65)	16.5(0.52)	23.3(4.90)
1986	15.0(1.97)	14.4(1.95)	16.5(0.62)	23.9(3.78)
1987	9.09(1.71)	13.3(1.86)	12.4(1.88)	19.5(2.66)
1988	13.4(1.31)	16.3(1.13)	16.1(4.01)	26.5(3.90)
1989	10.4(2.25)	12.3(2.65)	15.5(6.02)	25.9(7.56)
Belowground litter				
1985	17.3(2.22)a	16.5(2.88)	15.3(1.15)a	23.3(4.16)
1986	5.17(3.93)b	12.0(6.47)	5.75(2.88)a	22.1(11.8)
1987	15.1(7.14)a	16.8(6.31)	20.6(8.81)a	29.6(3.30)
1988	19.1(5.41)a	23.8(11.2)	16.4(8.35)a	49.8(1.64)
1989	40.4(9.43)c	29.2(13.0)	42.1(5.77)b	38.3(2.51)
Algae and invertebrates				
Algal uptake				
1985	24.8(11.3)	18.1(2.57)a	7.90(1.04)	7.20(1.64)
1986	9.94(3.90)	3.37(1.02)b	5.49(2.20)	5.51(1.50)
1987	17.7(8.61)	6.00(1.05)b	6.82(3.60)	7.70(1.30)
1988	14.6(11.6)	8.34(2.49)b	8.49(5.53)	3.48(1.43)
1989	10.5(2.92)	9.59(3.46)b	10.6(4.94)	4.66(1.51)
Algal litter fall				
1985	2.93(1.62)a	3.40(0.26)	0.31(0.26)	2.23(0.18)
1986	6.65(2.02)a	4.78(1.46)	2.58(2.50)	3.92(0.82)
1987	3.86(3.43)a	2.59(1.29)	4.11(1.10)	0
1988	11.8(5.65)a	4.23(2.13)	8.10(4.33)	5.91(2.04)
1989	23.4(3.65)b	8.10(4.38)	5.46(4.62)	7.73(1.35)

Continued

384

Table A2.3. (*Continued*)

Flux	Low (*n* = 4)	Medium (*n* = 3)	High (*n* = 3)	Reference (*n* = 2)
Invertebrate uptake				
1985	2.01(0.19)a	2.24(0.24)a	1.98(0.10)a	0.82(0.49)
1986	0.79(0.06)b	1.01(0.34)b	0.53(0.13)b	0.53(0.13)
1987	0.21(0.09)b	0.26(0.20)b	0.50(0.05)b	0.23(0.06)
1988	0.41(0.07)b	0.28(0.16)b	0.62(0.23)b	0.25(0.16)
1989	0.14(0.08)b	0.31(0.21)b	0.07(0.04)b	0.16(0.05)
Invertebrate loss				
1985	0.06(0.06)a	0.14(0.14)a	0.15(0.12)a	0.55(0.41)
1986	0.46(0.20)a	0.55(0.23)a	0.50(0.15)ab	0.28(0.03)
1987	0.87(0.09)b	1.31(0.20)b	0.76(0.11)b	0.23(0.06)
1988	0.92(0.23)b	1.08(0.34)ab	0.81(0.12)b	0.27(0.03)
1989	0.56(0.06)b	0.49(0.20)a	0.63(0.09)b	0.16(0.05)
Dissolved nutrients				
Mineralization SWDO to SWDI				
1985	1.63(0.88)a	0.18(0.18)a	2.07(2.07)a	4.62(0.48)
1986	4.70(0.73)a	6.04(0.20)b	11.6(1.75)b	8.91(1.14)
1987	3.73(1.61)a	3.80(1.94)ab	1.38(0.29)a	5.48(1.15)
1988	12.5(1.61)b	5.50(1.46)b	11.5(4.47)b	18.3(4.15)
1989	6.49(1.00)c	5.15(1.73)b	3.63(1.85)a	8.15(1.74)
Mineralization PWDO to PWDI				
1985	6.69(2.30)	11.4(3.61)	13.1(4.67)	7.33(4.74)
1986	6.31(2.31)	9.56(1.44)	8.95(2.28)	5.00(2.84)
1987	6.73(2.26)	2.51(0.42)	4.05(2.03)	5.35(3.68)
1988	4.45(2.03)	5.08(4.53)	2.89(1.06)	14.5(7.50)
1989	6.75(1.15)	4.64(1.93)	3.01(1.38)	8.00(7.35)
Pore–surface water				
1985	4.20(7.00)a	2.33(3.22)a	−9.87(1.64)	−15.9(3.88)
1986	−6.55(3.11)a	−16.9(1.58)b	−21.5(2.34)	−23.2(0.60)
1987	−5.17(4.35)a	−15.9(0.94)b	−13.5(6.06)	−22.8(2.48)
1988	−18.1(5.83)b	−15.9(2.04)b	−20.2(7.70)	−31.5(1.74)
1989	−14.5(4.38)b	−10.4(6.52)b	−12.5(9.74)	−29.6(7.56)
Litter leaching / sediment sink				
Dissolved inorganic to SW				
1985	3.63(3.63)	0.10(0.10)	−0.84(1.06)	0
1986	1.30(0.85)	0	1.98(2.58)	0
1987	4.67(3.56)	1.03(0.64)	2.35(1.69)	0.04(0.04)
1988	6.81(5.25)	4.04(3.67)	0.46(0.46)	0
1989	4.59(2.20)	3.91(3.67)	8.64(5.40)	0
Dissolved organic to SW				
1985	15.7(6.22)a	8.59(1.53)a	23.2(4.21)a	1.59(1.50)
1986	3.81(2.04)b	0.76(0.76)b	0b	7.12(0.13)
1987	1.52(0.82)b	3.06(1.52)b	2.72(0.51)b	2.81(1.44)
1988	5.05(1.75)b	3.24(1.38)b	7.71(3.25)b	0
1989	1.47(0.76)b	0.43(0.25)b	2.55(0.25)b	1.62(3.32)

Continued

385

Flux	Low (*n* = 4)	Medium (*n* = 3)	High (*n* = 3)	Reference (*n* = 2)
Dissolved organic to PW				
1985	5.51(2.45)	5.48(1.19)	2.04(1.01)	6.81(5.18)
1986	5.61(1.66)	8.78(0.83)	5.06(1.87)	5.68(2.26)
1987	6.30(0.52)	3.68(1.48)	4.32(1.97)	6.34(1.15)
1988	6.82(2.36)	8.71(3.40)	7.40(1.01)	12.0(0.35)
1989	4.73(2.11)	6.31(1.66)	5.57(3.89)	19.9(17.8)
Dissolved inorganic to PW				
1985	36.0(6.50)	27.6(9.96)	25.7(5.74)a	47.5(9.00)
1986	38.4(11.4)	28.2(11.0)	38.2(4.88)a	66.2(15.1)
1987	50.1(11.0)	45.6(18.9)	30.3(6.25)a	74.2(18.0)
1988	45.6(15.1)	22.3(15.5)	40.0(16.4)a	43.7(49.1)
1989	31.3(13.8)	14.6(15.5)	5.25(12.5)b	55.9(27.8)
Aboveground litter to sediments				
1985	23.3(5.27)	30.4(1.55)	23.8(4.76)	33.9(0.78)
1986	25.8(0.85)	29.2(0.80)	26.3(2.39)	29.7(0.97)
1987	21.6(3.12)	26.1(0.73)	24.9(1.48)	29.2(0.33)
1988	25.4(2.88)	23.5(3.45)	23.0(4.57)	30.0(0.55)
1989	23.1(4.47)	23.3(4.31)	18.7(6.58)	31.2(1.96)
Belowground litter to sediments				
1985	27.0(4.85)	25.2(2.99)	25.5(3.24)	33.3(12.6)
1986	23.3(3.05)	19.5(1.26)	19.6(5.15)	33.3(9.79)
1987	19.9(2.97)	21.0(3.75)	18.7(4.66)	38.4(11.1)
1988	21.0(3.18)	21.3(1.14)	16.9(4.67)	38.9(5.70)
1989	22.1(4.84)	22.4(2.96)	19.7(5.54)	28.8(12.6)

Note: Within a column within a treatment means with different letters are significantly different ($P < 0.05$). Surface water dissolved organic (SWDO), surface water dissolved inorganic (SWDI), pore water dissolved organic (PWDO), pore water dissolved inorganic (PWDI), surface water (SW), pore water (PW).

Table A2.4.   Mean (SE) annual (1 Oct.-30 Sept.) phosphorus (P) fluxes (kg/ha/yr) in the three flooding treatments in the diked experimental cells and the reference cells from 1985 to 1989

Flux	Low (n = 4)	Medium (n = 3)	High (n = 3)	Reference (n = 2)
Macrophytes				
Uptake				
1985	16.0(1.26)	15.4(2.22)	17.8(1.42)	21.6(3.70)
1986	10.0(1.39)	11.1(1.46)	11.8(1.96)	20.5(3.49)
1987	10.8(1.00)	13.1(1.58)	15.2(3.75)	21.4(0.54)
1988	8.12(1.54)	10.3(2.29)	14.1(5.19)	21.6(5.86)
1989	10.3(1.92)	11.3(5.36)	13.2(5.76)	25.4(9.37)
Translocation				
1985	5.67(0.33)	5.60(0.77)	6.43(0.52)	6.81(0.49)
1986	3.65(0.49)	3.89(0.50)	4.10(0.75)	7.44(0.87)
1987	4.08(0.37)	4.71(0.39)	5.28(1.08)	4.95(0.97)
1988	3.02(0.51)	3.68(0.84)	4.72(1.71)	5.30(1.25)
1989	3.64(0.76)	3.99(1.81)	4.55(1.99)	7.99(2.20)
Aboveground leaching				
1985	2.50(0.14)	2.47(0.34)	2.83(0.23)	3.00(0.22)
1986	1.51(0.22)	1.84(0.23)	1.96(0.31)	3.40(0.36)
1987	1.84(0.16)	2.15(0.16)	2.40(0.48)	2.36(0.39)
1988	1.43(0.23)	1.73(0.36)	2.20(0.77)	2.44(0.51)
1989	1.66(0.33)	1.83(0.80)	2.11(0.91)	3.60(0.98)
Belowground leaching				
1985	2.07(0.20)	1.96(0.29)	2.28(0.18)	2.97(0.64)
1986	1.26(0.17)	1.60(0.21)	1.77(0.16)	2.79(0.10)
1987	1.22(0.13)	1.62(0.14)	1.85(0.40)	2.94(0.31)
1988	1.51(0.18)	1.43(0.22)	1.90(0.65)	3.26(0.74)
1989	1.23(0.19)	1.42(0.58)	1.79(0.73)	3.48(1.26)
Aboveground litter-fall				
1985	3.17(0.18)	3.13(0.43)	3.59(0.29)	3.80(0.27)
1986	2.91(0.17)	2.88(0.40)	3.30(0.27)	3.50(0.25)
1987	2.13(0.25)	2.15(0.27)	2.29(0.37)	3.97(0.41)
1988	2.15(0.19)	2.50(0.19)	2.80(0.56)	2.76(0.45)
1989	1.66(0.27)	2.02(0.42)	2.56(0.89)	2.84(0.60)
Belowground litter-fall				
1985	8.27(0.79)a	7.84(1.16)	9.13(0.72)	11.8(2.56)
1986	5.03(0.68)b	6.41(0.85)	7.06(0.62)	11.1(0.41)
1987	4.88(0.52)b	6.46(0.57)	7.41(1.61)	11.7(1.23)
1988	4.44(0.72)b	5.71(0.88)	7.60(2.59)	13.0(2.98)
1989	4.93(0.74)b	5.69(2.30)	7.16(2.93)	13.9(5.03)

Continued

Table A2.4. *(Continued)*

Flux	Low (n = 4)	Medium (n = 3)	High (n = 3)	Reference (n = 2)
Algae and invertebrates				
Algal uptake				
1985	3.04(1.19)	2.74(0.41)a	0.98(0.15)	0.63(0.10)
1986	1.69(0.84)	0.44(0.21)b	0.78(0.36)	0.69(0.10)
1987	1.92(0.93)	0.62(0.15)b	0.83(0.43)	0.75(0.13)
1988	1.69(1.43)	0.95(0.38)b	1.08(0.67)	0.13(0.18)
1989	0.97(0.26)	1.13(0.47)b	1.15(0.44)	0.38(0.10)
Algal litter-fall				
1985	0.39(0.19)a	0.50(0.04)	0.09(0.05)	0.35(0.02)
1986	1.05(0.25)a	0.73(0.23)	0.40(0.38)	0.64(0.02)
1987	0.65(0.56)a	0.35(0.22)	0.51(0.20)	0.40(0.10)
1988	0.90(0.39)a	0.45(0.25)	1.09(0.54)	0.23(0.02)
1989	1.98(0.89)b	0.87(0.62)	0.69(0.56)	0.79(0.02)
Invertebrate uptake, algae/litter combined				
1985	0.20(0.02)a	0.22(0.02)a	0.20(0.01)a	0.08(0.05)
1986	0.08(0.01)b	0.08(0.03)b	0.05(0.01)b	0.05(0.01)
1987	0.01(0.01)c	0.03(0.02)b	0.01(0.01)b	0.02(0.01)
1988	0.04(0.03)c	0.03(0.02)b	0.06(0.02)b	0.03(0.02)
1989	0.01(0.01)c	0.03(0.02)b	0.01(0.01)b	0.02(0.01)
Invertebrate loss				
1985	0.01(0.01)a	0.01(0.01)a	0.02(0.01)a	0.06(0.04)
1986	0.05(0.02)a	0.06(0.02)a	0.05(0.02)ab	0.03(0.01)
1987	0.09(0.01)b	0.13(0.02)b	0.08(0.01)b	0.02(0.01)
1988	0.09(0.02)b	0.11(0.03)ab	0.08(0.01)b	0.03(0.01)
1989	0.06(0.01)b	0.05(0.02)a	0.06(0.01)ab	0.02(0.01)
Dissolved nutrients				
Mineralization SWDO to SWDI				
1985	0.17(0.06)a	0.11(0.01)a	0.29(0.24)a	0.04(0.02)
1986	0.04(0.02)b	0.09(0.03)a	0.06(0.03)b	0.46(0.46)
1987	0.22(0.02)a	0.27(0.02)ab	1.56(1.27)a	1.37(0.82)
1988	0.21(0.05)a	0.20(0.02)b	0.24(0.05)a	0.81(0.31)
1989	0.16(0.03)a	0.14(0.01)a	0.07(0.07)b	0.45(0.18)
Mineralization PWDO to PWDI				
1985	0.41(0.25)	0.15(0.05)	0.89(0.47)	0.07(0.07)
1986	0.41(0.13)	0.52(0.21)	0.32(0.11)	0.44(0.14)
1987	0.63(0.05)	0.43(0.11)	0.51(0.18)	0.29(0.18)
1988	1.31(0.57)	0.63(0.06)	0.33(0.20)	0.44(0.18)
1989	0.54(0.19)	0.46(0.19)	0.52(0.23)	0.22(0.02)
Pore–surface water exchange				
1985	−2.07(0.70)a	−2.80(0.76)a	−3.87(1.18)	−3.74(0.08)
1986	−0.85(0.63)a	−2.78(0.29)b	−2.75(0.22)	−4.67(0.45)

Continued

Table A2.4. (*Continued*)

Flux	Low (n = 4)	Medium (n = 3)	High (n = 3)	Reference (n = 2)
1987	−1.49(0.51)a	−2.91(0.11)b	−4.31(0.62)	−3.89(1.19)
1988	−1.19(0.91)b	−2.25(0.42)b	−1.95(1.01)	−4.53(0.26)
1989	−2.20(0.57)b	−2.04(1.14)b	−1.41(0.86)	−4.81(0.61)
Litter leaching / sediment sink				
Dissolved inorganic to SW				
1985	0.32(0.32)	0	0.38(0.24)	0
1986	0.06(0.06)	0	0.03(0.03)	0
1987	0.57(0.47)	0.11(0.07)	0.15(0.15)	0
1988	0.45(0.45)	0.44(0.37)	0.04(0.04)	0
1989	0.69(0.30)	0.15(0.15)	0.48(0.16)	0
Dissolved organic to SW				
1985	0.61(0.26)a	0.37(0.03)a	1.02(0.34)a	0.12(0.05)
1986	0.10(0.01)b	0.14(0.03)b	0.16(0.03)b	0.44(0.06)
1987	0b	0.02(0.02)b	0.10(0.07)b	0.60(0.60)
1988	0.03(0.03)b	0.10(0.02)b	0.10(0.07)b	0.09(0.01)
1989	0.01(0.01)b	0.05(0.03)b	0.23(0.07)b	0.18(0.15)
Dissolved organic to PW				
1985	0.46(0.34)	0.39(0.13)	0.99(0.59)	0.17(0.03)
1986	0.42(0.07)	0.53(0.26)	0.28(0.10)	0.29(0.03)
1987	0.96(0.19)	0.48(0.12)	0.34(0.14)	0.33(0.14)
1988	1.05(0.41)	0.63(0.19)	0.57(0.22)	0.59(0.28)
1989	0.58(0.20)	0.73(0.25)	0.87(0.40)	0.62(0.30)
Dissolved inorganic to PW				
1985	12.7(1.46)a	12.5(1.74)a	12.7(1.06)a	15.4(3.79)
1986	4.37(1.74)b	6.39(1.06)b	7.76(1.90)a	12.1(2.41)
1987	7.80(1.28)b	7.84(2.29)b	8.27(3.13)a	14.3(1.83)
1988	3.76(1.83)b	7.36(1.45)b	9.89(4.48)a	15.2(4.82)
1989	5.96(1.09)b	6.98(3.88)b	9.03(4.88)b	17.4(7.50)
Aboveground litter to sediments				
1985	3.07(0.47)	3.63(0.06)	2.71(0.68)a	3.94(0.13)
1986	3.39(0.02)	3.72(0.16)	3.86(0.40)ab	3.80(0.08)
1987	2.93(0.41)	3.81(0.15)	4.73(0.16)ab	4.48(0.01)
1988	3.15(0.43)	3.96(0.31)	5.43(0.28)b	4.46(0.02)
1989	3.43(0.54)	4.50(0.20)	5.17(0.22)b	4.54(0.04)
Belowground litter to sediments				
1985	3.93(0.78)	3.91(0.63)	2.54(0.75)a	3.47(0.52)
1986	4.58(0.69)	4.33(0.35)	3.81(0.35)ab	4.69(0.05)
1987	4.26(0.63)	4.57(0.35)	4.11(0.38)ab	5.52(0.24)
1988	3.65(0.70)	4.78(0.13)	4.60(0.07)b	6.43(0.20)
1989	4.33(0.61)	4.55(0.19)	4.40(0.35)b	6.08(2.24)

Note: Within a column within a treatment means with different letters are significantly different ($P < 0.05$). Surface water dissolved organic (SWDO), surface water dissolved inorganic (SWDI), pore water dissolved organic (PWDO), pore water dissolved inorganic (PWDI), surface water (SW), pore water (PW).

Table A2.5. Mean (SE) annual (1 Oct.-30 Sept.) carbon (C) fluxes (kg/ha/yr) in the three flooding treatments in the diked experimental cells and the reference cells from 1985 to 1989

Flux	Low (*n* = 4)	Medium (*n* = 3)	High (*n* = 3)	Reference (*n* = 2)
Macrophytes				
C fixation				
1985	1,650(199)	1,733(342)	1,714(30.1)	3,018(482)
1986	1,506(362)	1,700(42.7)	2,145(102)	2,805(821)
1987	1,721(349)	2,054(654)	1,907(702)	4,568(678)
1988	1,962(357)	1,090(139)	2,042(332)	3,794(1723)
1989	1,569(240)	989(554)	1,362(634)	2,491(1417)
Translocation				
1985	713(88.5)	732(172)	660(57.1)	1,134(179)
1986	714(184)	740(87.1)	950(43.4)	1,285(426)
1987	836(269)	980(397)	690(317)	2,656(312)
1988	972(176)	444(52.6)	929(83.1)	1,741(917)
1989	636(88.1)	424(292)	477(245)	891(450)
Aboveground leaching				
1985	432(54.3)	462(78.4)	486(20.7)	869(139)
1986	316(66.7)	406(72.4)	412(67.1)	713(159)
1987	401(56.2)	463(101)	550(169)	1,120(239)
1988	251(47.3)	280(73.7)	379(132)	936(339)
1989	405(87.0)	262(143)	406(176)	752(438)
Belowground leaching				
1985	142(17.7)	146(34.5)	132(11.4)	226(35.8)
1986	58.3(32.3)	120(62.9)	56.3(9.65)	155(77.5)
1987	107(51.5)	49.7(33.7)	162.1(51.9)	287(93.1)
1988	134(38.1)	149(77.9)	109(60.8)	302(44.5)
1989	232(60.0)	186(93.8)	231(45.5)	241(4.13)
Aboveground litter-fall				
1985	504(63.4)	529(96.4)	567(24.1)	1014(162)
1986	512(59.9)	539(91.4)	555(34.8)	1,123(45.5)
1987	369(77.9)	473(84.4)	481(78.2)	832(185)
1988	468(65.6)	540(118)	642(198)	1,307(279)
1989	293(55.1)	327(86.0)	442(155)	1,092(395)
Belowground litter-fall				
1985	571(70.8)	586(138)	528(45.6)a	907(143)
1986	333(129)	480(251)	225(38.6)a	621(310)
1987	431(206)	299(134)	648(207)a	1,251(372)
1988	537(152)	599(311)	438(243)a	1,010(178)
1989	931(240)	746(375)	1,327(182)b	966(16.5)

Continued

Table A2.5. (*Continued*)

Flux	Low (*n* = 4)	Medium (*n* = 3)	High (*n* = 3)	Reference (*n* = 2)
Algae and invertebrates				
Algal uptake				
1985	318(125)	259(43.7)a	99.9(16.6)	102(28.0)
1986	178(89.3)	44.5(22.4)b	81.7(37.8)	72.1(3.07)
1987	204(99.1)	65.2(16.3)b	88.8(46.1)	79.0(13.8)
1988	179(152)	100(39.7)b	113(71.0)	52.4(3.80)
1989	103(27.7)	117(49.7)b	122(46.9)	40.6(5.88)
Algal litter fall				
1985	43.0(20.3)	56.9(4.00)	10.6(5.69)	37.8(0.86)
1986	113(26.5)	78.8(24.3)	42.8(40.7)	68.2(0.87)
1987	68.4(59.4)	37.5(23.0)	54.1(21.0)	40.8(7.66)
1988	96.4(41.6)	48.9(26.6)	116(57.7)	44.7(4.77)
1989	316(200)	92.7(65.7)	73.4(59.1)	84.2(12.3)
Invertebrate uptake, algae/litter combined				
1985	9.37(0.88)a	10.4(1.11)a	9.22(0.20)a	3.80(2.29)
1986	3.66(0.28)b	4.71(1.59)b	2.47(0.58)b	2.46(0.44)
1987	1.01(0.01)b	1.24(0.92)b	1.23(0.63)b	1.07(0.28)
1988	1.89(1.16)b	1.31)0.73)b	2.88(1.07)b	1.15(0.76)
1989	0.66(0.38)b	1.48(0.98)b	0.34(0.38)b	0.75(0.25)
Invertebrate loss				
1985	0.28(0.27)a	0.65(0.65)a	0.69(0.56)a	2.58(1.90)
1986	2.15(0.90)b	2.58(1.10)b	2.33(0.72)b	1.29(0.10)
1987	4.07(0.41)c	6.09(0.90)c	3.51(0.51)b	1.07(0.28)
1988	4.28(1.09)c	5.06(1.58)c	3.76(0.56)b	1.23(0.18)
1989	2.60(0.24)b	2.29(0.94)b	2.92(0.42)b	0.73(0.23)
Dissolved nutrients				
Mineralization SWDO to SWDI				
1985	24.2(16.8)	0(0)a	24.0(4.01)	95.2(30.2)
1986	30.0(19.2)	15.9(7.99)b	28.9(19.1)	131.2(7.51)
1987	85.5(42.3)	69.4(14.7)c	70.3(34.6)	120(43.3)
1988	97.3(46.0)	51.1(20.5)c	68.9(56.3)	266(71.5)
1989	69.1(28.6)	36.4(19.9)c	48.7(25.9)	137(56.9)
Mineralization PWDO to PWDI				
1985	3.11(1.80)a	4.70(4.70)a	8.63(6.33)a	64.9(4.99)
1986	31.5(18.4)b	47.3(25.3)b	64.7(37.1)b	71.0(11.0)
1987	95.3(28.0)c	206(32.6)c	277(90.8)c	209(70.9)
1988	50.2(25.2)b	78.4(46.9)b	33.7(29.0)ab	174(142)
1989	73.9(24.3)bc	75.8(53.6)b	28.1(23.2)ab	38.0(8.01)

Continued

Table A2.5. (*Continued*)

Flux	Low (*n* = 4)	Medium (*n* = 3)	High (*n* = 3)	Reference (*n* = 2)
Pore–surface water exchange				
1985	137(139)a	30.8(164)a	−6.62(47.4)a	−686(88.5)
1986	−134(108)b	−438(50.8)b	−452(81.0)b	−721(77.2)
1987	−284(92.3)b	−456(19.6)b	−318(127)b	−937(100)
1988	−232(123)b	−155(114)b	−480(179)b	−944(363)
1989	−366(159)b	−111(68.9)b	−403(154)b	−858(367)
Litter leaching / sediment sink				
Dissolved inorganic to SW				
1985	109(39.8)	205(14.2)a	186(49.4)a	0
1986	105(28.1)	208(20.6)a	169(39.3)a	0
1987	147(71.5)	76.8(33.5)b	41.9(19.0)b	0
1988	65.1(38.9)	60.8(30.8)b	33.3(3.36)b	0
1989	145(29.4)	217(60.6)a	236(65.4)a	0
Dissolved organic to SW				
1985	340(149)	254(38.8)	491(82.7)a	98.6(22.9)
1986	128(49.5)	141(37.1)	120(23.1)b	84.7(4.98)
1987	85.0(36.5)	76.7(18.5)	107(64.7)b	105(37.6)
1988	75.4(39.9)	132(36.5)	52.1(29.0)b	31.9(2.82)
1989	124(75.1)	152(45.7)	117(32.8)b	72.9(59.1)
Dissolved organic to PW				
1985	75.6(21.0)	156(15.1)	120(26.5)	58.6(13.1)
1986	210(82.3)	197(31.3)	127(17.6)	160(104)
1987	95.9(6.28)	103(46.7)	119(85.3)	91.1(27.1)
1988	79.1(34.4)	95.1(32.3)	47.4(10.8)	122(23.9)
1989	120(39.9)	92.5(28.4)	58.1(30.6)	384(288)
Dissolved inorganic to PW				
1985	478(225)a	556(168)a	562(111)a	−637(75.1)
1986	−174(189)b	−633(249)b	−661(78.8)b	−959(142)
1987	−252(159)b	−539(244)b	−696(178)b	−1,079(101)
1988	−396(176)b	−272(365)b	−591(222)b	−1,457(131)
1989	−718(128)b	−453(134)b	−929(111)b	−955(263)
Aboveground litter to sediments				
1985	408(63.1)	315(55.1)	279(12.0)	732(35.7)
1986	413(91.7)	335(35.7)	360(69.9)	781(44.8)
1987	409(95.4)	547(35.4)	537(93.1)	787(24.2)
1988	497(85.5)	566(29.3)	624(31.0)	873(37.4)
1989	347(102)	351(88.0)	493(79.3)	915(102)

Continued

Table A2.5. (*Continued*)

Flux	Low (*n* = 4)	Medium (*n* = 3)	High (*n* = 3)	Reference (*n* = 2)
Belowground litter to sediments				
1985	112(28.9)	52.2(13.1)a	40.5(11.2)a	303(34.8)
1986	107(33.3)	60.4(4.11)a	71.3(38.9)a	154(121)
1987	163(60.8)	146(48.1)ab	134(43.7)ab	401(131)
1988	230(29.3)	240(64.4)b	206(41.8)b	655(153)
1989	255(40.0)	251(58.3)b	335(63.1)b	454(146)

Note: Within a column within a treatment means with different letters are significantly different ($P < 0.05$). Surface water dissolved organic (SWDO), surface water dissolved inorganic (SWDI), pore water dissolved organic (PWDO), pore water dissolved inorganic (PWDI), surface water (SW), pore water (PW).

Appendix 3

MERP Publications and Graduate Studies

GENERAL PUBLICATIONS

1. van der Valk, A.G. 1981. Succession in wetlands: a Gleasonian approach. Ecology 62:688-696.
2. van der Valk, A.G. 1982. Succession in temperate North American wetlands. In *Wetlands: Ecology and Management*. (Eds.) B. Gopal, R.E. Turner, R.G. Wetzel, and D.F. Whigham, pp.169-179. Jaipur, India: National Institute of Ecology.
3. Pederson, R.L. 1981. Seed bank characteristics of the Delta Marsh: applications for wetland management. In *Selected Proceedings of the Midwest Conference on Wetland Values and Management*. (Ed.) B. Richardson, pp.61-69. St. Paul: Minnesota Water Planning Board.
4. Kadlec, J.A. 1983. Water budgets for small diked marshes. Water Resources Bulletin 19:223-229.
5. Wrubleski, D.A., and D.M. Rosenberg. 1984. Overestimates of Chironomidae (Diptera) abundance from emergence traps with polystyrene floats. American Midland Naturalist 111:195-197.
6. Batt, B.D.J., P.J. Caldwell, C.B. Davis, J.A. Kadlec, R.M. Kaminski, H.R. Murkin, and A.G. van der Valk. 1983. The Delta Waterfowl Research Station - Ducks Unlimited Canada Marsh Ecology Research Program. In *First Western Hemisphere Waterfowl and Waterbird Symposium*. (Ed.) H. Boyd, pp.19-23. Edmonton: Canadian Wildlife Service.
7. Murkin, H.R., P.A. Abbott, and J.A. Kadlec. 1983. A comparison of activity traps and sweep nets for sampling nektonic invertebrates in wetlands. Freshwater Invertebrate Biology 2:99-106.
8. Kadlec, J.A. 1986. Effects of flooding on dissolved and suspended nutrients in small diked marshes. Canadian Journal of Fisheries and Aquatic Sciences 43:1999-2008.
9. Murkin, H.R., and J.A. Kadlec. 1986. Relationships between waterfowl and macroinvertebrate densities in a northern prairie marsh. Journal of Wildlife Management 50:212-217.
10. Murkin, H.R., B.D.J. Batt, P.J. Caldwell, C.B. Davis, J.A. Kadlec, and A.G. van der Valk. 1984. Perspectives on the Delta Waterfowl Research Station - Ducks Unlimited Canada Marsh Ecology Research Program. Transactions of the North American Wildlife and Natural Resources Conference 49:253-261.
11. Nelson, J.W., and J.A. Kadlec. 1984. A conceptual approach to relating habitat structure and macroinvertebrate production in freshwater wetlands. Transactions of the North American Wildlife and Natural Resources Conference 49:262-270.
12. Pederson, R.L., and A.G. van der Valk. 1984. Vegetation change and seed banks in marshes: ecological and management implications. Transactions of the North American Wildlife and Natural Resources Conference 49:271-280.
13. Murkin, H.R., and B.D.J. Batt. 1987. The interactions of vertebrates and invertebrates in peatlands and marshes. In *Aquatic Insects of Peatlands and Marshes in Canada*. (Eds.) D.M.

Rosenberg and H.V. Danks, pp.15-30. Memoirs of Entomological Society of Canada 140.

14. Murkin, H.R., and J.A. Kadlec. 1986. Responses by benthic macroinvertebrates to prolonged flooding of marsh habitat. Canadian Journal of Zoology 64:65-72.

15. van der Valk, A.G. 1986. The impact of litter and annual plants on recruitment of species from the seed bank of a lacustrine marsh. Aquatic Botany 24:13-26.

16. Wrubleski, D.A., and S.S. Roback. 1987. Two species of *Procladius* (Diptera: Chironomidae) from a northern prairie marsh: descriptions, phenologies and mating behaviour. Journal of the North American Benthological Society 6:198-212.

17. Neckles, H.A., J.A. Nelson, and R.L. Pederson. 1985. Management of whitetop (*Scholochloa festucacea*) marshes for livestock forage and wildlife. Delta Waterfowl and Wetlands Research Station Technical Bulletin 1.

18. Kadlec, J.A. 1986. Input–output nutrient budgets for small diked marshes. Canadian Journal of Fisheries and Aquatic Sciences 43:2009-2016.

19. Galinato, M.I., and A.G. van der Valk. 1986. Seed germination traits of annuals and emergents during drawdown in the Delta Marsh, Manitoba, Canada. Aquatic Botany 26:89-102.

20. Wrubleski, D.A. 1987. Chironomidae (Diptera) of peatlands and marshes in Canada. In *Aquatic Insects of Peatlands and Marshes in Canada*. (Eds.) D.M. Rosenberg and H.V. Danks, pp.141-161. Memoirs of the Entomological Society of Canada 140.

21. Murkin, H.R. 1989. The basis for food chains in prairie wetlands. In *Northern Prairie Wetlands*. (Ed.) A.G. van der Valk, pp.316-338. Ames: Iowa State University Press.

22. Wrubleski, D.A. 1989. The effect of waterfowl feeding activity on a chironomid (Diptera: Chironomidae) community. In *Freshwater Wetlands and Wildlife*. (Eds.) R.R. Sharitz and J.W. Gibbons, pp.691-696. USDOE Symposium Series 61, Oak Ridge: USDOE Office of Scientific and Technical Information.

23. Kadlec, J.A. 1989. Effects of deep flooding and drawdown on freshwater marsh sediments. In *Freshwater Wetlands and Wildlife*. (Eds.) R.R. Sharitz and J.W. Gibbons, pp.127-143. USDOE Symposium Series 61, Oak Ridge: USDOE Office of Scientific and Technical Information.

24. Neckles, H.A., and R.L. Wetzel. 1989. Effects of forage harvest in seasonally flooded prairie marshes: simulation model experiments. In *Freshwater Wetlands and Wildlife*. (Eds.) R.R. Sharitz and J.W. Gibbons, pp.127-143. USDOE Symposium Series 61, Oak Ridge: USDOE Office of Scientific and Technical Information.

25. van der Valk, A.G., C.H. Welling, and R.L. Pederson. 1989. Vegetation change in a freshwater wetland: a test of a priori predictions. In *Freshwater Wetlands and Wildlife*. (Eds.) R.R. Sharitz and J.W. Gibbons, pp.207-217. USDOE Symposium Series 61, Oak Ridge: USDOE Office of Scientific and Technical Information.

26. Murkin, H.R., R.M. Kaminski, and R.D. Titman. 1989. Responses by nesting red-winged black-birds to manipulated cattail habitat. In *Freshwater Wetlands and Wildlife*. (Eds.) R.R. Sharitz and J.W. Gibbons, pp.673-680. USDOE Symposium Series 61, Oak Ridge: USDOE Office of Scientific and Technical Information.

27. Peterson, L.P., H.R. Murkin, and D.A. Wrubleski. 1989. Waterfowl predation on benthic macroinvertebrates during fall drawdown of a northern prairie marsh. In *Freshwater Wetlands and Wildlife*. (Eds.) R.R. Sharitz and J.W. Gibbons, pp.681-689. DOE Symposium Series No. 61, Oak Ridge: USDOE Office of Scientific and Technical Information.

28. Nelson, J.W., J.A. Kadlec, and H.R. Murkin. 1990. Seasonal comparisons of weight loss for two types of *Typha glauca* Godr. leaf litter. Aquatic Botany 37:299-314.

29. Murkin, H.R., and D.A. Wrubleski. 1988. Aquatic invertebrates of freshwater marshes: function

and ecology. In *The Ecology and Management of Wetlands. Volume 1: Ecology of Wetlands.* (Eds.) D.D. Hook, W.H. McKee, H.K. Smith, J. Gregory, V.G. Burrell, M.R. DeVoe, R.E. Sojka, S. Gilbert, R. Banks, L.H. Stolzy, C. Brooks, T.D. Matthews, and T.H. Shear,, pp.239-249. Portland: Timber Press.

30. Kadlec, J.A. 1987. Nutrient dynamics in wetlands. In *Proceedings of Conference on Applications of Aquatic Plants for Water Treatment and Resource Recovery.* (Eds.) K.R. Reddy and W.H. Smith, pp.393-419. Orlando: Magnolia Publishing, Inc.

31. Pederson, R.L., and L.M. Smith. 1988. Implications of wetland seed bank research: a review of Great Basin and prairie marsh studies. In *Interdisciplinary Approaches to Freshwater Wetlands Research.* (Ed.) D.A. Wilcox, pp.81-95. East Lansing: Michigan State University Press.

32. Welling, C.H., R.L. Pederson, and A.G. van der Valk. 1988. Recruitment from the seed bank and the development of zonation of emergent vegetation during drawdown in a prairie marsh. Journal of Ecology 76:483-496.

33. Gurney, S.E., and G.G.C. Robinson. 1988. The influence of water level manipulation on metaphyton production in a temperate freshwater marsh. Verhandlungen Internationale Vereinigung für Theoretische und Angewandte Limnologie 23:1032-1040.

34. Welling, C.H., R.L. Pederson, and A.G. van der Valk. 1988. Temporal patterns in recruitment from the seed bank during drawdowns in a prairie wetland. Journal of Applied Ecology 25:999-1007.

35. Hosseini, S.M., and A.G. van der Valk. 1989. The impact of prolonged above-normal flooding on metaphyton in a freshwater marsh. In *Freshwater Wetlands and Wildlife.* (Eds.) R.R. Sharitz and J.W. Gibbons, pp.317-324. USDOE Symposium Series 61, Oak Ridge: USDOE Office of Scientific and Technical Information.

36. Murkin, H.R., J.A. Kadlec, and E.J. Murkin. 1991. Effects of prolonged flooding on nektonic invertebrates in small diked marshes. Canadian Journal of Fisheries and Aquatic Sciences 48:2355-2364.

37. Neill, C. 1994. Primary production and management of seasonally flooded prairie marshes harvested for wild hay. Canadian Journal of Botany 72:801-807.

38. van der Valk, A.G., and C.H. Welling. 1988. The development of zonation in freshwater wetlands: an experimental approach. In *Diversity and Pattern in Plant Communities.* (Eds.) H.J. During, M.J.A. Werger, and H.J. Willems, pp.145-158. The Hague, Netherlands: SPB Publishers.

39. Hosseini, S.M., and A.G. van der Valk. 1989. Primary productivity and biomass of periphyton and phytoplankton in flooded freshwater marshes. In *Freshwater Wetlands and Wildlife.* (Eds.) R.R. Sharitz and J.W. Gibbons, pp.303-315. USDOE Symposium Series 61, Oak Ridge: USDOE Office of Scientific and Technical Information.

40. van der Valk, A.G., and R.L. Pederson. 1989. Seed banks and the management and restoration of natural vegetation. In *The Ecology of Seed Banks.* (Eds.) M.A. Leck, V.T. Parker, and R.L. Simpson, pp.329-346. New York: Academic Press.

41. Wienhold, C.E., and A.G. van der Valk. 1989. The impact of duration of drainage on seed banks of northern prairie wetlands. Canadian Journal of Botany 67:1878-1884.

42. van der Valk, A.G. 1988. From community ecology to vegetation management: providing a scientific basis for management. Transactions of the North American Wildlife and Natural Resources Conference 53:463-470.

43. Murkin, H.R., A.G. van der Valk, and C.B. Davis. 1989. Decomposition of four dominant macrophytes in the Delta Marsh. Wildlife Society Bulletin 17:215-221.

44. van der Valk, A.G. (Ed.) 1989. *Northern Prairie Wetlands*. Ames: Iowa State University Press.

45. van der Valk, A.G., B.D.J. Batt, H.R. Murkin, P.J. Caldwell, and J.A. Kadlec. 1988. The Marsh Ecology Research Program (MERP): the organization and administration of a long-term mesocosm study. In *Proceedings of the Oceans '88 Conference*, pp.46-51. Baltimore, Maryland.

46. van der Valk, A.G., R.L. Pederson, and C.B. Davis. 1992. Restoration and creation of freshwater wetlands using seed banks. Wetlands Ecology and Management 1:191-197.

47. McKee, K.L., I.A. Mendelssohn, and D.M. Burdick. 1989. Effect of long-term flooding on root metabolic response in five freshwater marsh plant species. Canadian Journal of Botany 67:3446-3452.

48. Neckles, H.A., H.R. Murkin, and J.A. Cooper. 1990. Influences of seasonal flooding on macroinvertebrate abundance in wetland habitats. Freshwater Biology 23:311-322.

49. Wrubleski, D.A., H.R. Murkin, A.G. van der Valk, and J.W. Nelson. 1997. Decomposition of emergent macrophyte roots and rhizomes in a northern prairie marsh. Aquatic Botany 58:121-134.

50. Neill, C. 1990. Effects of nutrients and water levels on emergent macrophyte biomass in a prairie marsh. Canadian Journal of Botany 68:1007-1014.

51. Neill, C. 1990. Effects of nutrients and water levels on species composition in prairie whitetop (*Scolochloa festucacea*) marshes. Canadian Journal of Botany 68:1015-1020.

52. Wrubleski, D.A., and L.C.M. Ross. 1989. Diel periodicities of adult emergence of Chironomidae and Trichoptera from the Delta Marsh, Manitoba, Canada. Journal of Freshwater Ecology 5:163-169.

53. Neill, C. 1990. Nutrient limitation of hardstem bulrush (*Scirpus acutus* Muhl.) in a Manitoba Interlake Region Marsh. Wetlands 10:69-76.

54. Murkin, E.J., and H.R. Murkin. 1989. Marsh Ecology Research Program: long-term monitoring procedures manual. Delta Waterfowl and Wetlands Research Station Technical Bulletin 2.

55. Neill, C. 1993. Growth and resource allocation of whitetop (*Scolochloa festucacea*) along a water depth gradient. Aquatic Botany 46:235-246.

56. Nelson, J.W., J.A. Kadlec, and H.R. Murkin. 1990. Responses by macro-invertebrates to cattail litter quality and timing of litter submergence in a northern prairie marsh. Wetlands 10:47-60.

57. Wrubleski, D.A., and D.M. Rosenberg. 1990. The Chironomidae (Diptera) of Bone Pile Pond, Delta Marsh, Manitoba, Canada. Wetlands 10:243-275.

58. van der Valk, A.G., J.M. Rhymer, and H.R. Murkin. 1991. Water level and litter decomposition of four emergent species in a prairie wetland. Wetlands 11:1-16.

59. Merendino, M.T., L.M. Smith, H.R. Murkin, and R.L. Pederson. 1990. The response of prairie wetland vegetation to seasonality of drawdown. Wildlife Society Bulletin 18:245-251.

60. Clark, W.R. 1990. Compensation in furbearer populations: current data compared with a review of concepts. Transactions of the North American Wildlife and Natural Resources Conference 55:491-500.

61. Merendino, M.T., and L.M. Smith. 1991. Influence of drawdown date and subsequent reflood depth on wetland vegetation establishment. Wildlife Society Bulletin 19:143-150.

62. van der Valk, A.G. 1992. Response by wetland vegetation to a change in water level. *Wetland Management and Restoration. Swedish Environmental Protection Agency Report 3492*. (Eds.) C.M. Finlayson and T. Larsson, pp.7-16. Solna, Sweden: Swedish Environmental Protection Agency.

63. Wrubleski, D.A. The effect of submersed macrophytes on the benthic macroinvertebrate community of a freshwater marsh (submitted).

64. Murkin, E.J., H.R. Murkin, and R.D. Titman. 1992. Nektonic invertebrate abundance and distribution at the emergent vegetation - open water interface in the Delta Marsh, Manitoba, Canada. Wetlands 12:45-52.

65. van der Valk, A.G., and L. Squires. 1992. Indicators of flooding derived from aerial photography in northern prairie wetlands. In *Ecological Indicators: Volume 1*. (Eds.) D.H. McKenzie, D.E. Hyatt, and V.J. McDonald, pp.593-602. London, UK: Elsevier.

66. van der Valk, A.G. 1992. Establishment, colonization, and persistence. In *Plant Succession: Theory and Prediction*. (Eds.) D.C. Glenn-Lewin, R.K. Peet, and T.T. Veben, pp.60-102. New York: Chapman & Hall.

67. van der Valk, A.G., L. Squires, and C.H. Welling. 1994. Identifying the impacts of an increase in water level on wetland vegetation undergoing succession. Ecological Applications 4:525-534.

68. Campeau, S., H.R. Murkin, and R.D. Titman. The effect of fertilization on bulrush litter decomposition in a nutrient-limited wetland (submitted).

69. Neill, C., and J.C. Cornwell. 1992. Stable carbon, nitrogen, and sulfur isotopes in a prairie marsh food web. Wetlands 12:217-224.

70. Neill, C. 1992. Life history and population dynamics of whitetop (*Scolochloa festucacea*) shoots under different levels of flooding and nitrogen supply. Aquatic Botany 42:241-252.

71. Neill, C. 1992. Comparison of soil coring and ingrowth methods for measuring belowground production. Ecology 73:1918-1922.

72. Neill, C. 1993. Seasonal flooding, soil salinity and primary production in northern prairie marshes. Oecologia (Berl.) 95:499-505.

73. Campeau, S., H.R. Murkin, and R.D. Titman. 1994. The relative importance of algae and emergent plant litter to freshwater marsh invertebrates. Canadian Journal of Fisheries and Aquatic Sciences 51:681-692.

74. Squires, L., and A.G. van der Valk. 1992. Water depth tolerances of the dominant emergent macrophytes of the Delta Marsh, Manitoba. Canadian Journal of Botany 70:1860-1867.

75. Kadlec, J.A. 1993. Effect of depth of flooding on summer water budgets for small diked marshes. Wetlands 13:1-9.

76. Wrubleski, D.A., H.R. Murkin, A.G. van der Valk, and C.B. Davis. 1997. Decomposition of litter of three mudflat annual species in a northern prairie marsh during drawdown. Plant Ecology 129:141-148.

77. van der Valk, A.G. 1994. Effects of prolonged flooding on the distribution and biomass of emergent species along a freshwater wetland coenocline. Vegetatio 110:185-196.

78. Wrubleski, D.A. Responses of the Chironomidae (Diptera) to the experimental management of freshwater marshes (submitted).

79. Clark, W.R., and D.W. Kroeker. 1993. Population dynamics of muskrats in managed marshes at Delta, Manitoba. Canadian Journal of Zoology 71:1620-1628.

80. Seabloom, E.W., K.A. Moloney, and A.G. van der Valk. An experimental test of the importance of dispersal constraints in models of plant distributions along a dynamic environmental gradient (submitted).

81. Ross, L.C.M., and H.R. Murkin. 1993. The effect of above-normal flooding of a northern prairie marsh on *Agraylea multipunctata* Curtis (Trichoptera: Hydroptilidae). Journal of Freshwater Ecology 8:27-35.

82. Neckles, H.A., and C. Neill. 1994. Hydrologic control of litter decomposition in seasonally flooded prairie marshes. Hydrobiologia 286:155-165.

83. Clark, W.R. 1994. Habitat selection by muskrats in experimental marshes undergoing succession. Canadian Journal of Zoology 72:675-680.

84. de Swart, E.O.A.M., A.G. van der Valk, K.J. Koehler, and A. Barendregt. 1994. Experimental evaluation of realized niche models for predicting responses of plant species to a change in environmental conditions. Journal of Vegetation Science 5:541-552.

85. Neill, C. 1995. Seasonal flooding, nitrogen mineralization and nitrogen utilization in a prairie marsh. Biogeochemistry 30:171-189.

86. Robinson, G.G.C., S.E. Gurney, and L.G. Goldsborough. 1997. Response of benthic and planktonic algal biomass to experimental water-level manipulation in a prairie lakeshore wetland. Wetlands 17:167-181.

87. Robinson, G.G.C., S.E. Gurney, and L.G. Goldsborough. 1997. The primary productivity of benthic and planktonic algae in a prairie wetland under controlled water-level regimes. Wetlands 17:182-194.

88. Murkin, H.R., E.J. Murkin, and J.P. Ball. 1997. Avian habitat selection and prairie wetland dynamics: a ten-year experiment. Ecological Applications 7:1144-1159.

89. Cornwell, J.C., C. Neill, and J.C. Stevenson. 1995. Biogeochemical origin of Δ^{34}S isotopic signatures in a prairie marsh. Canadian Journal of Fisheries and Aquatic Sciences 52:1816-1820.

90. Murkin, H.R., and L.C.M. Ross. 1999. Northern prairie marshes (Delta Marsh, Manitoba): macroinvertebrate responses to a simulated wet-dry cycle. In *Invertebrates in Freshwater Wetlands of North America: Ecology and Management*. (Eds.) D. Batzer, R.D. Rader, and S.A. Wissinger, pp. 543-569. New York: Wiley.

91. Wrubleski, D.A. 1999. Northern prairie marshes (Delta Marsh, Manitoba): Chironomidae (Diptera) responses to changing plant communities in newly flooded habitats. In *Invertebrates in Freshwater Wetlands of North America: Ecology and Management*. (Eds.) D. Batzer, R.D. Rader, and S.A. Wissinger, pp. 571-601. New York: Wiley.

92. Adamus, P.R. 1996. Bioindicators for assessing ecological integrity of prairie wetlands. EPA/600/R-96/082. U.S. Environmental Protection Agency, National Health and Environmental Effects Laboratory, Western Ecology Division, Corvallis, Oregon. 209pp.

93. van der Valk, A.G., and H.R. Murkin. Changes in nutrient pools during an experimentally simulated wet-dry cycle in the Delta Marsh Manitoba (submitted).

Graduate Theses and Dissertations

Bicknese, N.A. 1987. The role of invertebrates in the decomposition of fallen macrophyte litter. M.S. thesis, Iowa State University, Ames, Iowa.

Campeau, S. 1990. The relative importance of algae and detritus to freshwater wetland food chains. M.S. thesis, McGill University, Montreal, Quebec.

Galinato, M.I. 1985. Seed germination studies of dominant mudflat-annual and emergent species of the Delta Marsh. M.S. thesis, Iowa State University, Ames, Iowa.

Hosseini, S.M. 1986. The effect of water level fluctuations on productivity and biomass of algae assemblages. Ph.D. dissertation, Iowa State University, Ames, Iowa.

Kroeker, D.W. 1988. Population dynamics of muskrats in managed marshes at Delta, Manitoba. M.S. thesis, Iowa State University, Ames, Iowa.

Merendino, M.T. 1989. The response of vegetation to seasonality of drawdown and reflooded depth. M.S. thesis, Texas Tech University, Lubbock, Texas.

Murkin, H.R. 1983. Responses by aquatic macroinvertebrates to prolonged flooding of marsh habitat. Ph.D. dissertation, Utah State University, Logan, Utah.

Neckles, H.A. 1984. Plant and macroinvertebrate responses to water regime in a whitetop marsh. M.S. thesis, University of Minnesota, St. Paul, Minnesota.

Neill, C. 1988. Control of the primary productivity of emergent macrophytes in prairie marshes: effects of water depth and nutrient availability. M.S. thesis, University of Massachusetts, Amherst, Massachusetts.

Neill, C. 1992. Relationships between emergent plant production and seasonal flooding in prairie whitetop (*Scolochloa festucacea*) marshes. Ph.D. dissertation, University of Massachusetts, Amherst, Massachusetts.

Nelson, J.W. 1982. The effects of varying detrital nutrient concentrations on macroinvertebrate production. M.S. thesis, Utah State University, Logan, Utah.

Pederson, R.L. 1983. Abundance, distribution, and diversity of buried seed populations in the Delta Marsh, Manitoba, Canada. Ph.D. dissertation, Iowa State University, Ames, Iowa.

Squires, L. 1991. Water depth tolerances of emergent species in the Delta Marsh, Manitoba. M.S. thesis, Iowa State University, Ames, Iowa.

Welling, C.H. 1987. Reestablishment of perennial emergent macrophytes during drawdown of lacustrine marsh. M.S. thesis, Iowa State University, Ames, Iowa.

Wrubleski, D.A. 1984. Species composition, emergence phenologies, and relative abundances of Chironomidae (Diptera) from the Delta Marsh, Manitoba, Canada. M.S. thesis, University of Manitoba, Winnipeg, Manitoba.

Wrubleski, D.A. 1991. Chironomid recolonization of marsh drawdown surfaces following reflooding. Ph.D. dissertation, University of Alberta, Edmonton, Alberta.

Index

Index

PLACE IN RETURN BOX to remove this checkout from your record.
TO AVOID FINES return on or before date due.
MAY BE RECALLED with earlier due date if requested.

GULL LAKE LIBRARY

DATE DUE	DATE DUE	DATE DUE
APR 2 7 2004	_____	_____
_____	_____	_____
_____	_____	_____
_____	_____	_____
_____	_____	_____

Prairie Wetland Ecology